CAMBRIDGE TRACTS IN MATHEMATICS

Poincaré Duality Algebras, Macaulay's Dual Systems, and Steenrod Operations

Poincaré Duality Algebras, Macaulay's Dual Systems, and Steenrod Operations

DAGMAR M. MEYER AND LARRY SMITH

CAMBRIDGE
UNIVERSITY PRESS

CAMBRIDGE UNIVERSITY PRESS
Cambridge, New York, Melbourne, Madrid, Cape Town, Singapore, São Paulo

Cambridge University Press
The Edinburgh Building, Cambridge CB2 2RU, UK

Published in the United States of America by Cambridge University Press, New York

www.cambridge.org
Information on this title: www.cambridge.org/9780521850649

First published 2005

Printed in the United Kingdom at the University Press, Cambridge

A catalogue record for this book is available from the British Library

Library of Congress Cataloguing in Publication data

ISBN-13 978-0-521-85064-3 hardback
ISBN-10 0-521-85064-9 hardback

Contents

Introduction **1**

Part I. Poincaré duality quotients **9**

I.1 Poincaré duality, Gorenstein algebras, and irreducible ideals . 9

I.2 Properties of irreducible ideals 14

I.3 The ancestor ideals 17

I.4 Fundamental classes 18

I.5 Poincaré duality quotients of $\mathbb{F}[V]$ 21

I.6 Counting Poincaré duality quotients up to isomorphism 23

Part II. Macaulay's dual systems and Frobenius powers **31**

II.1 Divided power algebras and operations 32

II.2 Macaulay's dual principal systems 35

II.3 An illustrative example 38

II.4 Relation to the classical form problem 39

II.5 The $K \subset L$ Paradigm: a computational tool 40

II.6 Frobenius powers 44

Part III. Poincaré duality and the Steenrod algebra **53**

III.1 $\mathscr{P}^*$-Unstable Poincaré duality quotients of $\mathbb{F}_q[V]$ 54

III.2 The $\mathscr{P}^*$-Double Annihilator Theorem 58

III.3 Wu classes 60

III.4 $\mathscr{P}^*$-Indecomposables: Hit Problems 65

III.5 A dual interpretation of the $\mathscr{P}^*$-Double Annihilator Theorem . 67
III.6 Steenrod operations and Frobenius powers 71
III.7 Examples 75

Part IV. Dickson, symmetric, and other coinvariants 81

IV.1 Dickson coinvariants 81
IV.2 Wu classes of algebras of coinvariants 86
IV.3 The Macaulay dual of Dickson coinvariants mod 2 88
IV.4 Symmetric coinvariants 89

Part V. The Hit Problem mod 2 93

V.1 Powers of Dickson polynomials in 2 variables 94
V.2 $\mathscr{A}^*$-Indecomposable elements in $\mathbb{F}_2[x, y]$ 100
V.3 Powers of Dickson polynomials in 3 variables 105
V.4 Powers of Dickson polynomials in many variables 109
V.5 Powers of mod 2 Stiefel–Whitney classes in 3 variables 123

Part VI. Macaulay's inverse systems and applications 133

VI.1 Macaulay's inverse principal systems 134
VI.2 Catalecticant Matrices and Ancestor Ideals 136
VI.3 Regular ideals 142
VI.4 Lying over for irreducible ideals 146
VI.5 Change of rings and inverse polynomials 155
VI.6 Inverse systems and the Steenrod algebra 160
VI.7 Change of Rings and Wu Classes 165
VI.8 The Hit Problem for the Dickson and other algebras 170

References 177

Notation 185

Index 189

Introduction

TO SET THE STAGE let $\mathbb{F}$ be a field (in much of what follows $\mathbb{F}$ will be finite, e.g., $\mathbb{F}_2$, the field with 2 elements) and $n \in \mathbb{N}$ a positive integer. Denote by $V = \mathbb{F}^n$ the n-dimensional vector space over $\mathbb{F}$, and by $\mathbb{F}[V]$ the graded algebra of homogeneous polynomial functions on V. To be specific, $\mathbb{F}[V]$ is the symmetric algebra $S(V^*)$ on the vector space V^* dual to V. Since graded commutation rules play no role here we will grade this as an algebraist would, i.e., putting the linear forms in degree 1 no matter what the characteristic of the ground field $\mathbb{F}$. The homogeneous component of $\mathbb{F}[V]$ of degree d will be denoted by $\mathbb{F}[V]_d$. If we need a notation for a basis of V^* we will use $z_1, \ldots, z_n$; the corresponding basis for V will be denoted by $u_1, \ldots, u_n$. For up to three variables we will most often write x, y, z, respectively u, v, w for the variables and their duals. Recall that a graded vector space, algebra, or module is said to have **finite type** if the homogeneous components are all finite dimensional vector spaces.

DEFINITION: *Let H be a commutative graded connected algebra of finite type over the field $\mathbb{F}$. We say that H is a* **Poincaré duality algebra** *of* **formal dimension** *d if*

(i) $H_i = 0$ for $i > d$,

(ii) $\dim_{\mathbb{F}}(H_d) = 1$,

(iii) the pairing $H_i \otimes H_{d-i} \longrightarrow H_d$ given by multiplication is nonsingular, i.e., an element $a \in H_i$ is zero if and only if $a \cdot b = 0 \in H_d$ for all $b \in H_{d-i}$.

If H is a Poincaré duality algebra we write f-dim(H) for the formal dimension of H. If the formal dimension is d and $[H]$ in H_d is nonzero, then $[H]$ is referred to as a **fundamental class** for H. Fundamental classes are determined only up to multiplication by a nonzero element of $\mathbb{F}$.

The notion of Poincaré duality comes from the study of closed manifolds in algebraic topology, and goes back at least to H. Poincaré; see e.g. [77] Section 69. Apart from the graded commutation rules the cohomology of a closed oriented manifold with field coefficients is the prototypical example of a Poincaré duality algebra. However, Poincaré duality algebras also appear quite naturally in invariant theory as rings of coinvariants of groups whose rings of invariants are polynomial algebras. Indeed, the less than complete understanding of the role of Poincaré duality algebras in invariant theory is part of the motivation for this study. We explain this next.

In characteristic zero, or more generally in the nonmodular case, i.e., where the order $|G|$ of G is invertible in the ground field, it is well known (see e.g. [80] or [87] Section 7.4) that pseudoreflection groups (better said, pseudoreflection representations) are characterized by the fact that their invariant rings are polynomial algebras. This is known to fail in the modular case: the ring of invariants of a reflection group need not be a polynomial algebra (see e.g. [100] or [87] Section 7.4 Example 4).

If $\rho : G \hookrightarrow \mathrm{GL}(n, \mathbb{F})$ is a representation of a finite group G over the field $\mathbb{F}$ for which the ring of invariants $\mathbb{F}[V]^G$ is a polynomial algebra then the ring of coinvariants $\mathbb{F}[V]_G$ is a Poincaré duality quotient of $\mathbb{F}[V]$ ([87] Theorem 6.5.1). Such a ring of coinvariants is therefore a very special type of Poincaré duality algebra, viz., a complete intersection. A theorem of R. Steinberg [98] (as formulated by R. Kane [40], [41] Chapter VII) says: if $\rho : G \hookrightarrow \mathrm{GL}(n, \mathbb{F})$ is a representation of a finite group over a field $\mathbb{F}$ *of characteristic zero*, then the ring of coinvariants $\mathbb{F}[V]_G$ is a Poincaré duality algebra if and only if G is a pseudoreflection group. Although Steinberg's proof, as well as Kane's, makes central use of the fact that the characteristic of $\mathbb{F}$ is zero and not just relatively prime to the order of the group, as would seem more natural, T.-C. Lin [47] has recently removed the need for this extra assumption and shown the result holds in the nonmodular case.

It is not known what the best extension of Steinberg's theorem to fields of nonzero characteristic might look like. Several variations of its hypotheses and conclusion are possible. There is an ad hoc, characteristic free, proof of the result as originally stated if one restricts to the case of two variables [92]. Examples in the modular case show that in more variables the original statement can be false. For example, consider the tautological representation of the alternating group A_n over a field of characteristic p less than or equal to n. Then $\mathbb{F}[z_1, \ldots, z_n]_{A_n}$ is a Poincaré duality algebra, but the ring of invariants is not a polynomial algebra (see [25], Section 11, [83] Section 5, and [93] Section 2). It is also not clear that in the modular case a ring of coinvariants which is a Poincaré duality algebra must be a complete intersection, though this is again correct for $n = 2$ by [102]. Other places where Poincaré duality algebras appear in connection with invariant theory may be found in [19] and [41].

An important feature of an algebra over a field of characteristic $p \neq 0$ is the operation of raising an element to the p-th power. This is loosely referred to as the Frobenius map. The Steenrod operations and the Steenrod algebra represent one way to organize information hidden in the Frobenius homomorphism provided the ground field is finite. Let $\mathbb{F}_q$ be the Galois field with $q = p^\nu$ elements, $V = \mathbb{F}_q^n$, and define[1]

$$\mathcal{P}(\xi) : \mathbb{F}_q[V] \longrightarrow \mathbb{F}_q[V][[\xi]]$$

by the rules

(i) $\mathcal{P}(\xi)$ is $\mathbb{F}$-linear,
(ii) $\mathcal{P}(\xi)(v) = v + v^q\xi$ for $v \in V^*$,
(iii) $\mathcal{P}(\xi)(u \cdot w) = \mathcal{P}(\xi)(u) \cdot \mathcal{P}(\xi)(w)$ for $u,\ w \in \mathbb{F}_q[V]$,
(iv) $\mathcal{P}(\xi)(1) = 1$.

$\mathcal{P}(\xi)$ is a ring homomorphism of degree 0 if we agree ξ has degree $(1-q)$. By separating out homogeneous components we obtain $\mathbb{F}_q$-linear maps

$$\mathcal{P}^i : \mathbb{F}_q[V] \longrightarrow \mathbb{F}_q[V]$$

by the requirement

$$\mathcal{P}(\xi)(f) = \sum_{i=0}^{\infty} \mathcal{P}^i(f)\xi^i.$$

The maps $\mathcal{P}^i$ may be assembled into an algebra called the Steenrod algebra $\mathscr{P}^*$ of the Galois field $\mathbb{F}_q$ (see e.g. [87] Chapters 10 and 11). The strength of these operations lies in the fact that they commute with the action of $\mathrm{GL}(V)$ on $\mathbb{F}_q[V]$ and satisfy the unstability conditions

$$\mathcal{P}^i(f) = \begin{cases} f^q & i = \deg(f) \\ 0 & i > \deg(f) \end{cases}$$

for all $f \in \mathbb{F}_q[V]$. The first unstability condition is a *nontriviality* condition, the second a *triviality* condition, and together they impose a very rigid restriction on $\mathbb{F}_q[V]^G$ as well as on which Poincaré duality quotients of $\mathbb{F}_q[V]$ can arise as $\mathbb{F}_q[V]_G$ for some representation $\rho : G \hookrightarrow \mathrm{GL}(n, \mathbb{F}_q)$. For example, using ideas from algebraic topology one can define invariants of such quotients called Wu classes. In Appendix B of [60] it is shown that the Wu classes (see Section IV.2) of a ring of coinvariants which is a Poincaré duality algebra are always trivial. Not all Poincaré duality quotients with an unstable action of the Steenrod operations have this property.

[1] If A is a ring then $A[[\xi]]$ denotes the ring of formal power series over A in the variable ξ.

Steinberg's paper [98] contains a number of other results on rings of coinvariants that are Poincaré duality algebras, such as how to construct a fundamental class. Again, as they stand, the proofs work only in characteristic zero. Some of these results have been extended to nonzero characteristic [91], but by no means all. This unsatisfactory state of affairs suggested that a systematic study of Poincaré duality quotients of $\mathbb{F}_q[V]$ admitting Steenrod operations satisfying the unstability conditions might be fruitful. We begin such a study here.

The first step in our program led us to Macaulay's theory of irreducible ideals in polynomial algebras. Due to the enormous changes in terminology since [49] was written, this theory is not as accessible as it might be. We present a treatment in current parlance and extend it to encompass the extra structure of a Steenrod algebra action when the ground field is finite. In doing so we uncover a very unexpected connection between $\mathscr{P}^*$-Poincaré duality quotients of $\mathbb{F}_q[V]$ and the so called *Hit Problems,*[2] e.g., the determination of the $\mathscr{P}^*$-indecomposable elements in $\mathbb{F}_q[V]$. This relationship is particularly appealing if formulated along the lines of M. C. Crabb and J. R. Hubbuck [16]. Applications to this problem based on what we have learned about $\mathscr{P}^*$-Poincaré duality algebras appear at various places in Parts III and VI, and in detail in Part V, where we work out a number of special cases.

There are many other motivations for studying Poincaré duality quotients of $\mathbb{F}[V]$. Let us just mention one more: it has its origins in algebraic topology and connects up with certain problems in invariant theory which we have not discussed here (see [44]). We paraphrase from the introduction to [1]. *"Recently,* in studying the coinvariants of reflection groups, *I had occasion to consider the formulae of Thom and Wu* [111] . . . *although these formulae are simple and attractive, I did not feel that they gave me that complete understanding that I sought."* In [1] J. F. Adams proves these formulae by constructing a universal example that is no longer a Poincaré duality algebra[3] but is an inverse limit of rings of coinvariants (see e.g. [10] or [44]) and verifying certain other formulae in this new object. We would like to understand why the formulae of Thom and Wu are *also* consequences of

[2] We use the expression *Hit Problem(s)* as in [108] Section 7: *quite generally, if M is a graded module over the positively graded algebra A over the field $\mathbb{F}$, we say that $u \in M$ is* **hit** *if there are elements $u_1, \ldots, u_k \in M$ and $a_1, \ldots, a_k \in A$ with* $\deg(a_1), \ldots, \deg(a_k) > 0$ *with $u = a_1u_1 + \cdots + a_ku_k$. The elements of M that are hit form the A-submodule $\overline{A} \cdot M$, and the quotient $M/\overline{A} \cdot M$ the module of A-indecomposable elements $\mathbb{F} \otimes_A M$.* *Hit Problem(s)* refer to the characterization of elements of M that are hit or not, e.g., in the case of the Steenrod algebra $\mathscr{P}^*$ acting on $\mathbb{F}_q[z_1, \ldots, z_n]$ finding conditions on a monomial that assure it is $\mathscr{P}^*$-indecomposable.

[3] In fact it isn't even Noetherian.

their validity in the Poincaré duality quotients of $\mathbb{F}_q[V]$ admitting an unstable Steenrod algebra action. The action of the Steenrod algebra in these cases is completely formal, it being a consequence of unstability and the Cartan formula. Why should these determine the formulae for arbitrary $\mathscr{P}^*$-unstable Poincaré duality algebras?

THE MANUSCRIPT divides naturally into several parts. The material in Part I is largely expository. There we explain the connection between Poincaré duality algebras, Gorenstein algebras, and irreducible ideals. Several different characterizations of irreducible ideals are given and a variety of methods to construct them are presented. We specialize and refine these results and constructions to the main case of interest for us, namely the Poincaré duality quotient algebras of $\mathbb{F}[V]$. A method is developed for counting the number of isomorphism classes of such quotients over a Galois field. It is based on invariant theory and is applied to count the number of such quotients of $\mathbb{F}_2[x, y]$.

Part II reformulates a number of results from [49] in modern language. In Section II.2 we present Macaulay's concept of inverse systems (which are called dual systems here) in the language of Hopf algebras and derive a number of results that will be useful in later sections. This is then illustrated with examples and connected with the classical form problem of ninteenth century invariant theory. A fundamental tool for making computations, the $K \subset L$ paradigm, appears in Section II.5. Section II.6 contains a result first proved in the ungraded case by R. Y. Sharp. Namely, a Frobenius power of an irreducible ideal in a regular local ring is again irreducible. Careful study of his proof has allowed us to give a proof adapted to the graded case. We use this to construct new Poincaré duality quotients from existing ones. As a bonus, the proof in the graded case yields formulae for a fundamental class and a generator for the dual principal system of the new Poincaré duality quotient.

In Part III we restrict the ground field to a Galois field. Here the Frobenius homomorphism provides us with an additional structure which we organize[4] into the Steenrod algebra $\mathscr{P}^*$ of $\mathbb{F}_q$ whose elements are called Steenrod operations. Section III.1 introduces the Steenrod algebra $\mathscr{P}^*$ and in Section III.2 we rework Macaulay's Double Annihilator Theorem

[4] This is by no means the only way to extract information from the Frobenius homomorphism: see [73] and [71] for a different approach altogether.

(see Section II.2) in this enhanced context. Wu classes are invariants of the action of the Steenrod algebra on a Poincaré duality algebra; they are introduced in Section III.3. In the case of a Poincaré duality quotient of $\mathbb{F}_q[V]$ their vanishing is related by means of Macaulay's $\mathscr{P}^*$-Double Annihilator Theorem to the problem of computing the invariants $\Gamma(V)^{\mathscr{P}^*}$ of the Steenrod algebra acting[5] on the dual divided power algebra $\Gamma(V)$. These results culminate in a surprising connection with a *Hit Problem*, namely of finding a minimal set of generators of $\mathbb{F}_q[V]$ as a $\mathscr{P}^*$-module.

In Part IV we investigate the structure of algebras of coinvariants that are Poincaré duality quotients of $\mathbb{F}_q[z_1, \ldots, z_n]$. A ring of coinvariants over a finite field is an unstable algebra over the Steenrod algebra. If a ring of invariants is a polynomial algebra then the corresponding ring of coinvariants is a Poincaré duality algebra, so potentially has nonzero Wu classes. The Dickson and symmetric coinvariants are such algebras and they are examined in detail. We determine fundamental classes, Macaulay duals, and by explicit computations show that the Wu classes are trivial. The relation of these algebras to the *Hit Problem* is explained here, with detailed computations carried out in Part V. In [60] S. A. Mitchell showed that a complete intersection algebra with unstable Steenrod algebra action that is a quotient of $\mathbb{F}_q[V]$ has trivial Wu classes. We present a variant of his proof in Section IV.2 that exploits our computations with the Dickson coinvariants and the Adams and Wilkerson/Neusel Imbedding Theorem, [66] Theorem 7.4.4.

Part V contains some detailed applications to the *Hit Problem* for $\mathbb{F}_2[V]$. We have confined ourselves to the choice of $\mathbb{F}_2$ as ground field to keep the gymnastics with binomial coefficients mod p within reasonable bounds. Section V.1 examines ideals generated by powers of the Dickson polynomials $\mathbf{d}_{2,0}$, $\mathbf{d}_{2,1} \in \mathbb{F}_2[x, y]$ where we determine which of these ideals are $\mathscr{A}^*$-invariant, and precisely which amongst those have a Poincaré duality quotient with trivial Wu classes. In Section V.2 we show that representatives for the fundamental classes of the latter, together with the so-called spikes (see [82]), provide a vector space basis for the $\mathscr{A}^*$-indecomposable elements of $\mathbb{F}_2[x, y]$. Some topological questions arising from these results are considered in [56]. Section V.3 extends this line of investigation to three variables. Our work with ideals generated by powers of Dickson polynomials culminates in a complete list of such ideals in an arbitrary number of variables that are $\mathscr{A}^*$-invariant and determines which of these

[5] If one introduces the **total Steenrod operator** $\mathcal{P} = 1 + \mathcal{P}^1 + \cdots + \mathcal{P}^k + \cdots$, then $\mathcal{P}$ induces an *ungraded* action of the integers $\mathbb{Z}$ on $\Gamma(V)$ and $\Gamma(V)^{\mathscr{P}^*}$ is the subalgebra of elements invariant under this action.

have trivial Wu classes. We close this part by commencing a study of the ideals generated by powers of Stiefel–Whitney classes; these results are not as complete as for the case of powers of Dickson polynomials. Section V.5 examines ideals generated by powers of the Stiefel–Whitney classes $w_2, w_3, w_4 \in \mathbb{F}_2[x, y, z]$, which provide some interesting examples of $\mathscr{A}^*$-indecomposable monomials.

Part VI grew out of a simple observation concerning the relation of the *Hit Problems* for $\mathbb{F}_q[z_1, \ldots, z_n]$ and $\mathbf{D}(n)$. We decided to develop a more general context in which to present this observation. The results in Part VI center around a discussion of the classical *lying over* relation, not for prime ideals, but for irreducible and regular ideals. We first introduce Macaulay's *inverse systems* in their original form using inverse polynomials. This we use to explain the *catalecticant matrices* which provide a tool to make computations. We illustrate this in a pair of examples in Section VI.2. We use the inverse system formulation of Macaulay's theory, together with our results on lying over for irreducible and regular ideals to prove a variety of *change of rings* results (see Sections VI.5 and VI.7). This allows us to better study the relation between two Poincaré duality quotients $\mathbb{F}[V]/I$ and $\mathbb{F}[V]/J$ under the hypothesis that $I \subset J$. We obtain a number of results relating fundamental classes and generators for the inverse principal systems. In the special case that $\mathbb{F} = \mathbb{F}_q$ and the ideals are closed under the action of the Steenrod algebra we enhance these results to include Steenrod operations, and apply them to the *Hit Problems* for the Dickson and other algebras.

For unexplained terminology or notation please see the index of notation, [68], or [87].

Part I
Poincaré duality quotients

IN THIS part we collect a certain amount of background material. There is basically not much new here, but these results are scattered throughout the literature, and where they do appear, they rarely do so in the form we need. We emphasize at the outset that, unless explicitly mentioned to the contrary, all algebras considered here are commutative, graded, connected algebras of finite type over a field. The component of degree k of a graded object X is denoted by X_k. We start by reviewing some definitions and then develop the relation between Gorenstein algebras, Poincaré duality algebras, and irreducible ideals. The study of irreducible ideals seems to have fallen out of favor with time so it is difficult to locate adequate references for some results. We therefore include a fair number of elementary proofs.

Once the basic objects of study have been introduced and their elementary properties developed we use them to present several different ways to construct Poincaré duality algebras that are quotients of a polynomial algebra. In the final section of this part we examine the problem of counting such Poincaré duality quotients up to isomorphism in the case of a finite ground field. This leads to an interesting invariant theoretic problem.

I.1 Poincaré duality, Gorenstein algebras, and irreducible ideals

If A is a commutative graded connected algebra over a field $\mathbb{F}$ we denote by $\overline{A}$ the **augmentation ideal** of A: this is the ideal of all homogeneous elements of strictly positive degree. This notation becomes a bit cumbersome for $\mathbb{F}[x, y, z]$, viz., $\overline{\mathbb{F}[x, y, z]}$, so if the algebra A is clear from the context we introduce the alternative notation $\mathfrak{m}$ for its augmentation ideal.

Note that this is the unique *graded* maximal ideal in A. If H is a commutative graded connected algebra over a field $\mathbb{F}$ and $[H] \in H$ is an element of degree d, then it is easy to see that H is a Poincaré duality algebra of formal dimension d with fundamental class $[H]$ if and only if $H_i = 0$ for $i > d$, $\operatorname{Ann}_H([H]) = \overline{H}$, and $\operatorname{Ann}_H(\overline{H}) = \mathbb{F} \cdot [H]$.

The study of Poincaré duality algebras is permeated with *double duality* results of one form or another. Here is one of the most basic of these (see e.g. [18] page 416 or [63]).

LEMMA I.1.1: *Let H be a Poincaré duality algebra over the field $\mathbb{F}$ and $I \subset H$ an ideal. Then* $\operatorname{Ann}_H(\operatorname{Ann}_H(I)) = I$.

PROOF: Let the formal dimension of H be d and choose a fundamental class $[H] \in H_d$. Define a bilinear pairing

$$< - \mid - > : H \times H \longrightarrow \mathbb{F}$$

by

$$<a \mid b> = \begin{cases} 0 & \text{if } \deg(a) + \deg(b) \neq d \\ \lambda & \text{if } \deg(a) + \deg(b) = d \text{ and } a \cdot b = \lambda[H]. \end{cases}$$

Note that this pairing is symmetric, middle associative, and nondegenerate, i.e.,

$$\begin{aligned} <a \mid b> &= <b \mid a> \quad \forall\, a,\, b \in H \\ <a \cdot c \mid b> &= <a \mid c \cdot b> \quad \forall\, a,\, c,\, b \in H \\ <a \mid b> = 0 \quad \forall\, b \in H &\iff a = 0. \end{aligned}$$

We claim that $\operatorname{Ann}_H(I) = \{h \in H \mid <h \mid I> = 0\}$. To see this suppose that $u \in \operatorname{Ann}_H(I)$ and $w \in I$. If $\deg(u) + \deg(w) \neq d$ then $<u \mid w> = 0$ by definition. If $\deg(u) + \deg(w) = d$, then since u annihilates I, we have $u \cdot w = 0 = 0 \cdot [H]$, so $<u \mid w> = 0$. Therefore it follows that $\operatorname{Ann}_H(I) \subseteq \{h \in H \mid <h \mid I> = 0\}$. On the other hand, if $u \in \{h \in H \mid <h \mid I> = 0\}$, and $w \in I$, then for any $x \in H_{d-(\deg(u)+\deg(w))}$ we have $w \cdot x \in I$, so

$$0 = <u \mid w \cdot x> = <u \cdot w \mid x>.$$

Hence $u \cdot w$ annihilates $H_{d-(\deg(u)+\deg(w))}$ so by Poincaré duality $u \cdot w = 0$, therefore $u \in \operatorname{Ann}_H(I)$. This establishes the claim. From this the lemma follows using elementary facts about nondegenerate bilinear forms (see for example [35] Chapter V Section 3). □

The tensor product $H = H' \otimes H''$ of two Poincaré duality algebras H' and H'' is again a Poincaré duality algebra: if $[H'] \in H'_{d'}$ and $[H''] \in H''_{d''}$ are fundamental classes, then $[H'] \otimes [H''] \in H_{d'+d''}$ is a fundamental class for

H. This follows directly from the definitions, and shows in addition that $\text{f-dim}(H' \otimes H'') = \text{f-dim}(H') + \text{f-dim}(H'')$.

If two Poincaré duality algebras H' and H'' have the same formal dimension d, one can define their **connected sum** $H'\#H''$ in the following way:

$$(H'\#H'')_k = \begin{cases} \mathbb{F} \cdot [H'\#H''] & \text{if } k = d \\ H'_k \oplus H''_k & \text{if } 0 < k < d \\ 1 \cdot \mathbb{F} & \text{if } k = 0. \end{cases}$$

The products of two elements in either H' or H'' are as before, modulo the identification of the three fundamental classes $[H']$, $[H'\#H'']$, $[H'']$. The product of an element of H' of positive degree and of H'' of positive degree is zero. The operation # turns the isomorphism classes of Poincaré duality algebras of a fixed formal dimension d over a fixed ground field $\mathbb{F}$ into a commutative torsion free monoid. The Poincaré duality algebra[1] $H^*(S^d;\mathbb{F})$ serves as a two sided unit: this is the algebra H with $H_0 = \mathbb{F} = H_d$ and all other homogeneous components trivial. Already at this point a number of unsolved problems appear. Here is one such.

PROBLEM I.1.2: *What are the* **indecomposable** *Poincaré duality algebras with respect to the connected sum operation, i.e., what generates the monoid of isomorphism classes of Poincaré duality algebras over a fixed ground field and of a fixed formal dimension under the operation of connected sum? What is the Grothendieck group of this monoid?*

One might hope for a simple answer, such as: the complete intersections provide generators. However this is not the case: the Poincaré duality algebra $H = H^*(S^2 \times S^2;\mathbb{F})$ cannot be written as a nontrivial connected sum. This is best seen on the basis of the accompanying graphic for this algebra. In the graphic the two generators of the algebra, which appear in degree 2, are x' and x'', whose squares are zero, but whose product is a fundamental class. Up to a change of basis a nontrivial connected sum decomposition $H = H'\#H''$ would have to put x' in H' say, and x'' in H''. But then both $(x')^2$ and $(x'')^2$ would have to be fundamental classes of H' and H'' respectively, so in H the squares of x' and x'' would become a fundamental class, contrary to the fact that they are zero in H.

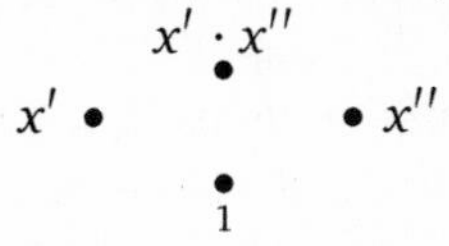

Moreover complete intersections need not be indecomposable: for example if $\mathbb{CP}(2)$ is the complex projective plane and $\mathbb{CP}(2)\#\mathbb{CP}(2)$ the connected

[1] If X is a topological space $H^*(X;\mathbb{F})$ denotes the cohomology of X with coefficients in $\mathbb{F}$. S^d denotes the d-dimensional sphere.

sum of two copies of it, then the cohomology algebra $H^*(\mathbb{CP}(2)\#\mathbb{CP}(2);\mathbb{F})$ is a complete intersection, viz., $\mathbb{F}[x, y]/(x^2 + y^2, xy)$, but decomposable since it is isomorphic to $H^*(\mathbb{CP}(2);\mathbb{F})\#H^*(\mathbb{CP}(2);\mathbb{F})$.

A **Poincaré duality quotient of** $\mathbb{F}[V]$ is just that: a quotient algebra H of $\mathbb{F}[V]$ that is a Poincaré duality algebra. We use the notation $\mathbb{F}[V] \xrightarrow{\pi} H$ if we need to make the quotient map explicit. The dimension $\dim_{\mathbb{F}}(H_1)$ of the component of H of degree one is called the **rank** of H and is denoted by $\operatorname{rank}(H)$. If the ideal $\ker(\pi)$ contains no linear forms this is just the dimension of V as a vector space.

There are many different ways to construct and classify Poincaré duality quotients of $\mathbb{F}[V]$. For example the connected sum of two Poincaré duality quotients $H' = \mathbb{F}[V']/I'$ and $H'' = \mathbb{F}[V'']/I''$ is a Poincaré duality quotient of $\mathbb{F}[V' \oplus V'']$ so Problem I.1.2 can be reformulated for the restricted class of Poincaré duality quotients of $\mathbb{F}[V]$ as V ranges over the finite dimensional $\mathbb{F}$-vector spaces. See Example 2 in Section VI.2 for more about this.

Since we are discussing quotient algebras, one could place the emphasis on the kernel of the quotient map, instead of on the quotient itself. Of these two extremes, we begin with the former, which is the most ideal theoretic, and work towards a description of Poincaré duality quotients of $\mathbb{F}[V]$ via Macaulay's *dual systems* ([49] Chapter IV) in the language of Hopf algebra duality and double annihilators in Part II.

As is often the case it is convenient to set things up in a bit more generality than we actually make use of here.

We recall that an ideal I in a commutative graded connected algebra A is called **irreducible** if whenever $I = I' \cap I''$ for ideals I', I'', then either $I = I'$ or $I = I''$. If A is Noetherian, a **parameter ideal** for A is an ideal generated by a system of parameters. A parameter ideal is $\overline{A}$-primary.

LEMMA I.1.3: *Let A be a commutative graded connected algebra of finite type over a field and $I \subset A$ an ideal. If A/I satisfies Poincaré duality then I is irreducible and $\overline{A}$-primary.*

PROOF: If A/I satisfies Poincaré duality then it is totally finite. Hence the elements of positive degree in A/I must be nilpotent which implies $\sqrt{I} = \overline{A}$, so I is $\overline{A}$-primary. Let $[A]$ be a fundamental class for A. If J is a nonzero ideal of A/I and $u \neq 0 \in J$ then there is a w in A/I with $w \cdot u = [A]$, so $[A] \in J$. Since $[A]$ belongs to every nonzero ideal of A/I the zero ideal of A/I cannot be written as an intersection of two or more nonzero ideals. This lifts to A to say that I is not the intersection of proper ideals containing it, so I is irreducible. □

A **regular ideal** is one that is generated by a regular sequence. Cohen–Macaulay algebras are characterized by the fact that their parameter ideals are regular; the Gorenstein algebras are a special class of Cohen–Macaulay algebras.

DEFINITION: *A commutative graded connected Noetherian algebra over a field is called* **Gorenstein** *if it is Cohen–Macaulay and every parameter ideal is irreducible.*

REMARK: It is actually enough to check one parameter ideal is irreducible, in other words, if A is a commutative graded connected algebra over a field that is Cohen–Macaulay and one parameter ideal is irreducible, then every parameter ideal is irreducible (see e.g. the classic paper of H. Bass [6] or [68] Corollary 5.7.4).

The reason why Gorenstein algebras are relevant for us is clarified by the following elementary results.

PROPOSITION I.1.4: *A commutative graded connected Noetherian Cohen–Macaulay algebra A over a field is Gorenstein if and only if for every parameter ideal $I \subset A$ the quotient algebra $H = A/I$ satisfies Poincaré duality.*

In point of fact, the proof of this proposition shows a bit more is true, so we reformulate the positive portion of this result as follows.

PROPOSITION I.1.5: *Let A be a commutative graded connected Noetherian algebra over a field $\mathbb{F}$. Then for every $\overline{A}$-primary irreducible ideal $I \subset A$ the quotient algebra $H = A/I$ satisfies Poincaré duality.*

PROOF: Suppose that $I \subset A$ is an $\overline{A}$-primary irreducible ideal. There is no loss in generality in assuming that $I \neq \overline{A}$, so $H \neq \mathbb{F}$. Since $I \subset A$ is irreducible, $(0) \subset H$ is irreducible. Let J be the intersection of all the nonzero ideals of H. Since (0) is irreducible in H, it follows that $J \neq (0)$, and is therefore the unique nonzero minimal ideal of H. The totalization[2] of H is finite dimensional because I contains a power of $\overline{A}$ since A is Noetherian and I is $\overline{A}$-primary. Let $d \in \mathbb{N}$ be the largest integer such that $H_d \neq 0$. Then $\dim_{\mathbb{F}}(H_d) = 1$ since otherwise H would have more than one nonzero minimal ideal. (If $x, y \in H_d$ are linearly independent then $(x), (y) \subset H$ are distinct nonzero minimal ideals in H.) Let $[H] \in H_d$ be a basis vector. Then $J = ([H])$. If $w \neq 0 \in H$ then the ideal (w) contains J since J is the unique nonzero minimal ideal in H. So there is an element $u \in H$ such that $uw = [H]$ and H satisfies Poincaré duality with fundamental class $[H]$. □

---

[2] The totalization of H is the ungraded algebra $\oplus_{i=0}^{\infty} H_i$.

PROOF OF PROPOSITION I.1.4: If A is Gorenstein then parameter ideals are irreducible $\overline{A}$-primary so the positive assertion follows from Proposition I.1.5. Conversely, if $I \subset A$ is a parameter ideal and $H = A/I$ satisfies Poincaré duality with fundamental class $[H]$, then the principal ideal generated by $[H]$ is the unique nonzero minimal ideal of H. Therefore $(0) \subset H$ is irreducible in H which implies $I \subset A$ is irreducible in A. □

So the study of Poincaré duality quotient algebras of a Gorenstein algebra A is the same as the study of the $\overline{A}$-primary irreducible ideals of A.

I.2 Properties of irreducible ideals

If A is a commutative graded connected Noetherian algebra over a field $\mathbb{F}$ and $I \subset A$ is an ideal, we denote by over(I) the set of all over ideals of I, i.e.,

$$\text{over}(I) = \left\{J \mid J \subset A \text{ is an ideal, and } I \subseteq J\right\}.$$

Note that a proper ideal I is irreducible if and only if it has a unique minimal proper over ideal, i.e., if and only if there is a unique minimal element in the set over$(I) \setminus \{I\}$.

If I is an $\overline{A}$-primary irreducible ideal then the correspondence Ξ on ideals defined by

$$\Xi(J) = (I : J)$$

induces an involution[3]

$$\Xi : \text{over}(I) \longrightarrow \text{over}(I)$$

with the following properties:

$$\begin{aligned} J' \subseteq J'' \;&\Rightarrow\; \Xi(J'') \subseteq \Xi(J') \\ \Xi(J' \cap J'') &= \Xi(J') + \Xi(J'') \\ \Xi(J' + J'') &= \Xi(J') \cap \Xi(J'') \\ \left(\Xi(J') : J''\right) &= \Xi(J' \cdot J'') = \left(\Xi(J'') : J'\right) \\ \Xi(J' : J'') &= J'' \cdot \Xi(J') \end{aligned}$$

whenever J', $J'' \in$ over(I). Moreover, $J \in$ over(I) is irreducible if and only if $\Xi(J)$ is principal modulo I, i.e., $\Xi(J) = (a) + I$ for some element $a \in A$.

[3] According to W. Krull ([43] page 32) and W. Gröbner ([28] footnote 15 in Section III) the fact that Ξ is an involution in this generality is due to Emmy Noether, who felt her proof was too long and complicated to publish. The first published proof would appear to be [28] and a modern reference is [114] Chapter IV Theorems 34 and 35. Prior to this the result had been established for homogeneous ideals in polynomial rings by F. S. Macaulay (see [48] page 114). A proof in the graded case is much easier than the one in [114] so for the sake of completeness we include one in the text.

Following K. Kuhnigk [44] we call such an element $a \in A$ a **transition element** for J over I or a **transition invariant** for J over I. Such an invariant becomes unique in A/I up to a nonzero scalar. For the sake of easy reference in later sections we record this as a theorem. The proof yields a number of bonuses in the graded context appearing in the corollaries that follow.

THEOREM I.2.1: *Suppose that A is a commutative graded connected Noetherian algebra over a field and I is an $\overline{A}$-primary irreducible ideal. For $J \in \mathrm{over}(I)$ denote by $\Xi(J)$ the ideal $(I : J)$. Then $\Xi : \mathrm{over}(I) \rightleftarrows$ is an involution, and an ideal $J \in \mathrm{over}(I)$ is itself irreducible if and only if $\Xi(J)$ is principal over I.*

PROOF: Since I is an $\overline{A}$-primary irreducible ideal in A the ideal (0) is an $\overline{A/I}$-primary irreducible ideal in A/I. So if we pass down to A/I we may suppose without loss of generality that A is a Poincaré duality algebra and $I = (0)$. In this new context we need to prove the following three facts.

(i) The map Ξ defined on the ideals of A by $\Xi(J) = (0 : J) = \mathrm{Ann}_A(J)$ is an involution.
(ii) If $J \subset A$ is an irreducible ideal then $\mathrm{Ann}_A(J)$ is a principal ideal.
(iii) If $0 \neq a \in A$ then $\mathrm{Ann}_A(a) \subset A$ is an irreducible ideal.

The first statement is just Lemma I.1.1, so we turn to the second statement. Let $a_1, \ldots, a_m \in A$ be a minimal set of generators for the ideal $\mathrm{Ann}_A(J)$. By minimality none of the principal ideals $(a_1), \ldots, (a_m) \subset A$ is contained in one of the others. Since the map Ξ reverses inclusions none of the annihilators $\mathrm{Ann}_A(a_1), \ldots, \mathrm{Ann}_A(a_m) \subset A$ is contained in one of the others also. So, using that $\Xi(-) = \mathrm{Ann}_A(-)$ is an involution, we obtain

$$J = \mathrm{Ann}_A(\mathrm{Ann}_A(J)) = \mathrm{Ann}_A(a_1, \ldots, a_m) = \mathrm{Ann}_A(a_1) \cap \cdots \cap \mathrm{Ann}_A(a_m).$$

Since J is irreducible we must have $m = 1$ which means that $\mathrm{Ann}_A(J)$ is a principal ideal.

Choose a fundamental class $[A] \in A$. For $a \neq 0 \in A$ let $a^\vee$ be a Poincaré dual to a, i.e., $aa^\vee = [A]$. Set $B = A/\mathrm{Ann}_A(a)$ and let $[B]$ be the image of $a^\vee$ in B and $d = \deg([B])$. Suppose that $0 \neq b \in B$. Lift b to $\widehat{b} \in A$ and note that $\widehat{b} \notin \mathrm{Ann}_A(a)$, so $\widehat{b}a \neq 0 \in A$. In particular, this implies that $\deg(b) \leq d$ since $A_k = 0$ for $k > d + \deg(a)$. So there is an element $c \in A$ such that $c(\widehat{b}a) = [A] = aa^\vee$. Let $b^\#$ be the image of c in B. Since

$$0 = c\widehat{b}a - a^\vee a = (c\widehat{b} - a^\vee)a$$

we see that $c\widehat{b} - a^\vee \in \mathrm{Ann}_A(a)$ and hence $b^\# b = [B]$ in B, proving that B is a Poincaré duality algebra with fundamental class $[B]$. Therefore $\mathrm{Ann}_A(a) \subset A$ is an irreducible ideal by I.1.3. □

If A is a polynomial algebra over a field then $\mathfrak{m}$-primary regular ideals are irreducible. If $I \subset J$ are two such ideals then an explicit formula for a transition element for J over I is given in Proposition VI.3.1.

An $\overline{A}$-primary ideal $I \subset A$ in a Noetherian ring A contains a system of parameters for A, and also the ideal that they generate. Therefore in a Gorenstein algebra, since Ξ is an involution, the $\overline{A}$-primary irreducible ideals are all of the form $(Q : a)$ for some parameter ideal $Q \subset A$ and element $a \in A$.

If H is a Poincaré duality algebra over a field then the zero ideal $(0) \subset H$ is $\overline{H}$-primary and irreducible, and $(0 : I) = \operatorname{Ann}_H(I)$ for any ideal $I \subset H$. Therefore Theorem I.2.1 has the following consequence for Poincaré duality algebras. Indeed, this was shown in the proof of Theorem I.2.1.

COROLLARY I.2.2: *Let H be a Poincaré duality algebra over a field. Then taking annihilator ideals sets up a bijective correspondence between nonzero proper irreducible ideals $I \subset H$ and nonzero proper principal ideals $(h) \subset H$. In other words, a nonzero proper ideal $I \subset H$ is irreducible if and only if $\operatorname{Ann}_H(I) \subset H$ is a nonzero proper principal ideal.* □

Lemma I.1.1 allows us to reformulate Corollary I.2.2 using the bilinear form $\langle - \mid - \rangle$ given by taking products into the top degree of the Poincaré duality algebra H. Here is one way to do so conceived as an analog of *dualizing a line bundle* in topology. It records additional information about the formal dimension.

COROLLARY I.2.3: *Let H be a Poincaré duality algebra over a field and $0 \neq u \in H$. Then $H/\operatorname{Ann}_H(u)$ is a Poincaré duality algebra of formal dimension $\operatorname{f-dim}(H) - \deg(u)$. If $u^\vee \in H$ is a Poincaré dual for u, then the image of $u^\vee$ in $H/\operatorname{Ann}_H(u)$ is a fundamental class for $H/\operatorname{Ann}_H(u)$.* □

Rephrased in terms of transition elements this corollary takes the following form.

COROLLARY I.2.4: *Let A be a commutative graded connected Noetherian algebra A over a field. If $I \subset J$ are $\overline{A}$-primary irreducible ideals in A and $(I : J) = (a) + I$, so a is a transition invariant for J over I, then $\deg(a) = \operatorname{f-dim}(A/I) - \operatorname{f-dim}(A/J)$.* □

One final result that belongs to this circle of ideas will be of use in the sequel.

COROLLARY I.2.5: *Let H be a Poincaré duality algebra over a field and $0 \neq u \in H$. If $h \in H$ projects in $H/\mathrm{Ann}_H(u)$ to a fundamental class then $uh \in H$ represents a fundamental class of H.*

PROOF: Let $u^\vee$ be a Poincaré dual to u. The element $u^\vee - \lambda h$ lies in $\mathrm{Ann}_H(u)$ for a suitable $\lambda \in \mathbb{F}^\times$ so $0 = u(u^\vee - \lambda h)$ and hence $uu^\vee = \lambda uh$. The result follows from the fact that $uu^\vee$ represents a fundamental class of H. □

I.3 The ancestor ideals

The $\overline{A}$-primary irreducible ideals with Poincaré duality quotients of formal dimension d are determined by their degree d homogeneous components. To explain this we introduce a construction due to A. Iarrobino and V. Kanev [34] (see also [24]).

DEFINITION: *Let A be a commutative graded connected algebra over the field $\mathbb{F}$ and $W \subset A_d$ a vector subspace of the homogeneous component A_d of A of degree d. The* **little ancestor ideal** *of W, denoted by $\mathfrak{a}(W)$, is the ideal generated by the sets*

$$W,\ [W : A_1],\ \ldots,\ [W : A_{d-1}]$$

where, for vector subspaces $X \subset A_i$ and $Y \subset A_j$ with $j \leq i$, $[X : Y]$ denotes the set of all elements $a \in A_{i-j}$ such that $a \cdot Y \subseteq X$. The ideal $\mathfrak{a}(W) + A_{\geq d+1}$, where $A_{\geq k}$ denotes the ideal of all elements whose degree is greater than or equal to k, is called the **big ancestor ideal** *of W and is denoted by $\mathfrak{A}(W)$.*

The big ancestor ideal is an $\overline{A}$-primary extension of $\mathfrak{a}(W)$. In the special case $A = \mathbb{F}[V]$ note that $\mathbb{F}[V]_{\geq k} = \left(\overline{\mathbb{F}[V]}\right)^k$ so $\mathfrak{A} = \mathfrak{a} + \left(\overline{\mathbb{F}[V]}\right)^{d+1}$ in this case.

If H', H'' are Poincaré duality algebras over the field $\mathbb{F}$ of formal dimension d a map of algebras $f : H' \longrightarrow H''$ is said to have **degree one** if it maps a fundamental class to a fundamental class. Such a map must be a monomorphism as we next show.

LEMMA I.3.1: *Let H' and H'' be Poincaré duality algebras of the same formal dimension d and $\varphi : H' \longrightarrow H''$ a map of algebras. If $\varphi_d : H'_d \longrightarrow H''_d$ is an epimorphism, so φ has degree one, then φ is monic.*

PROOF: Suppose not, and let $u \neq 0 \in \ker(\varphi)$. Choose fundamental classes $[H'] \in H'_d$ and $[H''] \in H''_d$ such that $\varphi([H']) = [H'']$, which is possible since φ is an epimorphism in degree d. Let $u^\vee \in H'$ be such that $uu^\vee = [H']$. Then

$$0 = \varphi(u)\varphi(u^\vee) = \varphi(uu^\vee) = \varphi([H']) = [H'']$$

which is a contradiction. Therefore no such u can exist and φ is monic as claimed. □

PROPOSITION I.3.2: *Let A be a commutative graded connected Noetherian algebra over the field $\mathbb{F}$ and $I \subset A$ an $\overline{A}$-primary irreducible ideal with Poincaré duality quotient $H = A/I$ of formal dimension d. Then $I_d \subset A_d$ is a codimension one subspace and $I = \mathfrak{A}(I_d)$. Conversely, if $W \subset A_d$ is a codimension one subspace, then $\mathfrak{A}(W) \subset A$ is an $\overline{A}$-primary irreducible ideal with Poincaré duality quotient $A/\mathfrak{A}(W)$ of formal dimension d.*

PROOF: Suppose that $W \subset A_d$ is a codimension one vector subspace. Then the homogeneous component of $A/\mathfrak{A}(W)$ of degree d is one dimensional. Choose an element $a \in A_d \setminus W$. Then the residue class of a in $A/\mathfrak{A}(W)$ is a basis for the homogeneous component of degree d. Call this residue class $[A]$. Let $u \neq 0$ in $A/\mathfrak{A}(W)$ and choose $b \in A$ a representative for u. Since $u \neq 0$ it follows that $b \neq \mathfrak{A}(W)$ so there must exist an element $b^\vee$ of degree $d - \deg(b)$ such that $bb^\vee \notin W$. If $u^\vee$ denotes the residue class of $b^\vee$ then $uu^\vee = \lambda[A]$ for some nonzero $\lambda \in \mathbb{F}$, and hence $A/\mathfrak{A}(W)$ is a Poincaré duality algebra, so $\mathfrak{A}(W)$ is an irreducible ideal by Theorem I.2.1.

Conversely, suppose A/I is a Poincaré duality quotient of A of formal dimension d. Then $I_d \subset A_d$ is a codimension one subspace since A/I in degree d is one dimensional. It follows from the definition of the big ancestor ideal that $\mathfrak{A}(I_d)$ projects to zero in A/I. Hence $\mathfrak{A}(I_d) \subseteq I$, so there is an induced epimorphism $A/\mathfrak{A}(I_d) \longrightarrow A/I$. By Lemma I.3.1 this map is also a monomorphism, so we conclude $\mathfrak{A}(I_d) = I$. □

COROLLARY I.3.3: *Let A be a commutative graded connected Noetherian algebra over a field and I', $I'' \subset A$ be $\overline{A}$-primary and irreducible ideals in A. Let $d' = \text{f-dim}(A/I')$ and $d'' = \text{f-dim}(A/I'')$. Then $I' = I''$ if and only if $d' = d''$ and $I'_{d'} = I''_{d''}$.*

PROOF: Proposition I.3.2 implies $I' = \mathfrak{A}(I'_{d'})$ and $I'' = \mathfrak{A}(I''_{d''})$, so the result follows. □

I.4 Fundamental classes

In the study of Poincaré duality quotients of a commutative graded connected algebra A it is also possible to place the emphasis on a fundamental class of the Poincaré duality quotient rather than on the kernel of the quotient map as we have done up to this point. In fact one might wonder whether there are any restrictions on the representatives in A for a fundamental class of a Poincaré duality quotient of A. There are none, apart from the obvious requirement that the element be nonzero, as the next result shows.

PROPOSITION I.4.1: *Let A be a commutative graded connected Noetherian algebra over a field and $a \neq 0 \in A_d$. If I is maximal among all ideals not containing a then I is an $\overline{A}$-primary irreducible ideal and a represents a fundamental class of the Poincaré duality quotient A/I.*

PROOF: From the maximality of I we see that I contains all elements of A of degree strictly greater than d and that $I_d \subset A_d$ is a complementary subspace for $\mathrm{Span}_{\mathbb{F}}\{a\}$. So I_d has codimension one as a vector subspace of A_d. For an element $c \in A$ denote by $[c]$ its residue class in A/I. If $b \notin I$, then $I + (b)$ is strictly larger than I, so must contain a by the maximality of I. This means that $a = h + bb^{\vee}$ for some $h \in I$ and $b^{\vee}$ with $\deg(b) + \deg(b^{\vee}) = d$. In the quotient algebra A/I this says $[b] \cdot [b^{\vee}] = [a]$ so A/I is a Poincaré duality algebra with fundamental class $[a]$. □

Fix a graded connected commutative Noetherean algebra A over a field and a nonzero element $a \in A_d$. Propositions I.3.2 and I.4.1 say that the set of Poincaré duality quotients of A that admit a as a representative for their fundamental class are in bijective correspondence with the codimension one subspaces $W \subset A_d$ such that $W + \mathrm{Span}_{\mathbb{F}}\{a\} = A_d$. The set of all codimension one subspaces of A_d forms the **projective space** $\mathbb{P}(A_d)$ of the vector space A_d. So we can reformulate Proposition I.3.2 in the following way.

PROPOSITION I.4.2: *Let A be a commutative graded connected Noetherian algebra over a field and $d \in \mathbb{N}$. Then there is a bijective correspondence between Poincaré duality quotients A of formal dimension d and the points of the projective space $\mathbb{P}(A_d)$.*

PROOF: The desired bijection quite simply associates to a codimension one subspace $W \subset A_d$, first the ideal $\mathfrak{A}(W)$, and then the quotient $A/\mathfrak{A}(W)$. □

If what one wants is a moduli space for the *isomorphism classes* of Poincaré duality quotients of formal dimension d of a fixed commutative graded connected algebra A over the field $\mathbb{F}$, one needs to pass to an orbit space of the projective space. For the special case of $A = \mathbb{F}[V]$ this is made explicit in Section I.5 and exploited in Section I.6 to count the number of isomorphism classes if the ground field is a Galois field.

The complete intersection algebras arising as quotients of $\mathbb{F}[V]$ present an attractive class of Poincaré duality algebras. We collect some of their properties in the next example.

EXAMPLE 1: Suppose H is a Poincaré duality quotient of $\mathbb{F}[V]$ which is a complete intersection, viz., $H = \mathbb{F}[V]/(f_1, \ldots, f_n)$ where $f_1, \ldots, f_n \in \mathbb{F}[V]$ is a system of parameters. Since $\mathbb{F}[V]$ is Cohen–Macaulay $f_1, \ldots, f_n$ are a regular sequence in $\mathbb{F}[V]$, so algebraically independent, and $\mathbb{F}[V]$ is a free finitely generated $\mathbb{F}[f_1, \ldots, f_n]$-module (see e.g. [87] Section 6.2). Therefore the Poincaré series $P(H, t)$ is given by

$$P(H, t) = \frac{P(\mathbb{F}[V], t)}{P(\mathbb{F}[f_1, \ldots, f_n], t)} = \frac{\dfrac{1}{(1-t)^n}}{\dfrac{1}{\prod_{i=1}^{n}(1 - t^{\deg(f_i)})}}$$
$$= \prod_{i=1}^{n}(1 + t + \cdots + t^{\deg(f_i)-1}).$$

From this we see that the formal dimension of H is $\sum_{i=1}^{n}(\deg(f_i) - 1)$. In addition [86] (or see [87] Proposition 6.5.1) provides a means to compute a fundamental class for H. Namely, if we write

$$f_i = \sum_{j=1}^{n} h_{i,j} z_j \quad i = 1, \ldots, n,$$

where $z_1, \ldots, z_n \in V^*$ is a basis for the linear forms, then

$$[H] = \det(h_{i,j})$$

represents a fundamental class for H. If the characteristic of $\mathbb{F}$ does not divide the product $\deg(f_1) \cdots \deg(f_n)$ then the formula of Euler

$$\deg(f) \cdot f = \sum_{j=1}^{n} \frac{\partial f}{\partial z_j} z_j$$

shows that we may in fact take the Jacobian determinant

$$\det\left(\frac{\partial f_i}{\partial z_j}\right)$$

as a representative for a fundamental class. Simple examples show this can fail if the characteristic of $\mathbb{F}$ divides the product $\deg(f_1) \cdots \deg(f_n)$. For example take $f = z^p \in \mathbb{F}[z]$ where $\mathbb{F}$ has characteristic p. It can even fail for a ring of invariants. For example it fails for the usual algebra of symmetric functions $\mathbb{F}[z_1, \ldots, z_n]^{\Sigma_n}$ if the characteristic of the ground field $\mathbb{F}$ divides $|\Sigma_n| = n!$. In this modular situation the Jacobian determinant is the discriminant which belongs to the ideal generated by the elementary symmetric polynomials (see [25], [83], or [93]). For a related result see Section II.5 Example 3.

See Proposition VI.3.1 for an ideal theoretic generalization of this example and Section II.5 for another way to compute a fundamental class.

I.5 Poincaré duality quotients of $\mathbb{F}[V]$

We are principally interested in Poincaré duality quotient algebras H of $\mathbb{F}[V]$. If H has formal dimension d and $[H] \in H$ is a fundamental class then we may use $[H]$ to identify H_d with $\mathbb{F}$, so the quotient map $\mathbb{F}[V]_d \longrightarrow H_d$ defines a linear form $\gamma : \mathbb{F}[V]_d \longrightarrow \mathbb{F}$. The following lemma allows us to classify the Poincaré duality quotients H of $\mathbb{F}[V]$ up to isomorphism from this viewpoint.

LEMMA I.5.1: *Suppose that* $\mathbb{F}[V'] \xrightarrow{\pi'} H'$ *and* $\mathbb{F}[V''] \xrightarrow{\pi''} H''$ *are Poincaré duality quotient algebras of formal dimension* d*. Choose fundamental classes for* H' *and* H'' *and let* $\gamma', \gamma'' : \mathbb{F}[V]_d \longrightarrow \mathbb{F}$ *be the associated linear forms. If* $\varphi : H' \longrightarrow H''$ *is a map of algebras such that* $\varphi([H']) = [H'']$ *then there exists a map* $\Phi : V'' \longrightarrow V'$ *of vector spaces such that the diagram*

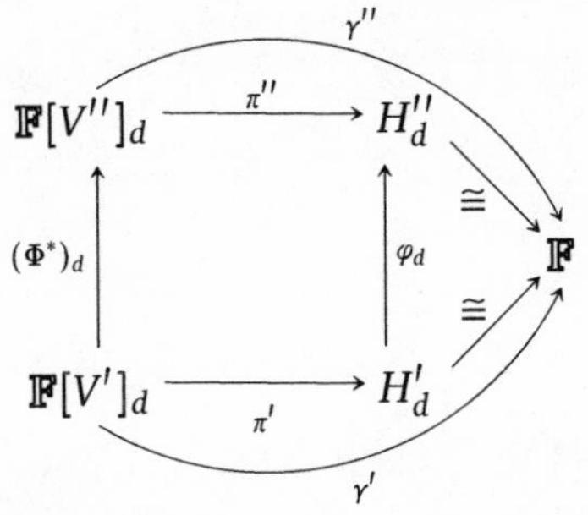

commutes. If $\dim_{\mathbb{F}}(V') = \dim_{\mathbb{F}}(V'')$ *and* φ *is an isomorphism, then we may choose* Φ *to be an isomorphism. If* $\operatorname{rank}(H') = \dim_{\mathbb{F}}(V') = \dim_{\mathbb{F}}(V'') = \operatorname{rank}(H'')$ *then* Φ *is unique.*

PROOF: The existence of a linear map completing the diagram

$$\begin{array}{ccc} V''^{*} = \mathbb{F}[V'']_1 & \xrightarrow{\pi''} & H''_1 \\ \vdots\uparrow & & \uparrow \varphi_1 \\ V'^{*} = \mathbb{F}[V']_1 & \xrightarrow{\pi'} & H'_1 \end{array}$$

is elementary linear algebra. Its dual has all the asserted properties. □

We next show how a streamlined version of the big ancestor ideal can be used to construct all examples of Poincaré duality quotient algebras of $\mathbb{F}[V]$. As with any such general construction it gives rather little insight into the structure of any particular algebra, and raises more questions than it answers.

Let d be a natural number and $\gamma : \mathbb{F}[V]_d \longrightarrow \mathbb{F}$ a nonzero linear map defined on the degree d component of $\mathbb{F}[V]$. We associate to γ an ideal $I(\gamma) \subset \mathbb{F}[V]$ as follows:

$$f \in I(\gamma) \Longleftrightarrow \forall\, h \in \mathbb{F}[V]_{d-\deg(f)} \text{ one has } \gamma(f \cdot h) = 0.$$

Let $H(\gamma) = \mathbb{F}[V]/I(\gamma)$ be the quotient algebra. Notice that the definition puts all the elements in $\mathbb{F}[V]$ of degree $> d$ into $I(\gamma)$, and the elements of $\mathbb{F}[V]_d$ that are not in $I(\gamma)$ are the forms f of degree d for which $\gamma(f) \neq 0$. A form of degree k, with $k < d$, lies in $I(\gamma)$ if and only if its product with all forms of degree $d-k$ lies in $\ker\{\gamma : \mathbb{F}[V]_d \longrightarrow \mathbb{F}\}$, so $I(\gamma) = \mathfrak{A}(\ker(\gamma))$, and from Proposition I.4.2 and Lemma I.5.1 one obtains the following.

PROPOSITION I.5.2: *Let $\mathbb{F}$ be a field, $n,\ d \in \mathbb{N}$ positive integers, $V = \mathbb{F}^n$, and $\gamma : \mathbb{F}[V]_d \longrightarrow \mathbb{F}$ a nonzero linear map. Then $H(\gamma)$ is a Poincaré duality quotient algebra of $\mathbb{F}[V]$. Every Poincaré duality quotient of $\mathbb{F}[V]$ of formal dimension d arises from some nonzero linear form $\gamma : \mathbb{F}[V]_d \longrightarrow \mathbb{F}$ in this way. Two nonzero linear forms $\gamma',\ \gamma'' : \mathbb{F}[V]_d \longrightarrow \mathbb{F}$ define the same ideal $I(\gamma)$ if and only if they are nonzero scalar multiples of each other. They define isomorphic quotient algebras if and only if they are in the same* $\mathrm{GL}(n, \mathbb{F})$*-orbit of the action of* $\mathrm{GL}(n, \mathbb{F})$ *on the space of linear forms on* $\mathbb{F}[V]_d$. □

So to construct a Poincaré duality quotient H of $\mathbb{F}[V]$ of formal dimension d is the same thing as to give a nonzero linear form $\mathbb{F}[V]_d \xrightarrow{\gamma} \mathbb{F}$ which defines the corresponding ideal $I(\gamma)$ that is the kernel of the induced epimorphism $\mathbb{F}[V] \longrightarrow H$. By way of example, if the maps $\mathbb{F}[V']_d \xrightarrow{\gamma'} \mathbb{F} \xleftarrow{\gamma''} \mathbb{F}[V'']_d$ define the Poincaré quotients H' of $\mathbb{F}[V']$ and H'' of $\mathbb{F}[V'']$, then the algebra $H' \# H''$ is a Poincaré quotient of $\mathbb{F}[V' \oplus V''] \cong \mathbb{F}[V'] \otimes \mathbb{F}[V'']$. It is defined by the linear form $\gamma' \# \gamma''$ where

$$(\gamma' \# \gamma'')(f' \otimes f'') = \begin{cases} \gamma'(f') \otimes f'' & \text{if } \deg(f') = d \text{ and } \deg(f'') = 0 \\ f' \otimes \gamma''(f'') & \text{if } \deg(f') = 0 \text{ and } \deg(f'') = d \\ 0 & \text{otherwise.} \end{cases}$$

Let H be a Poincaré duality quotient of $\mathbb{F}[V]$ of formal dimension d defined by the linear form $\gamma : \mathbb{F}[V]_d \longrightarrow \mathbb{F}$. So $H = \mathbb{F}[V]/I(\gamma)$. By Lemma I.5.1 any other Poincaré duality quotient of $\mathbb{F}[V]$ that is isomorphic to H must be of the form $\mathbb{F}[V]/\Phi^*(I(\gamma))$ for some isomorphism $\Phi : V \longrightarrow V$. The number of distinct such quotients is the index of the isotropy subgroup $\mathrm{GL}(V)_{I(\gamma)}$ of the ideal $I(\gamma) \subset \mathbb{F}[V]$ for the natural action of $\mathrm{GL}(V)$ on $\mathbb{F}[V]$, i.e.,

$$\mathrm{GL}(V)_{I(\gamma)} = \{\Phi \in \mathrm{GL}(V) \mid \Phi(I) = I\}$$

is the subgroup that stabilizes $I(\gamma)$ setwise.

Here is a simple example to illustrate some of the concepts introduced up to this point.

EXAMPLE 1 (F. S. Macaulay [49] Section 71): Consider the linear function on the quadratic forms in n variables defined by specifying its value on the monomial basis $\{z_i z_j \mid i \leq j\}$ as follows

$$\mathbb{F}[z_1, \ldots, z_n]_2 \xrightarrow{\gamma_n} \mathbb{F} \qquad \gamma_n(z_i z_j) = \begin{cases} 1 & \text{if } i = j \\ 0 & \text{otherwise.} \end{cases}$$

If the characteristic of $\mathbb{F}$ is not 2, then γ_n can be thought of as half the Laplacian $\frac{1}{2}[\frac{\partial^2}{\partial z_1^2} + \cdots + \frac{\partial^2}{\partial z_n^2}]$ regarded as a map of quadratic forms to field elements. One easily sees that

$$I(\gamma_n) = (z_1^2 - z_2^2, \ldots, z_1^2 - z_n^2, z_i z_j \mid i < j),$$

which means that $H(\gamma_n) \cong \mathbb{F}[z_1]/(z_1^3) \# \cdots \# \mathbb{F}[z_n]/(z_n^3)$. This calculation can be done with the catalecticant matrix method described in Section VI.2. See in particular Example 1 in that section, as well as [34] Example 1.20, and [24] Lecture 9. The algebra $H(\gamma_n)$ is a complete intersection for $n = 1$ or 2, but not for $n \geq 3$. These are Macaulay's original examples of $\mathfrak{m}$-primary irreducible ideals that are not regular ([49] Section 71).

I.6 Counting Poincaré duality quotients up to isomorphism

Over a finite field $\mathbb{F}_q$ there can be only a finite number of Poincaré duality quotients of $\mathbb{F}_q[V]$ of a given formal dimension up to isomorphism. So instead of asking for a moduli space as we did in Section I.5 one could ask for a count. As an application of Proposition I.5.2 we describe an invariant theoretic procedure to compute the number of nonisomorphic Poincaré duality quotients of $\mathbb{F}_q[z_1, \ldots, z_n]$ of formal dimension d, where $\mathbb{F}_q$ is the Galois field with q elements. This leads to a number of interesting problems in invariant theory which we are able to solve in a few cases. We present the solution for $n = 2 = q$ and comment on a couple of other *small* cases.

The starting point is the Cauchy–Frobenius lemma whose proof is an elementary counting argument (see e.g. [96] volume II page 404). For a finite set Y write $|Y|$ for the number of elements in Y. If the group G acts on a set X then X/G is the set of G-orbits and is called the **orbit space**. If $g \in G$ then X^g denotes the subset of X whose elements are fixed by g.

LEMMA I.6.1 (A. Cauchy, G. Frobenius): *Let G be a finite group and X a finite G set. Then* $|X/G| = \frac{1}{|G|} \sum_{g \in G} |X^g|$. □

Proposition I.5.2 tells us that to count the number of isomorphism classes of Poincaré duality quotients of $\mathbb{F}[z_1, \ldots, z_n]$ of formal dimension d we must count the number of $\mathrm{GL}(n, \mathbb{F}_q)$ orbits on the space of linear forms on $\mathbb{F}_q[z_1, \ldots, z_n]_d$. As a first step we show that this is equivalent to counting orbits on $\mathbb{F}_q[z_1, \ldots, z_n]_d$ itself.

LEMMA I.6.2: *Let $\mathbb{F}_q$ be a finite field and $g \in \mathrm{GL}(n, \mathbb{F}_q)$. Denote by $g^{\mathrm{tr}} \in \mathrm{GL}(n, \mathbb{F}_q)$ the transpose of g and set $V = \mathbb{F}_q^n$. Then* $\dim_{\mathbb{F}_q}(V^g) = \dim_{\mathbb{F}_q}(V^{g^{\mathrm{tr}}})$.

PROOF: Let $\mathbb{F}_{\bar{q}} \mid \mathbb{F}_q$ be a field extension that contains all the eigenvalues of g. Since for any $h \in \mathrm{GL}(n, \mathbb{F}_q)$ one has $\dim_{\mathbb{F}_q}(V^h) = \dim_{\mathbb{F}_{\bar{q}}}((V \otimes_{\mathbb{F}_q} \mathbb{F}_{\bar{q}})^h)$ we may start over replacing $\mathbb{F}_q$ by $\mathbb{F}_{\bar{q}}$. Then the characteristic polynomial of g splits into linear factors, and g can be put into Jordan normal form. Let $\mathbf{J} \in \mathrm{GL}(n, \mathbb{F}_q)$ be the Jordan form of g. Then g and $\mathbf{J}$ are similar, as are g^{tr} and $\mathbf{J}^{\mathrm{tr}}$. Since

$$\dim_{\mathbb{F}_q}(V^g) = \dim_{\mathbb{F}_q}(V^{\mathbf{J}})$$

and

$$\dim_{\mathbb{F}_q}(V^{g^{\mathrm{tr}}}) = \dim_{\mathbb{F}_q}(V^{\mathbf{J}^{\mathrm{tr}}})$$

it will be enough to show that

$$\dim_{\mathbb{F}_q}(V^{\mathbf{J}}) = \dim_{\mathbb{F}_q}(V^{\mathbf{J}^{\mathrm{tr}}}).$$

The matrix $\mathbf{J}$ is a block diagonal matrix,

$$\mathbf{J} = \begin{bmatrix} \mathbf{J}_1 & & 0 \\ & \ddots & \\ 0 & & \mathbf{J}_s \end{bmatrix}$$

where each block $\mathbf{J}_i$ has the form

$$\mathbf{J}_i = \begin{bmatrix} \lambda_i & \varepsilon_{i,1} & 0 & \cdots \\ 0 & \ddots & & 0 \\ 0 & \cdots & \lambda_i & \varepsilon_{i,k} \\ 0 & \cdots & 0 & \lambda_i \end{bmatrix}$$

where $\lambda_i \in \mathbb{F}_q^{\times}$ is an eigenvalue of $\mathbf{J}$ and $\varepsilon_{i,j} \in \{0, 1\}$. Inspection shows that

$\dim_{\mathbb{F}_q}(V^{\mathbf{J}})$ is the cardinality of the set

$$\{(i, j) \mid \lambda_i = 1 \text{ and } \varepsilon_{i,j} = 0\} \bigcup \{i \mid \lambda_i = 1\}.$$

Passing from $\mathbf{J}$ to $\mathbf{J}^{\mathrm{tr}}$ does not change this count, which becomes $\dim_{\mathbb{F}_q}(V^{\mathbf{J}^{\mathrm{tr}}})$. □

LEMMA I.6.3: *Let $\mathbb{F}_q$ be a finite field and $g \in \mathrm{GL}(n, \mathbb{F}_q)$. Set $V = \mathbb{F}_q^n$ and let V^* be the dual vector space. Then $\dim_{\mathbb{F}_q}(V^g) = \dim_{\mathbb{F}_q}((V^*)^g)$.*

PROOF: Let $E_1, \ldots, E_n \in V$ be the standard basis and $E_1^*, \ldots, E_n^* \in V^*$ the corresponding dual basis. The matrix of g acting on V^* with respect to the basis $E_1^*, \ldots, E_n^*$ is just g^{tr} so the result follows from Lemma I.6.2. □

PROPOSITION I.6.4: *Let $\rho : G \hookrightarrow \mathrm{GL}(n, \mathbb{F}_q)$ be a representation of a finite group over the finite field $\mathbb{F}_q$. Set $V = \mathbb{F}_q^n$ and let V^* be the dual vector space. Then $|V/G| = |V^*/G|$.*

PROOF: Both V and V^* are finite G-sets so we may use the Cauchy–Frobenius lemma, Lemma I.6.1 to count the number of orbits, viz.,

$$|V/G| = \frac{1}{|G|}\sum_{g\in G} |V^g|$$

$$|V^*/G| = \frac{1}{|G|}\sum_{g\in G} |(V^*)^g|.$$

The number of elements in $\mathbb{F}_q$ is q, so

$$|V^g| = q^{\dim_{\mathbb{F}_q}(V^g)}$$

$$|(V^*)^g| = q^{\dim_{\mathbb{F}_q}((V^*)^g)}$$

for any $g \in G$, and it is enough to show that

$$\dim_{\mathbb{F}_q}(V^g) = \dim_{\mathbb{F}_q}((V^*)^g).$$

This is the content of Lemma I.6.3. □

EXAMPLE 1: Let $\rho : G \hookrightarrow \mathrm{GL}(n, \mathbb{F}_q)$ be a representation of a finite group over the finite field $\mathbb{F}_q$, and $V = \mathbb{F}_q^n$ with V^* the dual vector space. Although Proposition I.6.4 tells us that the numbers of orbits of G on V and V^* are the same, it does *not* tell us anything about their structure. It certainly should not, as the orbit structures can be quite different as we show next. Choose

$$G = \left\{ \begin{bmatrix} 1 & 0 & * \\ 0 & 1 & * \\ 0 & 0 & 1 \end{bmatrix} \text{ where } * \in \mathbb{F}_2 \right\} < \mathrm{GL}(3, \mathbb{F}_2).$$

Abstractly G is $\mathbb{Z}/2 \oplus \mathbb{Z}/2$. The embedding into $\mathrm{GL}(3, \mathbb{F}_2)$ makes it a transvection group, i.e., a group in which the nonidentity elements are transvections (see e.g. [87] Section 8.2 or [68] Section 6.2). If $x, y, z \in V$ denote the standard basis vectors, then the 8 elements of $V = \mathbb{F}_2^3$ decompose into 5 orbits as follows.

Orbit	1	2	3	4	5
Elements	0	x	y	$x+y$	$z,\ z+x,\ z+y,\ z+x+y$
Cardinality	1	1	1	1	4

TABLE I.6.1: Orbits of G on V

By contrast, if $x^*, y^*, z^* \in V^*$ is the dual basis, then the 8 elements of V^* decompose also into 5 orbits, but as follows.

Orbit	1	2	3	4	5
Elements	0	z^*	$y^*,\ y^*+z^*$	$x^*,\ x^*+z^*$	$x^*+y^*,\ x^*+y^*+z^*$
Cardinality	1	1	2	2	2

TABLE I.6.2: Orbits of G on V^*

So although the numbers of orbits are the same, as they should be, their cardinalities are not.

The preceding discussion leads naturally to the following result.

THEOREM I.6.5: *The number of orbits of* $\mathrm{GL}(n, \mathbb{F}_q)$ *acting on the homogeneous polynomials of degree* d *in* $\mathbb{F}_q[z_1, \ldots, z_n]$ *is the same as the number of orbits of* $\mathrm{GL}(n, \mathbb{F}_q)$ *acting on the space of linear forms on* $\mathbb{F}_q[z_1, \ldots, z_n]_d$. *Therefore the number of isomorphism classes of Poincaré duality quotient algebras of* $\mathbb{F}_q[z_1, \ldots, z_n]$ *of formal dimension* d *is equal to* $\left|\mathbb{F}_q[z_1, \ldots, z_n]_d/\mathrm{GL}(n, \mathbb{F}_q)\right| - 1$. □

So to count the number of isomorphism classes of Poincaré duality quotients of $\mathbb{F}_q[z_1, \ldots, z_n]$ of formal dimension d we need to count the number of $\mathrm{GL}(n, \mathbb{F}_q)$-orbits on $\mathbb{F}[z_1, \ldots, z_n]_d$. Lemma I.6.1 gives us a formula for the number of $\mathrm{GL}(n, \mathbb{F}_q)$-orbits on $\mathbb{F}_q[z_1, \ldots, z_n]_d$, viz.,

$$\left|\mathbb{F}_q[z_1, \ldots, z_n]_d/\mathrm{GL}(n, \mathbb{F}_q)\right| = \frac{1}{|\mathrm{GL}(n, \mathbb{F}_q)|} \sum_{g \in \mathrm{GL}(n, \mathbb{F}_q)} \left|\mathbb{F}_q[z_1, \ldots, z_n]_d^g\right|.$$

The number of fixed points of $g \in \mathrm{GL}(n, \mathbb{F}_q)$ on $\mathbb{F}_q[z_1, \ldots, z_n]_d$ can be computed by exponentiating the dimension of the fixed point set in degree d of the ring of invariants of the subgroup generated by g. The preceding sum can be further simplified[4] by summing over one representative for each conjugacy class of elements of $\mathrm{GL}(n, \mathbb{F}_q)$. To use this method one needs to compute

(1) a transversal $g_1, \ldots, g_k$ for the conjugacy classes of $\mathrm{GL}(n, \mathbb{F}_q)$,
(2) for each conjugacy class, the number of elements it contains, and
(3) for each g_i the Poincaré series of the ring of invariants of the cyclic group C_i generated by g_i.

Here is an example to show how this works. The group $\mathrm{GL}(2, \mathbb{F}_2)$ is the symmetric group Σ_3. The conjugacy classes of elements in standard permutation group notation are listed in the following table along with other useful information.

Conjugacy Class	χ_1	χ_2	χ_3
Elements	id	(12), (23), (13)	(123), (132)
Number of Elements	1	3	2
Order of Elements	1	2	3
Generic Representative	id	$\tau = (12)$	$\xi = (123)$

TABLE I.6.3: $\mathrm{GL}(2, \mathbb{F}_2)$

From this we derive the following formula for the number of isomorphism classes of Poincaré duality quotients of $\mathbb{F}_2[x, y]$ of formal dimension d:

$$(\because)\qquad \frac{1}{6}\left[2^{d+1} + 3\cdot 2^{\dim_{\mathbb{F}_2}(\mathbb{F}_2[x,y]_d^{\tau})} + 2\cdot 2^{\dim_{\mathbb{F}_2}(\mathbb{F}_2[x,y]_d^{\xi})}\right] - 1.$$

So we need to compute the dimensions of $\mathbb{F}_2[x, y]_d^{\tau}$ and $\mathbb{F}_2[x, y]_d^{\xi}$.

Let us handle the involution first. Since τ permutes x and y the ring of invariants is $\mathbb{F}_2[x + y, xy]$ and this has Poincaré series

$$P(\mathbb{F}_2[x + y, xy], t) = \frac{1}{(1-t)(1-t^2)} = \sum_{i=0}^{\infty}(i + 1)t^{2i} + \sum_{i=0}^{\infty}(i + 1)t^{2i+1}.$$

Therefore we obtain

$$\dim_{\mathbb{F}_2}(\mathbb{F}_2[x, y]_d^{\tau}) = \left[\frac{d}{2}\right] + 1.$$

[4] Other ways of rearranging this sum may be found in [42]. For example, one could employ Burnside's Lemma, [42] Lemma 3.1.4, and sum over the conjugacy classes of subgroups instead.

The three cycle ξ can be lifted to characteristic zero. The Poincaré series of the rings of invariants in characteristic 0 and 2 are the same (see e.g. [68] Section 3.1, in particular Proposition 3.1.2). In characteristic 0 the computations may be made with Molien's theorem ([87] Section 4.3). Here are the details of this method.

We first pass from the ground field $\mathbb{F}_2$ to the quadratic extension $\mathbb{F}_4$ which contains a third root of unity ζ. Over $\mathbb{F}_4$ we can diagonalize ξ obtaining the matrix

$$\begin{bmatrix} \zeta & 0 \\ 0 & \zeta^{-1} \end{bmatrix}.$$

To obtain a Brauer lift identify ζ with the complex third root of unity $\frac{-1+\sqrt{-3}}{2}$. Then the matrix

$$\mathbf{A} = \begin{bmatrix} \zeta & 0 \\ 0 & \zeta^{-1} \end{bmatrix} \in \mathrm{GL}(2, \mathbb{Z}(\zeta))$$

provides a lift to characteristic 0 of $\xi \in \mathrm{GL}(2, \mathbb{F}_4)$. Namely, $2 \in \mathbb{Z}(\zeta)$ is prime and $\mathbb{Z}(\zeta)/(2) \cong \mathbb{F}_4$, so passing to the quotient field the matrix $\mathbf{A}$ becomes conjugate to ξ. The ring of invariants of the lifted action to $\mathbb{C}[x, y]$ of ξ is easily computed (see e.g. [87] Chapter 4 Section 3 Example 3). One finds

$$\mathbb{C}[x, y]^{\xi} = \mathbb{C}[x^3, y^3] \oplus \mathbb{C}[x^3, y^3] \cdot xy \oplus \mathbb{C}[x^3, y^3] \cdot (xy)^2.$$

For the Poincaré series of $\mathbb{C}[x, y]^{\xi}$ we therefore have

$$\begin{aligned} P(\mathbb{C}[x, y]^{\xi}, t) &= \frac{1 + t^2 + t^4}{(1-t^3)^2} \\ &= \sum_{i=0}^{\infty}(i+1)t^{3i} + \sum_{I=0}^{\infty}(i+1)t^{3i+2} + \sum_{i=0}^{\infty}(i+1)t^{3i+4}. \end{aligned}$$

Hence

$$\dim_{\mathbb{F}_2}(\mathbb{F}_2[x, y]_d^{\xi}) = \begin{cases} \left[\frac{d}{3}\right] + 1 & \text{for } d \equiv 0, 2 \bmod 3 \\ \left[\frac{d}{3}\right] & \text{for } d \equiv 1 \bmod 3. \end{cases}$$

Putting all this information into the formula (∵) and applying Proposition I.5.2 yields the following result.

PROPOSITION I.6.6: *The number of nonisomorphic Poincaré duality quotient algebras of $\mathbb{F}_2[x, y]$ of formal dimension d is*

$$\frac{1}{6}\left[2^{d+1} + 3 \cdot 2^{\left[\frac{d}{2}\right]+1} + 2 \cdot 2^{\left[\frac{d}{3}\right]+1}\right] - 1 \quad \text{for } d \equiv 0, 2 \bmod 3$$

$$\frac{1}{6}\left[2^{d+1} + 3 \cdot 2^{\left[\frac{d}{2}\right]+1} + 2 \cdot 2^{\left[\frac{d}{3}\right]}\right] - 1 \quad \text{for } d \equiv 1 \bmod 3. \quad \square$$

For example, if the formal dimension is $d = 8$, we find there are

$$\frac{1}{6}\left[2^9 + 3 \cdot 2^{4+1} + 2 \cdot 2^{2+1}\right] - 1 = \frac{1}{6}[512 + 96 + 16] - 1 = \frac{624}{6} - 1 = 103$$

distinct nonisomorphic Poincaré duality quotients of $\mathbb{F}_2[x, y]$.

A similar analysis can be carried out for other low rank examples over small fields, such as $\mathrm{GL}(3, \mathbb{F}_2)$, $\mathrm{GL}(2, \mathbb{F}_3)$ and $\mathrm{GL}(2, \mathbb{F}_5)$. In each of these cases it is not the p-Sylow subgroup that causes the most trouble, but the *large* cyclic subgroups. What makes the computations possible is that in these cases there is a Hall subgroup whose index is the order of the p-Sylow subgroup and there are not too many conjugacy classes of elements to worry about. For $\mathrm{GL}(3, \mathbb{F}_2)$ we obtain after a tedious computation a recursion of length 36 for the number of Poincaré duality quotients of a given formal dimension. In other words, if $\Upsilon(d)$ denotes the number of isomorphism classes of Poincaré duality quotients of $\mathbb{F}_2[x, y, z]$ of formal dimension d, and $d \geq 36$, then there is a formula expressing $\Upsilon(d)$ in terms of $\Upsilon(d-36)$. The first 36 values $\Upsilon(0), \dots, \Upsilon(35)$ were computed individually and tabulated. There is little new to be learned from these computations, so we do not include them. We do think however that the problem of counting the number of $\mathrm{GL}(n, \mathbb{F}_q)$ orbits on $\mathbb{F}_q[z_1, \dots, z_n]$ is an interesting problem worthy of more attention.

Part II
Macaulay's dual systems and Frobenius powers

WE BEGIN this part with an exposition of Macaulay's theory of irreducible ideals in polynomial algebras, [49] Part IV. Curiously enough, despite the passage of time, the original treatment is the closest to what we require, but the mathematical terminology used by Macaulay is no longer current. Of the several possible presentations of F. S. Macaulay's theory in current terminology we have chosen a Hopf algebra approach, as it is best adapted to our study of Poincaré duality quotients of $\mathbb{F}[V]$. Macaulay used the term *inverse system* but we will use the phrase *dual system* (see Section II.2) which seems more appropriate to the context of Hopf algebra duality. We use the theory of dual systems to provide a computational tool, the $K \subset L$ paradigm, that allows us to deduce information about $\mathbb{F}[V]/L$ from information about $\mathbb{F}[V]/K$, where $K \subset L$ are irreducible ideals that are $\overline{\mathbb{F}[V]}$-primary. For alternative expositions of Macaulay's theory see [50], Macaulay's own *translation* of [49] into the language of *modern algebra* à la Emmy Noether [69]. Other contemporary treatments may be found in [34], [24], [21], and [27]. In Section VI.1, we explain Macaulay's original version using *inverse polynomials*.

In some sense the parameter ideals in $\mathbb{F}[V]$ provide the simplest examples of $\overline{\mathbb{F}[V]}$-primary irreducible ideals. From such an ideal $Q = (f_1, \ldots, f_n)$ one can create new $\overline{\mathbb{F}[V]}$-primary irreducible ideals by taking powers of the generators. The methods employed in Example 1 of Section I.4 allow one to compute the formal dimension, fundamental classes, and so on for the new Poincaré duality quotient from the original data. In the case where the ground field $\mathbb{F}$ has characteristic p it turns out that the Frobenius power $I^{[p]} = (f^p \mid f \in I)$ of an arbitrary irreducible ideal I is again irreducible. This was first shown to us by R. Y. Sharp in the ungraded case (see Section II.6).

We use the $K \subset L$ paradigm to deduce a number of additional relations between the Poincaré duality quotients $\mathbb{F}[V]/I$ and $\mathbb{F}[V]/I^{[p]}$. In particular we relate generators of their Macaulay duals and their fundamental classes by means of simple formulae (see Theorem II.6.6). We also show that $\mathbb{F}[V]/I^{[p]}$ can be expressed using a graded version of the Frobenius functor of [73] shown to us by N. Nossem in terms of $\mathbb{F}[V]/I$ and $\mathbb{F}[V]/(V)^{[p]}$ (see Proposition II.6.7).

II.1 Divided power algebras and operations

We devote this section to some preliminaries needed for our discussion of Macaulay's theory of $\overline{\mathbb{F}[V]}$-primary irreducible ideals. References for this material are: [49] Chapter IV, [21] pages 526–527 and Exercise 21.7, as well as [24]. Our intention is to present Macaulay's theory in the language of Hopf algebra duality. To do so we first define a Hopf algebra structure on $\mathbb{F}[V]$ and review the structure of its dual Hopf algebra, the divided power algebra $\Gamma(V)$.

The polynomial algebra $\mathbb{F}[V]$ can be made into a Hopf algebra as in [59] by demanding that the coproduct[1] $\nabla : \mathbb{F}[V] \longrightarrow \mathbb{F}[V] \otimes \mathbb{F}[V]$ be a map of algebras and satisfy $\nabla(z) = z \otimes 1 + 1 \otimes z$ for any linear form $z \in V^*$. The dual Hopf algebra is denoted by $\Gamma(V)$ and the canonical pairing $\Gamma(V)_i \times \mathbb{F}[V]_i \longrightarrow \mathbb{F}$ by $<- \mid ->$. The algebra $\Gamma(V)$ is a **divided power algebra** (see [13] Exposé 7) and its multiplication may be described as follows: let $z_1, \ldots, z_n \in V^*$ be a basis for the space of linear forms and denote the associated monomial basis for $\mathbb{F}[V]$ by $\{z^E = z_1^{e_1} \cdots z_n^{e_n} \mid E = (e_1, \ldots, e_n) \in \mathbb{N}_0^n\}$. Let $\{\gamma_E \in \Gamma(V) \mid E \in \mathbb{N}_0^n\}$ be the dual basis for $\Gamma(V)$, so that $<\gamma_E \mid z^F> = \delta_{E,F}$. The multiplication of $\Gamma(V)$ is determined by the rule

$$\gamma_{E'} \cdot \gamma_{E''} = \binom{E' + E''}{E'} \gamma_{E'+E''},$$

where

$$\binom{A+B}{B} = \frac{(A+B)!}{A!B!} = \binom{A+B}{A},$$

and

$$C! = c_1! \cdots c_n! \qquad \forall\, C \in \mathbb{N}_0^n.$$

The structure of the divided power algebra is dependent on the characteristic of the ground field $\mathbb{F}$. If $\mathbb{F}$ has characteristic zero then a simple change of basis shows that $\Gamma(V)$ is a polynomial algebra on the generators

[1] The terminology comultiplication or diagonal is also used for the coproduct.

$\gamma_{\Delta_i} \in \Gamma(V)_1$, where for $i = 1, \ldots, n$, $\Delta_i = (\delta_{1,i}, \ldots, \delta_{n,i}) \in \mathbb{N}_0^n$. By contrast, if $\mathbb{F}$ has characteristic $p \neq 0$ then $\gamma^p = 0$ for all $\gamma \in \Gamma(V)$. Hence, in characteristic $p \neq 0$, $\overline{\Gamma(V)}$ is the nil radical of $\Gamma(V)$ and $\Gamma(V)$ is *not* Noetherian.

To make this a bit more explicit, let $z_1, \ldots, z_n \in V^*$ be a basis. Then we have $\mathbb{F}[V] \cong \mathbb{F}[z_1] \otimes \cdots \otimes \mathbb{F}[z_n]$ as Hopf algebras. If $u_1, \ldots, u_n \in \Gamma(V)_1$ is the basis dual to $z_1, \ldots, z_n$ then $\Gamma(V) \cong \Gamma(u_1) \otimes \cdots \otimes \Gamma(u_n)$. If the ground field has characteristic zero then, in the case of a single variable u one has in $\Gamma(u)$ the formula $\gamma_k = \frac{1}{k!}\gamma_1^k$, and hence one sees that $\Gamma(u)$ is a polynomial algebra (cf. the theorem of Samelson and Leray [59] Theorem 7.20 or [13] Exposé 7). If the ground field $\mathbb{F}$ has characteristic $p \neq 0$ then

$$\Gamma(u) \cong \frac{\mathbb{F}[u]}{(u^p)} \otimes \frac{\mathbb{F}[u^p]}{(u^{p^2})} \otimes \cdots \otimes \frac{\mathbb{F}[u^{p^k}]}{(u^{p^{k+1}})} \otimes \cdots$$

as an algebra and the diagonal is determined on the algebra generators by $\nabla(u^{p^k}) = \mu^*(u^{p^k}) = \sum_{i+j=p^k} u^i \otimes u^j$. Taking tensor products leads to a description of $\Gamma(u_1, \ldots, u_n)$ as a large tensor product of truncated polynomial algebras, with the generators in degrees p^k for $k \in \mathbb{N}_0$, each generator being truncated at height p. For yet another description of the divided power algebra as an associated graded algebra of the polynomial algebra with respect to the filtration by the Frobenius powers of the maximal ideal see [13] and [84].

The algebra $\Gamma(V)$ supports a **system of divided power operations**, also called **$\boldsymbol{\gamma}$-operations**, $\gamma_0, \gamma_1, \ldots, \gamma_k, \ldots$. These are operators

$$\gamma_k : \Gamma(V)_i \longrightarrow \Gamma(V)_{ki} \qquad i > 0$$

satisfying the rules (see e.g. [13] Exposé 7 or [21] Proposition A.2.6) listed in Table II.1.1. In the table $u, v \in \Gamma(V)$ and $\lambda \in \mathbb{F}$.

These properties of the γ-operations are those of the operators $x \longmapsto \frac{1}{k!}x^k$ in the polynomial algebra $\mathbb{F}[x]$ for $\mathbb{F}$ a field of characteristic zero. If $z_1, \ldots, z_n \in V^*$ is a basis and $u_1, \ldots, u_n \in V$ the dual basis, then $\gamma_p(u_i)$ is just $\gamma_p(u_i)$, i.e., the dual to z_i^p.

REMARK: Perhaps the most obscure of these rules is the last, which implies that $\frac{(k'k'')!}{(k'!)^{k''}k''!}$ is an integer. The following argument of P. Zion extracted from [24] page 86 shows this to be the case.

LEMMA II.1.1: *Let $d, a \in \mathbb{N}$. Then $d!$ divides $\frac{(da)!}{(a!)^d}$.*

$$\begin{aligned}
\gamma_0(u) &= 1 \\
\gamma_1(u) &= u \\
\gamma_{k'}(u)\gamma_{k''}(u) &= \binom{k'+k''}{k'}\gamma_{k'+k''}(u) \\
\gamma_k(u+v) &= \sum_{i+j=k}\gamma_i(u)\gamma_j(v) \\
\gamma_k(\lambda u) &= \lambda^k\gamma_k(u) \\
\gamma_k(uv) &= k!\gamma_k(u)\gamma_k(v) \\
\gamma_{k''}(\gamma_{k'}(u)) &= \frac{(k'k'')!}{(k'!)^{k''}k''!}\gamma_{k'k''}(u) \quad \text{for } k' > 0
\end{aligned}$$

TABLE II.1.1: Properties of the γ-operators

PROOF (P. Zion): One notes that

$$\frac{(da)!}{(a!)^d} = \binom{da}{\underbrace{a\cdots a}_{\leftarrow\ d \rightarrow}} = \binom{da}{a}\binom{(d-1)a}{a}\cdots\binom{a}{a}$$

is an integer, so it is enough to show that d divides $\binom{da}{a}$ for any d, $a \in \mathbb{N}$. Since

$$\binom{da}{a} = \frac{(da)(da-1)\cdots(da-a+1)}{a(a-1)!} = \frac{da}{a}\binom{da-1}{a-1}$$

and $\binom{da-1}{a-1}$ is an integer we are done. □

If $u_1, \ldots, u_n \in \Gamma(V)_1$ is the vector space basis dual to $z_1, \ldots, z_n \in \mathbb{F}[V]_1$ then $\gamma_i(u_j) = \gamma_{i\Delta_j}$ where Δ_j is the index sequence with a 1 in the j-th position and 0 elsewhere. Put another way, $\gamma_i(u_j)$ is the dual of z_j^i with respect to the monomial basis of $\mathbb{F}[V]$. In particular, if $n = 1$, and we write $\Gamma(u)$ for the Hopf algebra dual to $\mathbb{F}[z]$, then $\gamma_k(u) = \gamma_{(k)}$ is dual to z^k.

Assuming the characteristic of the ground field is $p \neq 0$ and the p-adic expansion of k is $k = k_s p^s + \cdots + k_1 p + k_0$, then the product formula $\gamma_{k'}(u)\gamma_{k''}(u) = \binom{k'+k''}{k'}\gamma_{k'+k''}(u)$ in conjunction with the composition formula yields

$$\gamma_k(u) = \gamma_{k_0}(u)\gamma_{k_1}(\gamma_p(u))\cdots\gamma_{k_s}(\gamma_{p^s}(u))$$

as well as

$$\gamma_{p^i}(u) = \underbrace{\gamma_p(\cdots\gamma_p}_{\longleftarrow\ i \longrightarrow}(u)\cdots).$$

II.2 Macaulay's dual principal systems

This section is devoted to an exposition of Macaulay's theory of $\overline{\mathbb{F}[V]}$-primary irreducible ideals using the dual pair of Hopf algebras $\mathbb{F}[V]$ and $\Gamma(V)$. For a discussion of these ideas in Macaulay's original context see [49] Part IV or Section VI.1. Other sources for Macaulay's theory in modern language are [24] Lecture 9 and [34].

If $\gamma \in \Gamma(V)_d$ we may regard γ as a linear form defined on $\mathbb{F}[V]_d$ via the duality pairing, viz.,

$$\gamma(f) = <\gamma \mid f> \in \mathbb{F} \text{ for } f \in \mathbb{F}[V]_d.$$

We associate to this linear form the ideal $I(\gamma) \subset \mathbb{F}[V]$ defined by

$$f \in I(\gamma) \Longleftrightarrow \gamma(f \cdot h) = 0 \ \ \forall\, h \in \mathbb{F}[V]_{d-\deg(f)},$$

cf. Section I.5. By Proposition I.5.2, if γ is nonzero, then $I(\gamma) = \mathfrak{A}(\ker(\gamma))$, and hence by Proposition I.3.2 $I(\gamma) \subset \mathbb{F}[V]$ is an $\overline{\mathbb{F}[V]}$-primary irreducible ideal. Alternatively, reasoning as in the proof of Proposition I.4.1, one can prove this directly, as $I(\gamma)$ consists of all elements of $\mathbb{F}[V]$ for which there is no Poincaré dual element in the quotient algebra $\mathbb{F}[V]/I(\gamma)$. The following result in different language is due to F. S. Macaulay ([49] Section 72).

THEOREM II.2.1 (F. S. Macaulay): *Let $\mathbb{F}$ be a field, $n \in \mathbb{N}$, and set $V = \mathbb{F}^n$. The assignment $\gamma \rightsquigarrow I(\gamma)$ induces a correspondence between nonzero elements $\gamma \in \Gamma(V)_d$ and $\overline{\mathbb{F}[V]}$-primary irreducible ideals $I(\gamma) \subset \mathbb{F}[V]$ such that the corresponding Poincaré duality quotient algebra $\mathbb{F}[V]/I(\gamma)$ has formal dimension d. Two nonzero elements γ', $\gamma'' \in \Gamma(V)_d$ determine the same ideal I if and only if each is a nonzero multiple of the other: equivalently, they have the same kernel.*

PROOF: This follows from the fact that $I(\gamma) = \mathfrak{A}(I(\gamma)_d)$, that $\mathfrak{A}(I(\gamma)_d)$ is determined by $I(\gamma)_d = \ker(\gamma)$, and Proposition I.3.2. □

Macaulay's dual systems as they apply to the study of Poincaré duality quotients of $\mathbb{F}[V]$ ([49] Sections 69–72) are a reformulation of Theorem II.2.1 in ideal and module theoretic terms. To explain this we introduce an $\mathbb{F}[V]$-module structure on $\Gamma(V)$. The product of an element $f \in \mathbb{F}[V]$ with an element $\gamma \in \Gamma(V)$ is denoted by $f \cap \gamma$ and is defined by the rule

$$<f \cap \gamma \mid h> = <\gamma \mid f \cdot h> \qquad \forall\, f,\, h \in \mathbb{F}[V],\ \gamma \in \Gamma(V)$$

with $\deg(\gamma) = \deg(f) + \deg(h)$. Note that for $f \in \mathbb{F}[V]_i$ and $\gamma \in \Gamma(V)_j$ we have $f \cap \gamma \in \Gamma(V)_{j-i}$, i.e., the action of $\mathbb{F}[V]$ on $\Gamma(V)$ *lowers* the grading; it does not raise it. For this reason $\Gamma(V)$ is not a finitely generated $\mathbb{F}[V]$-module. Topologists should recognize this $\mathbb{F}[V]$-module structure on $\Gamma(V)$

as an analog of the cap-product for cohomology acting on homology, or a *stripping operation*. It is for this reason that we have chosen the notation $f \cap \gamma$ for the product of an element $f \in \mathbb{F}[V]$ with an element $\gamma \in \Gamma(V)$. The relation between the $\mathbb{F}[V]$-module structure and the algebra structure of $\Gamma(V)$ is expressed by the commutativity of the diagram

$$\begin{array}{ccc}
\mathbb{F}[V] \otimes \Gamma(V) \otimes \Gamma(V) & \xrightarrow{1 \otimes \nabla^*} & \mathbb{F}[V] \otimes \Gamma(V) \\
\downarrow{\scriptstyle \nabla \otimes 1 \otimes 1} & & \downarrow{\scriptstyle \cap} \\
\mathbb{F}[V] \otimes \mathbb{F}[V] \otimes \Gamma(V) \otimes \Gamma(V) & & \Gamma(V) \\
\downarrow{\scriptstyle 1 \otimes T \otimes 1} & & \uparrow{\scriptstyle \nabla^*} \\
\mathbb{F}[V] \otimes \Gamma(V) \otimes \mathbb{F}[V] \otimes \Gamma(V) & \xrightarrow{\cap \otimes \cap} & \Gamma(V) \otimes \Gamma(V)
\end{array}$$

where T is the twisting map. In terms of elements this says

$$f \cap (\gamma' \cdot \gamma'') = \sum (f_i' \cap \gamma') \cdot (f_i'' \cdot \gamma'')$$

if $f \in \mathbb{F}[V]$, $\gamma', \gamma'' \in \Gamma(V)$, and $\nabla(f) = \sum f_i' \otimes f_i''$. Apart from the curious grading $\Gamma(V)$ is an algebra over the Hopf algebra $\mathbb{F}[V]$.

Note that if $z_1, \ldots, z_n$ is a basis for V^* with $\{z^E \mid E \in \mathbb{N}_0^n\}$ the corresponding monomial basis for $\mathbb{F}[V]$ and $\{\gamma_E \mid E \in \mathbb{N}_0^n\}$ the dual basis for $\Gamma(V)$ then

$$z^E \cap \gamma_F = \begin{cases} \gamma_{F-E} & \text{if } F - E \in \mathbb{N}_0^n \\ 0 & \text{otherwise.} \end{cases}$$

For this reason the $\cap$-action of $\mathbb{F}[V]$ on $\Gamma(V)$ is often called the **contraction pairing** by analogy with the classical language of tensor algebra. We define the **support** of $\theta \in \Gamma(V)_d$ (with respect to the monomial basis $\{z^D \mid |D| = d\}$) to be

$$\operatorname{supp}(\theta) = \left\{ z^D \mid |D| = d \text{ and } \theta(z^D) \neq 0 \right\}.$$

Note, if we write $\Gamma(V) = \Gamma(u_1) \otimes \cdots \otimes \Gamma(u_n)$, then the basis element γ_F is a divided power monomial. For an index sequence $F = (f_1, \ldots, f_n)$ set

$$f_i = \sum_{j=0}^{\infty} \alpha_j(f_i) p^j$$

for $i = 1, \ldots, n$, where $0 \leq \alpha_j(f_i) < p$ for all $j \in \mathbb{N}_0$. Then

$$\gamma_F = \prod_{i=1}^{n} \prod_{j=0}^{\infty} \gamma_{\alpha_j(f_i) \cdot p^j}(u_i),$$

where $\gamma_a(u_i) = \boldsymbol{\gamma}_a(u_i) \in \Gamma(u_i)$ (the inner of the two products is of course finite).

If $\gamma \in \Gamma(V)$ we denote by $M(\gamma)$ the $\mathbb{F}[V]$-submodule of $\Gamma(V)$ generated by γ. Unraveling the definitions and referring to Theorem II.2.1 shows that the ideal $I(\gamma) \subset \mathbb{F}[V]$ is the annihilator $\mathrm{Ann}_{\mathbb{F}[V]}(\gamma)$ in $\mathbb{F}[V]$ of the element γ, or what is the same, of $M(\gamma)$, and that the submodule of $\Gamma(V)$ that is annihilated by $I(\gamma)$ is precisely $M(\gamma)$. We summarize this in the following theorem.

THEOREM II.2.2 (F. S. Macaulay): *There is a bijective correspondence between nonzero cyclic $\mathbb{F}[V]$-submodules of $\Gamma(V)$ and proper $\overline{\mathbb{F}[V]}$-primary irreducible ideals in $\mathbb{F}[V]$ given by associating to a nonzero cyclic submodule $M(\gamma) \subset \Gamma(V)$ its annihilator ideal $I(\gamma) = \mathrm{Ann}_{\mathbb{F}[V]}(M(\gamma)) \subset \mathbb{F}[V]$, and to a proper $\overline{\mathbb{F}[V]}$-primary irreducible ideal $I \subset \mathbb{F}[V]$ the submodule $\mathrm{Ann}_{\Gamma(V)}(I) \subset \Gamma(V)$ of elements annihilated by I.* □

We will refer to the correspondence given by this theorem as **Macaulay's double annihilator correspondence.** If $I \subset \mathbb{F}[V]$ is any ideal then we denote $\mathrm{Ann}_{\Gamma(V)}(I)$ by $I^{\perp}$ and call it the **dual system** to I. The minimum number of generators for $I^{\perp}$ turns out to be the dimension of the socle of $\mathbb{F}[V]/I$ as a vector space over $\mathbb{F}$. If $I \subset \mathbb{F}[V]$ is a $\overline{\mathbb{F}[V]}$-primary irreducible ideal, then $I^{\perp} \subset \Gamma(V)$ is a cyclic $\mathbb{F}[V]$-module and is called the dual **principal** system of I. A generator for this module is called a **Macaulay dual** for I and will be denoted by θ_I.

For example, if $I \subset \mathbb{F}[V]$ is a monomial ideal and $\mathcal{F}(I) = \{F \in \mathbb{N}_0^n \mid z^F \notin I\}$ then $I^{\perp} = \mathrm{Span}_{\mathbb{F}}\{\gamma_F \mid F \in \mathcal{F}(I)\}$. If I is $\overline{\mathbb{F}[V]}$-primary and irreducible then $\mathcal{F}(I)$ contains a unique element of maximal degree (see e.g. [57] Chapter 11), say M, and $\theta_I = \gamma_M$. For example, the monomial ideal $(x^2, y^4) \subset \mathbb{F}[x, y]$ is $\mathfrak{m}$-primary and irreducible. It contains all monomials of degree at least 5, and $\mathcal{F}(I) = \{1, x, y, xy, y^2, y^3, xy^2, xy^3\}$ which contains the unique maximal element xy^3. A Macaulay dual of (x^2, y^4) is $\gamma_1(u) \cdot \gamma_3(v)$, where $u, v \in V$ is the dual basis to $x, y \in V^*$.

Let $0 \neq \gamma \in \Gamma(V)_d$. From the double annihilator correspondence it follows that the map $\alpha_\gamma : \mathbb{F}[V] \longrightarrow \Gamma(V)$ defined by $\alpha_\gamma(f) = f \cap \gamma$ induces an isomorphism $\mathbb{F}[V]/I(\gamma) \longrightarrow M(\gamma)$. It sends the homogeneous component of degree k of $\mathbb{F}[V]/I(\gamma)$ to the homogeneous component of degree $d-k$ of $M(\gamma)$. Since $\mathbb{F}[V]/I(\gamma)$ satisfies Poincaré duality, and has formal dimension d its Poincaré polynomial $P(\mathbb{F}[V]/I(\gamma), t)$ is palindromic, i.e.,

$$P(\mathbb{F}[V]/I(\gamma), t) = t^d P\big(\mathbb{F}[V]/I(\gamma), 1/t\big)$$

and hence we have the following.

COROLLARY II.2.3: *Suppose $\mathbb{F}[V] \xrightarrow{\pi} H$ is a Poincaré duality quotient of $\mathbb{F}[V]$, then the Poincaré polynomials of H and $\ker(\pi)^{\perp}$ are equal, i.e., $P(H, t) = P(\ker(\pi)^{\perp}, t)$.* □

We leave the proof of the following fact to the reader: it is a good exercise in the use of the tools developed up to this point.

PROPOSITION II.2.4: *Suppose that* $H' = \mathbb{F}[V']/I'$ *and* $H'' = \mathbb{F}[V'']/I''$ *are Poincaré duality quotients of* V', V'' *respectively, and set*

$$H = H' \# H'' = \mathbb{F}[V' \oplus V'']/I.$$

If $\gamma' \in \Gamma(V')$ *and* $\gamma'' \in \Gamma(V'')$ *are Macaulay duals for* I' *respectively* I'', *then* $\gamma' + \gamma'' \in \Gamma(V' \oplus V'')$ *is a Macaulay dual for* I. □

II.3 An illustrative example

To illustrate the ideas introduced so far we will determine all the Poincaré duality quotients of $\mathbb{F}_2[x, y]$ of formal dimension 2 up to algebra isomorphism. Proposition I.6.6 tells us that there are three such examples, but gives us no information on what they might be. By Proposition II.4.1 we need to compute the orbits of $GL(2, \mathbb{F}_2)$ on the set of hyperplanes of $\mathbb{F}_2[x, y]_2$.

We tackle the dual problem. Let $u, v \in V$ be the basis dual to $x, y \in V^*$ and consider the action of $GL(2, \mathbb{F}_2)$ on the seven 1-dimensional subspaces of $\Gamma(u, v)_2$ instead. Recall that $GL(2, \mathbb{F}_2)$ is generated by the cyclic permutation $\xi : u \longrightarrow u + v \longrightarrow v$, which has order 3, and the involution $\tau : u \leftrightarrow v$, which has order 2. Since the divided power operations commute with the $GL(2, \mathbb{F}_2)$ action we find

$$\xi(\gamma_2(u)) = \gamma_2(\xi(u)) = \gamma_2(u + v) = \gamma_2(u) + \gamma_1(u)\gamma_1(v) + \gamma_2(v)$$

and

$$\tau(\gamma_2(u)) = \gamma_2(\tau(u)) = \gamma_2(v).$$

Using these formulae one finds there are three orbits for the action of $GL(2, \mathbb{F}_2)$ on the seven lines in $\Gamma(u, v)_2$. We list them in Table II.3.1.

Orbit	Elements
1	$\gamma_2(u)$, $\gamma_2(u) + \gamma_1(u)\gamma_1(v) + \gamma_2(v)$, $\gamma_2(v)$
2	$\gamma_2(u) + \gamma_2(v)$, $\gamma_2(u) + \gamma_1(u)\gamma_1(v)$, $\gamma_2(v) + \gamma_1(u)\gamma_1(v)$
3	$\gamma_1(u)\gamma_1(v)$

TABLE II.3.1: Orbits of $GL(2, \mathbb{F}_2)$ on $\Gamma(u, v)_2 \setminus \{0\}$

This means we have three cases to consider.

ORBIT 1: As orbit representative choose $\gamma_2(u)$. The kernel of the map $\gamma_2(u) : \mathbb{F}_2[x, y]_2 \longrightarrow \mathbb{F}_2$ is spanned by y^2 and xy, so the big ancestor ideal of $\ker(\gamma_2(u))$ is easily seen to contain the linear form y. Hence $\mathfrak{A}(\ker((\gamma_2(u)) = (y, x^3)$ and the Poincaré duality quotient algebra $\mathbb{F}_2[x, y]/I(\gamma_2(u))$ is isomorphic to $\mathbb{F}_2[x]/(x^3)$. This case is somewhat degenerate in the sense that it has only rank one instead of two.

ORBIT 2: A representative for the orbit is $\gamma_2(u) + \gamma_2(v)$. The kernel of the map $\mathbb{F}_2[x, y]_2 \longrightarrow \mathbb{F}_2$ induced by $\gamma_2(u) + \gamma_2(v)$ is the linear span of $x^2 + y^2 = (x + y)^2$ and xy. The big ancestor ideal contains no linear forms, so $I(\gamma_2(u) + \gamma_2(v)) = (xy, x^2 + y^2)$ and the Poincaré duality quotient algebra is isomorphic to $\mathbb{F}_2[x, y]/(xy, x^2 + y^2)$ which is the connected sum of two copies of the previous case.

ORBIT 3: Choose $\gamma_1(u)\gamma_1(v)$ to represent this orbit. The kernel of the map $\mathbb{F}_2[x, y]_2 \longrightarrow \mathbb{F}_2$ induced by $\gamma_1(u)\gamma_1(v)$ is the span of x^2 and y^2 which also generates the ideal $I(\gamma_1(u)\gamma_1(v))$. Hence the corresponding Poincaré duality quotient is isomorphic to $\mathbb{F}[x, y]/(x^2, y^2)$.

To make more extensive computations than these requires further tools. We describe one such in Section VI.2: the *catalecticant matrices* of [34]. This is the means by which we arrive at the results in Examples 1 and 2 of Section III.7. In Section II.5 we describe another method to assist in computations that is based on Theorem I.2.1.

II.4 Relation to the classical form problem

The results I.1.5, I.3.2, I.4.1, II.2.1, and II.2.2 each provide a means of describing Poincaré duality quotients of $\mathbb{F}[V]$. The passage from one description to another is by no means clear. Indeed, this is part of the raison-d'être of this manuscript, as more than one surprise occurs in this connection. For example, Lemma I.5.1 provides a connection between the classification of Poincaré duality quotient algebras of $\mathbb{F}[V]$ and the classical form problem, see e.g. [34] and [87] Section 2.5. To better formulate our results we first draw another conclusion from Lemma I.5.1. The proof is routine and is left to the reader.

PROPOSITION II.4.1: *There is a bijective correspondence between the isomorphism classes of Poincaré duality quotients of* $\mathbb{F}[V]$ *of formal dimension* d *with the orbits of* $\mathrm{GL}(V)$ *on* $\mathbb{P}(\mathbb{F}[V]_d)$. □

One interpretation of the classical form problem for forms $f \in \mathbb{F}[V]_d$ is to classify them up to linear change of variables. This amounts to describing the orbit space $\mathbb{P}(\mathbb{F}[V]_d)/\mathrm{PGL}(V)$, where $\mathbb{P}(\mathbb{F}[V]_d)$ denotes the projective space associated to the vector space $\mathbb{F}[V]_d$. The approach chosen in classical invariant theory (i.e., where $\mathbb{F} = \mathbb{C}$) is to describe the ring of $\mathbb{F}$-valued polynomial functions on $\mathbb{P}(\mathbb{F}[V]_d)/\mathrm{PGL}(V)$ and prove they separate the points (see e.g. [14]). The analog for the classification of Poincaré duality quotients of $\mathbb{F}[V]$ runs as follows. The proof is a translation of Lemma I.5.1 into a global form.

THEOREM II.4.2: *Let $V = \mathbb{F}^n$ be an n-dimensional vector space over the field $\mathbb{F}$ and $d \in \mathbb{N}$ a positive integer. The group $\mathrm{PGL}(n, \mathbb{F})$ acts on the projective space $\mathbb{P}(\Gamma(V)_d)$ of the homogeneous component $\Gamma(V)_d$ of degree d of the dual Hopf algebra $\Gamma(V)$ of $\mathbb{F}[V]$. There is a bijective correspondence between the orbits of $\mathrm{PGL}(n, \mathbb{F})$ on $\mathbb{P}(\Gamma(V)_d)$ and isomorphism classes of Poincaré duality quotients of $\mathbb{F}[V]$.* □

II.5 The $K \subset L$ Paradigm: a computational tool

In this section we introduce a tool, based on Theorem I.2.1, for making concrete computations with Macaulay dual principal systems. It applies to a pair of ideals K, L with $L \in \mathrm{over}(K)$ and both K and L being $\overline{\mathbb{F}[V]}$-primary and irreducible. We call it the $K \subset L$ **paradigm**. The motivation runs as follows: an $\overline{\mathbb{F}[V]}$-primary irreducible ideal I always contains the ideal $(z_1^k, \ldots, z_n^k)$ for a suitable large k. A generator of the dual principal system to the ideal $(z_1^k, \ldots, z_n^k)$ in $\mathbb{F}[z_1, \ldots, z_n]$ is $\gamma_{k-1}(u_1) \cdots \gamma_{k-1}(u_n)$. If one writes using Theorem I.2.1 $I = \big((z_1^k, \ldots, z_n^k) : h\big)$, then extensive computation with examples convinced us that $h \cap (\gamma_{k-1}(u_1) \cdots \gamma_{k-1}(u_n))$ is a generator for the dual principal system of I. Here is a proof of this fact.

THEOREM II.5.1: *Let $\mathbb{F}$ be a field and $K \subsetneq L$ be a proper $\overline{\mathbb{F}[V]}$-primary irreducible ideals in $\mathbb{F}[V]$. Write $(K : L)$ in the form $(h) + K$ for some $h \in \mathbb{F}[V]$. If θ_K is a Macaulay dual for K then $h \cap \theta_K$ is a Macaulay dual for L. Conversely, if $K \subset \mathbb{F}[V]$ is an $\overline{\mathbb{F}[V]}$-primary irreducible ideal and $h \in \mathbb{F}[V]$ does not belong to K then $(K : h) = L$ is a proper over ideal of K which is also $\overline{\mathbb{F}[V]}$-primary, irreducible, and $\theta_L = h \cap \theta_K$.*

PROOF: Let θ_L be a Macaulay dual for L. Note that $h \notin K$ for otherwise we would have

$$L = \big(K :(K : L)\big) = \big(K :((h) + K)\big) = (K : K) = \mathbb{F}[V]$$

contrary to hypothesis. Therefore $h \cap \theta_K \neq 0 \in \Gamma(V)$ since the annihilator of θ_K is K. Corollary I.2.3 applied to the map $\mathbb{F}[V]/K \longrightarrow \mathbb{F}[V]/L$ yields $\text{f-dim}(\mathbb{F}[V]/L) = \text{f-dim}(\mathbb{F}[V]/K) - \deg(h)$. Therefore

$$\deg(\theta_L) = \text{f-dim}(\mathbb{F}[V]/L) = \text{f-dim}(\mathbb{F}[V]/K) - \deg(h) = \deg(h \cap \theta_K).$$

Finally, if $f \in L$ then $hf \in K$ so f annihilates $h \cap \theta_K$, viz., $f \cap (h \cap \theta_K) = (fh) \cap \theta_K = 0$. Hence $h \cap \theta_K \in L^\perp$. Since $L^\perp$ is cyclic as an $\mathbb{F}[V]$-module, and both θ_L and $h \cap \theta_K$ are nonzero and have the same degree as a generator of $L^\perp$, they must be nonzero multiples of each other. The converse follows once one notes that $(K : L) = (h) + K$ since Ξ is an involution on the over ideals of K by Theorem I.2.1. □

In Theorem II.5.1 we have a powerful tool to find a generator for the Macaulay dual of an arbitrary $\overline{\mathbb{F}[V]}$-primary irreducible ideal of $\mathbb{F}[V]$ by comparing the ideal with an ideal generated by powers of the generators. In the special case where $\mathbb{F}$ is a Galois field we also provide an explicit formula in the next section relating Macaulay duals of an ideal and its Frobenius powers.

Here are some simple examples to illustrate the use of this theorem. More extensive computations with the $K \subset L$ paradigm appear in Parts III and IV as well as [44].

EXAMPLE 1: The Dickson polynomials[2] $\mathbf{d}_{2,0}, \mathbf{d}_{2,1} \in \mathbb{F}_q[x, y]$ in two variables over the Galois field $\mathbb{F}_q$ may be defined by means of the identity

$$\Phi(X) = \prod_{0 \neq z \in V^*} (X + z) = X^{q^2-1} + \mathbf{d}_{2,1} X^{q-1} + \mathbf{d}_{2,0}.$$

They generate a parameter ideal so the methods of Example 1 in Section I.4 apply here. Let $u, v \in V = \mathbb{F}_q^2$ be the standard basis and $x, y \in V^*$ the dual basis. Since $\Phi(x) = 0 = \Phi(y)$ it follows that

$$\begin{aligned} -x^{q^2-1} &= \mathbf{d}_{2,0} + x^{q-1} \cdot \mathbf{d}_{2,1} \\ -y^{q^2-1} &= \mathbf{d}_{2,0} + y^{q-1} \cdot \mathbf{d}_{2,1}, \end{aligned}$$

so $x^{q^2-1}, y^{q^2-1} \in (\mathbf{d}_{2,0}, \mathbf{d}_{2,1})$, or in terms of ideals $(x^{q^2-1}, y^{q^2-1}) \subset (\mathbf{d}_{2,0}, \mathbf{d}_{2,1})$. We apply a variant of Cramer's rule to the preceding pair of equations: namely, we regard them as a single matrix equation and multiply both sides of the equality by the transposed cofactor matrix of the coefficients (see e.g. [88] Section 14.3). One obtains

$$\begin{bmatrix} y^{q-1} & -x^{q-1} \\ -1 & 1 \end{bmatrix} \begin{bmatrix} -x^{q^2-1} \\ -y^{q^2-1} \end{bmatrix} = (y^{q-1} - x^{q-1}) \begin{bmatrix} \mathbf{d}_{2,0} \\ \mathbf{d}_{2,1} \end{bmatrix},$$

since the product of a matrix and its transposed cofactor matrix is the determinant of the matrix times the identity matrix. From this it follows that

$$y^{q-1} - x^{q-1} \in ((x^{q^2-1}, y^{q^2-1}) : (\mathbf{d}_{2,0}, \mathbf{d}_{2,1})).$$

[2] L. E. Dickson showed that $\mathbb{F}_q[x, y]^{\mathrm{GL}(2, \mathbb{F}_q)} = \mathbb{F}_q[\mathbf{d}_{2,0}, \mathbf{d}_{2,1}]$, see e.g. [20] or [87] Section 8.1.

If h is a transition element for $(\mathbf{d}_{2,0}, \mathbf{d}_{2,1})$ over (x^{q^2-1}, y^{q^2-1}) then h has degree $q-1$ by Corollary I.2.4. This is also the degree of $y^{q-1}-x^{q-1}$. Since $y^{q-1}-x^{q-1} \in (h) + (\mathbf{d}_{2,0}, \mathbf{d}_{2,1})$, $h \notin (\mathbf{d}_{2,0}, \mathbf{d}_{2,1})$, and a transition element for $(\mathbf{d}_{2,0}, \mathbf{d}_{2,1})$ over (x^{q^2-1}, y^{q^2-1}) is unique modulo $(\mathbf{d}_{2,0}, \mathbf{d}_{2,1})$, we may as well choose $h = y^{q-1}-x^{q-1}$.

The Macaulay dual of the monomial ideal (x^{q^2-1}, y^{q^2-1}) is generated by $\gamma_{q^2-2}(u)\gamma_{q^2-2}(v)$ (see Section II.2). Therefore by Theorem II.5.1 a generator of the dual principal system to the Dickson ideal $(\mathbf{d}_{2,0}, \mathbf{d}_{2,1}) \subset \mathbb{F}_q[x, y]$ is

$$(y^{q-1}-x^{q-1}) \cap \gamma_{q^2-2}(u)\gamma_{q^2-2}(v) = \gamma_{q^2-2}(u)\gamma_{q^2-q-1}(v) - \gamma_{q^2-q-1}(u)\gamma_{q^2-2}(v).$$

For the generalization of this formula to an arbitrary number of variables and any Galois field see [44].

Example 2: Suppose that $f_1, \ldots, f_n \in \mathbb{F}[z_1, \ldots, z_n]$ is a system of parameters and $a_1, \ldots, a_n \in \mathbb{N}$. Then

$$K = (f_1^{a_1}, \ldots, f_n^{a_n}) \subseteq (f_1, \ldots, f_n) = L$$

and since $f_1, \ldots, f_n$ is a regular sequence

$$(K : L) = (f_1^{a_1-1} \cdots f_n^{a_n-1}) + K.$$

So if θ_K is a Macaulay dual for K then $f_1^{a_1-1} \cdots f_n^{a_n-1} \cap \theta_K$ is a Macaulay dual for L. This result will be very useful in the sequel and will often be used without special mention.

Example 3 (K. Kuhnigk [44]): The elementary symmetric polynomials $e_1, \ldots, e_n \in \mathbb{F}[z_1, \ldots, z_n]$ generate an $\mathfrak{m}$-primary irreducible ideal in the algebra $\mathbb{F}[z_1, \ldots, z_n]$, where as before $\mathfrak{m} \subset \mathbb{F}[z_1, \ldots, z_n]$ is the augmentation ideal $(z_1, \ldots, z_n)$. One definition of $e_1, \ldots, e_n$ is via the identity

$$\prod_{i=1}^{n}(X + z_i) = \sum_{r+s=n} e_r X^s,$$

with the convention that $e_0 = 1$. Note that the left hand side is obviously zero for $X = -z_i$, $i = 1, \ldots, n$. If we write the resulting n equations as a single matrix equation and rearrange terms we find

$$(-1)^n \begin{bmatrix} -z_1^n \\ \vdots \\ -z_n^n \end{bmatrix} = \begin{bmatrix} (-z_1)^{n-1} & \cdots & -z_1 & 1 \\ \vdots & \vdots & \vdots & \vdots \\ (-z_n)^{n-1} & \cdots & -z_n & 1 \end{bmatrix} \begin{bmatrix} e_1 \\ \vdots \\ e_n \end{bmatrix}.$$

Therefore $z_1^n, \ldots, z_n^n \in (e_1, \ldots, e_n)$, so $(z_1^n, \ldots, z_n^n) \subset (e_1, \ldots, e_n)$.

Set

$$\nabla_n = \det \begin{bmatrix} (-z_1)^{n-1} & \cdots & -z_1 & 1 \\ \vdots & \vdots & \vdots & \vdots \\ (-z_n)^{n-1} & \cdots & -z_n & 1 \end{bmatrix} = (-1)^{\binom{n}{2}} \sum_{\sigma \in \Sigma_n} \operatorname{sgn}(\sigma) z_1^{\sigma(n-1)} \cdots z_n^{\sigma(0)},$$

where the symmetric group Σ_n acts on the set $\{0, 1, \ldots, n-1\}$ by permutation of the elements. ∇ is up to sign the discriminant and Cramer's rule implies as in Example 1 that

$$\nabla_n \in \big((z_1^n, \ldots, z_n^n) : (e_1, \ldots, e_n)\big).$$

Note ∇_n has degree $\binom{n}{2}$ which is also the degree of a transition element for $(e_1, \ldots, e_n)$ over $(z_1^n, \ldots, z_n^n)$. So again, as in Example 1, we obtain

$$\big((z_1^n, \ldots, z_n^n) : (e_1, \ldots, e_n)\big) = (\nabla_n) + (z_1^n, \ldots, z_n^n).$$

A generator for the dual principal system $(z_1^n, \ldots, z_n^n)^{\perp} \subset \Gamma(u_1, \ldots, u_n)$ is $\gamma_{n-1}(u_1) \cdots \gamma_{n-1}(u_n)$, so Theorem II.5.1 tells us that

$$\theta_{(e_1, \ldots, e_n)} = \nabla_n \cap \gamma_{n-1}(u_1) \cdots \gamma_{n-1}(u_n) = \sum_{\sigma \in \Sigma_n} \operatorname{sgn}(\sigma) \gamma_{\sigma(n-1)}(u_1) \cdots \gamma_{\sigma(0)}(u_n)$$

is a Macaulay dual for the ideal $(e_1, \ldots, e_n)$ generated by the elementary symmetric polynomials. The support of $\theta_{(e_1, \ldots, e_n)}$ on the set of monomials is $\{z_1^{\sigma(0)} \cdots z_n^{\sigma(n-1)} \mid \sigma \in \Sigma_n\}$ and hence the monomials in this set are precisely the monomials representing a fundamental class for $\mathbb{F}[z_1, \ldots, z_n]_{\Sigma_n}$.

As is well known when the ground field $\mathbb{F}$ is of characteristic zero, $\nabla_n \in \mathbb{F}[z_1, \ldots, z_n]$ represents a fundamental class of the algebra of coinvariants $\mathbb{F}[z_1, \ldots, z_n]_{\Sigma_n}$. The preceding computations have the surprising consequence that this is no longer the case for fields $\mathbb{F}$ whose characteristic divides the order of Σ_n. Namely if $\mathbb{F}$ has characteristic $p \leq n$ then

$$\theta_{(e_1, \ldots, e_n)}(\nabla_n) = \sum_{\Sigma_n} 1 = n! = 0 \in \mathbb{F}.$$

A result of this type was originally discovered by D. Glassbrenner [25] Section 11. The simple argument given here appears in [93].

Remark: In both Examples 1 and 3 we used Cramer's rule to find a transition element. This is a special case of the more general result that we prove in Proposition VI.3.1.

II.6 Frobenius powers

If $\mathbb{F}$ is a field of characteristic $p \neq 0$ and A is a commutative graded algebra over $\mathbb{F}$ then the **Frobenius power** $I^{[p]}$ of an ideal $I \subseteq A$ is defined to be the ideal generated by the p-th powers of the elements of I. Given generators $a_1, \ldots, a_m$ for I the elements $a_1^p, \ldots, a_m^p$ generate $I^{[p]}$. One could ask whether taking the Frobenius power preserves that an ideal I is $\overline{A}$-primary and irreducible. In the special case that A is Gorenstein and I is generated by a regular sequence this is clear since taking the p-th powers of the elements of a regular sequence yields another regular sequence.

If $A = \mathbb{F}[V]$ and $I \subset \mathbb{F}[V]$ is a parameter ideal then the method described in Example 1 of Section I.4 allows us to be even more precise. Namely, $\text{f-dim}(\mathbb{F}[V]/I^{[p]}) = n(p-1) + p \cdot \text{f-dim}(\mathbb{F}[V]/I)$, where $n = \dim_{\mathbb{F}}(V)$. Moreover, if $f \in \mathbb{F}[V]$ represents a fundamental class of $\mathbb{F}[V]/I$ then together with the use of the chain rule one sees that $(z_1 \cdots z_n)^{p-1} f^p$ represents a fundamental class for $\mathbb{F}[V]/I^{[p]}$.

It is natural to ask what happens if I is not a parameter ideal. We are indebted to N. Nossem for numerous discussions on Frobenius functors and assistance in computing an illuminating example (see Example 1 in this section) which made it plausible that taking Frobenius powers preserves irreducibility of $\mathfrak{m}$-primary ideals in a polynomial algebra. R. Y. Sharp provided us with a proof of this fact, and we thank him for his permission to use the result here. Once knowing the result was correct we constructed a proof adapted to the graded case which yields several additional properties of the Poincaré duality quotient $\mathbb{F}[V]/I^{[p]}$, in particular a representative for a fundamental class in terms of a fundamental class of $\mathbb{F}[V]/I$, and a generator for Macaulay's dual principal system for $I^{[p]}$ completely analagous to the case of parameter ideals. We begin with a pair of elementary lemmas from [64].

LEMMA II.6.1 ([64] Theorem 18.1 (1)): *Let $\varphi : R \longrightarrow S$ be a homomorphism of commutative rings that makes S into a flat R-module. If M is an R-module and $N', N'' \subseteq M$ submodules, then*

$$(N' \cap N'') \otimes_R S = (N' \otimes_R S) \cap (N'' \otimes_R S).$$

PROOF: The Noether isomorphism theorem says that for any ring A and any pair of submodules K', K'' of an A-module L we have

$$K''/(K' \cap K'') \cong (K' + K'')/K',$$

the isomorphism being induced by the natural inclusion $K'' \hookrightarrow K' + K''$.

This means that we have a commutative diagram

$$\begin{array}{ccccccccc} 0 & \longrightarrow & N' \cap N'' & \longrightarrow & N'' & \longrightarrow & N''/(N' \cap N'') & \longrightarrow & 0 \\ & & \downarrow & & \downarrow & & \cong \downarrow & & \\ 0 & \longrightarrow & N' & \longrightarrow & N' + N'' & \longrightarrow & (N' + N'')/N' & \longrightarrow & 0 \end{array}$$

where the vertical maps are induced by the inclusion. Applying the exact functor $- \otimes_R S$ leads to another diagram of exact rows from which we conclude

$$\frac{N'' \otimes_R S}{(N' \cap N'') \otimes_R S} \cong \frac{(N' \otimes_R S) + (N'' \otimes_R S)}{N' \otimes_R S} \cong \frac{N'' \otimes_R S}{(N' \otimes_R S) \cap (N'' \otimes_R S)}$$

by the Noether isomorphism theorem. □

NOTATION: *If R and S are rings, $\varphi : R \longrightarrow S$ a ring homomorphism, and $K \subseteq R$ an ideal, then we write KS for the ideal of S generated by $\varphi(K)$. Note that $KS = K \otimes_R S$ regarded as an ideal of $R \otimes_R S = S$. If $I, J \subseteq R$ are ideals then we write $(IS \underset{S}{:} JS)$ for the corresponding ideal quotient of JS against IS in S.*

LEMMA II.6.2 ([64] Theorem 18.1 (2)): *Let $\varphi : R \longrightarrow S$ be a homomorphism of commutative rings that makes S into a flat R-module. If $I, J \subseteq R$ are ideals with J finitely generated, then*

$$(I \underset{R}{:} J)S = (IS \underset{S}{:} JS).$$

PROOF: Since J is finitely generated it is enough to consider the case $J = (u)$, as the general case then follows by induction on the number of generators of J. The map $R \longrightarrow (u)$ sending $1 \in R$ to u induces an isomorphism

$$\eta_R : \frac{R}{(I \underset{R}{:} u)} \longrightarrow \frac{(u)}{I \cap (u)}.$$

Therefore by Lemma II.6.1 and the exactness of the functor $- \otimes_R S$ we obtain

$$\frac{S}{(I \underset{R}{:} u)S} \underset{\cong}{\xrightarrow{\eta_R \otimes_R S}} \frac{(u) \otimes_R S}{(I \cap (u)) \otimes_R S} \cong \frac{(u) \otimes_R S}{(I \otimes_R S) \cap ((u) \otimes_R S)} \underset{\cong}{\xleftarrow{\eta_S}} \frac{S}{(IS \underset{S}{:} uS)}$$

and the result follows. □

The next result would appear to be well known in the ungraded case (see e.g. [32] Proposition 4.3 and the references in its proof). For the sake of completeness we include a proof for the graded case. The restriction to polynomial algebras may seem artificial, but it is forced by the main result of [45] since the proof uses Lemma II.6.2. This lemma depends on the fact that the algebra A in question is flat (which in the graded case is equivalent to free) over itself along the Frobenius homomorphism $A \longrightarrow A$ given by $a \longmapsto a^p$.

PROPOSITION II.6.3: *Let $\mathbb{F}$ be a field of characteristic $p \neq 0$ and V a finite dimensional vector space over $\mathbb{F}$. Then for any two ideals $I, J \subseteq \mathbb{F}[V]$ we have*

$$(I^{[p]} : J^{[p]}) = (I : J)^{[p]}.$$

In particular, for any ideal $K \subseteq \mathbb{F}[V]$ and any $u \in \mathbb{F}[V]$ we have $(K : u)^{[p]} = (K^{[p]} : u^p)$.

PROOF: Introduce the algebra $\Phi(\mathbb{F}[V])$ which is a copy of $\mathbb{F}[V]$ but regraded as follows:

$$\Phi(\mathbb{F}[V])_i = \begin{cases} 0 & \text{if } i \not\equiv 0 \bmod p \\ \mathbb{F}[V]_{i/p} & \text{if } i \equiv 0 \bmod p. \end{cases}$$

For $f \in \mathbb{F}[V]_j$ write $\Phi(f)$ for the element of $\Phi(\mathbb{F}[V])_{jp}$ corresponding to f. Introduce the map $\lambda : \Phi(\mathbb{F}[V]) \longrightarrow \mathbb{F}[V]$ defined by $\lambda(\Phi(f)) = f^p$. The grading on $\Phi(\mathbb{F}[V])$ has been chosen to make this a map[3] of rings. Since λ is monic we may identify $\Phi(\mathbb{F}[V])$ with a subalgebra of $\mathbb{F}[V]$. In a polynomial algebra all ideals are finitely generated, and since $\mathbb{F}[V]$ is free over $\Phi(\mathbb{F}[V])$ along λ the hypotheses of Lemma II.6.2 are fullfilled. For any ideal $K \subseteq \mathbb{F}[V]$ one has $\Phi(K)$ is an ideal of $\Phi(\mathbb{F}[V])$ and $K^{[p]} = \lambda(\Phi(K))\mathbb{F}[V]$. So if $I, J \subseteq \mathbb{F}[V]$ are ideals, then

$$\begin{aligned}
(I^{[p]} \underset{\mathbb{F}[V]}{:} J^{[p]}) &= (\Phi(I) \cdot \mathbb{F}[V] \underset{\mathbb{F}[V]}{:} \Phi(J) \cdot \mathbb{F}[V]) && \text{(by the preceding discussion)} \\
&= (\Phi(I) \underset{\Phi(\mathbb{F}[V])}{:} \Phi(J)) \cdot \mathbb{F}[V]) && \text{(by Lemma II.6.2)} \\
&= \Phi((I \underset{\mathbb{F}[V]}{:} J)) \cdot \mathbb{F}[V] && \text{(by the definition of } \Phi) \\
&= (I \underset{\mathbb{F}[V]}{:} J)^{[p]} && \text{(by the preceding discussion)}
\end{aligned}$$

as required. □

COROLLARY II.6.4: *Let $\mathbb{F}$ be a field of characteristic $p \neq 0$, V a finite dimensional vector space over $\mathbb{F}$, and $I \subset J$ a pair of $\overline{\mathbb{F}[V]}$-primary irreducible ideals in $\mathbb{F}[V]$. If h is a transition element for J over I then h^p is a transition element for $J^{[p]}$ over $I^{[p]}$.*

PROOF: By definition of a transition element $(I : J) = (h) + I$ so by Proposition II.6.3 taking Frobenius powers gives $(h^p) + I^{[p]} = \big((h) + I\big)^{[p]} = (I : J)^{[p]} = (I^{[p]} : J^{[p]})$ as required. □

[3] In fact, apart from a possible Frobenius automorphism of the ground field, a map of algebras.

THEOREM II.6.5 (R. Y. Sharp): *Let $\mathbb{F}$ be a field of characteristic p and V a finite dimensional $\mathbb{F}$-vector space. If $I \subseteq \mathbb{F}[V]$ is an $\overline{\mathbb{F}[V]}$-primary irreducible ideal, then so is $I^{[p]}$.*

PROOF: Since $I \subseteq \mathbb{F}[V]$ is $\overline{\mathbb{F}[V]}$-primary I contains a parameter ideal, say Q, so $I \in \operatorname{over}(Q)$. By Theorem I.2.1 it follows that $(Q:I)$ is principal over Q, say

$$(Q:I) = (u) + Q.$$

Then by Proposition II.6.3

$$(Q^{[p]}:I^{[p]}) = (Q:I)^{[p]} = ((u) + Q)^{[p]} = (u^p) + Q^{[p]}.$$

The ideal $Q^{[p]}$ is a parameter ideal and $I^{[p]} \in \operatorname{over}(Q^{[p]})$ so again by Theorem I.2.1 it follows that $I^{[p]}$ is irreducible. Since $\mathbb{F}[V]/I^{[p]}$ is totally finite $I^{[p]}$ is also $\overline{\mathbb{F}[V]}$-primary. □

The converse of Theorem II.6.5 has been proven by N. Nossem (see [71] Lemma 2.5.1). Our proof of Theorem II.6.5 allows us to deduce much more about the nature of the relation between $\mathbb{F}[V]/I$ and $\mathbb{F}[V]/I^{[p]}$.

THEOREM II.6.6: *Let $\mathbb{F}$ be a field of characteristic p and V a finite dimensional $\mathbb{F}$-vector space. Let $e \in \mathbb{N}_0$ and set $q = p^e$. If $I \subseteq \mathbb{F}[V]$ is an $\overline{\mathbb{F}[V]}$-primary irreducible ideal, then so is $I^{[q]}$. If $z_1, \ldots, z_n \in V^*$ is a basis and z^D a monomial representing a fundamental class of $\mathbb{F}[V]/I$ then $(z_1 \cdots z_n)^{q-1} z^{qD}$ represents a fundamental class of $\mathbb{F}[V]/I^{[q]}$, so*

$$\text{f-dim}\big(F[V]/I^{[q]}\big) = n(q-1) + q \cdot \big(\text{f-dim}(\mathbb{F}[V]/I)\big).$$

If $u_1, \ldots, u_n \in V$ is a basis and $\theta_I \in \Gamma(V)$ is a generator of Macaulay's dual principal system for $\mathbb{F}[V]/I$ then $\gamma_{q-1}(u_1) \cdots \gamma_{q-1}(u_n)\Upsilon_q(\theta_I)$ is a generator for Macaulay's dual principal system for $\mathbb{F}[V]/I^{[q]}$.

PROOF: That $I^{[q]}$ is an $\overline{\mathbb{F}[V]}$-primary irreducible ideal follows by induction on e from R. Y. Sharp's theorem, Theorem II.6.5. Since I is $\overline{\mathbb{F}[V]}$-primary it contains the ideal $Q = (z_1^{q^s}, \ldots, z_n^{q^s})$ for $s \in \mathbb{N}$ suitably large. Then $Q^{[q]} = (z_1^{q^{s+1}}, \ldots, z_n^{q^{s+1}}) \subseteq I^{[q]}$. We therefore have epimorphisms

$$\mathbb{F}[V]/Q \longrightarrow \mathbb{F}[V]/I$$

and

$$\mathbb{F}[V]/Q^{[q]} \longrightarrow \mathbb{F}[V]/I^{[q]}.$$

By Theorem I.2.1 $(Q:I)$ is principal over Q, say $(Q:I) = (u) + Q$. Then by Proposition II.6.3 $(Q^{[q]}:I^{[q]}) = (u^q) + Q^{[q]}$. The monomial $z_1^{q^s-1} \cdots z_n^{q^s-1}$ represents a fundamental class of $\mathbb{F}[V]/Q$, so from Corollary I.2.5 we learn that

$$uz^D \overset{\circ}{=} z_1^{q^s-1} \cdots z_n^{q^s-1}$$

in $\mathbb{F}[V]/Q$, where $a \mathrel{\dot{=}} b$ denotes that a and b are nonzero scalar multiples of each other. Without loss of generality we may suppose uz^D and $z_1^{q^s-1} \cdots z_n^{q^s-1}$ are equal. Therefore

$$u^q z^{qD} z_1^{q-1} \cdots z_n^{q-1} \equiv z_1^{q^{s+1}-1} \cdots z_n^{q^{s+1}-1}$$

in $\mathbb{F}[V]/Q^{[q]}$, so by Corollary I.2.3 we conclude that $z^{qD} z_1^{q-1} \cdots z_n^{q-1}$ represents a fundamental class for $\mathbb{F}[V]/I^{[q]}$.

We turn next to the statement about Macaulay duals. From what we have just shown a generator of $(I^{[q]})^{\perp}$ has degree $n(q-1) + qd$, where d is the formal dimension of $\mathbb{F}[V]/I$. The element $\gamma_{q-1}(u_1) \cdots \gamma_{q-1}(u_n)\Upsilon_q(\theta_I) \in \Gamma(V)$ has degree $n(q-1) + qd$ and is nonzero, since regarded as a linear form on $\mathbb{F}[V]_{n(q-1)+qd}$ it evaluates nonzero on the representative of the fundamental class of $\mathbb{F}[V]/I^{[q]}$ just constructed. The dual system $(I^{[q]})^{\perp}$ is a cyclic $\mathbb{F}[V]$-module so it is therefore enough to show that $\gamma_{q-1}(u_1) \cdots \gamma_{q-1}(u_n)\Upsilon_q(\theta_I)$ belongs to $(I^{[q]})^{\perp}_{n(q-1)+qd}$. This means we must show that $\gamma_{q-1}(u_1) \cdots \gamma_{q-1}(u_n)\Upsilon_q(\theta_I)$ regarded as a linear form on $\mathbb{F}[V]_{n(q-1)+qd}$ is zero on the elements of $I^{[q]}_{n(q-1)+qd}$.

A typical element of $I^{[q]}$ is a sum of terms of the form $z^E f^q$ where $f \in I$ and $E \in \mathbb{N}_0^n$. The monomial z^E can be rewritten in the form $z^E = z^A z^{qB}$, where $A, B \in \mathbb{N}_0^n$ and $A = (a_1, \ldots, a_n)$ with $0 \le a_1, \ldots, a_n \le q-1$. Therefore $z^E f^q = z^A z^{qB} f^q = z^A (z^B f)^q$. The element $z^B f$ belongs to I as I is an ideal, so in fact the typical element of $I^{[q]}$ is a sum of terms of the form $z^A h^q$, where $h \in I$ and $A = (a_1, \ldots, a_n)$ with $0 \le a_1, \ldots, a_n \le q-1$. Let $z^A h^q$ be such a term of degree $n(q-1) + qd$. Since $\Gamma(V)$ is an algebra over the Hopf algebra $\mathbb{F}[V]$ (see Section II.2) we obtain

$$\begin{aligned}
&\left(z^A \cdot h^q\right) \cap \left(\gamma_{q-1}(u_1) \cdots \gamma_{q-1}(u_n)\Upsilon_q(\theta_I)\right) \\
&\quad = \left(z^A \cap \left(\gamma_{q-1}(u_1) \cdots \gamma_{q-1}(u_n)\right)\right) \cdot \left(h^q \cap \Upsilon_q(\theta_I)\right) \\
&\quad = \left(\gamma_{q-1-a_1}(u_1) \cdots \gamma_{q-1-a_n}(u_n)\right) \cdot \left(h^q \cap \Upsilon_q(\theta_I)\right) \\
&\quad = \begin{cases} \left(\gamma_{q-1-a_1}(u_1) \cdots \gamma_{q-1-a_n}(u_n)\right) \cdot \theta_I(h) & \text{if } \deg(h) = d \\ \left(\gamma_{q-1-a_1}(u_1) \cdots \gamma_{q-1-a_n}(u_n)\right) \cdot \Upsilon_q\left(h \cap \theta_I\right) & \text{if } \deg(h) \neq d. \end{cases}
\end{aligned}$$

Since $h \in I$ it annihilates θ_I under the $\cap$-product and θ_I evaluates to zero on h. Hence the element $\gamma_{q-1}(u_1) \cdots \gamma_{q-1}(u_n)\Upsilon_q(\theta_I)$ belongs to $(I^{[q]})^{\perp}_{n(q-1)+qd}$ as was to be shown. □

Definition: *If V is an n-dimensional vector space over the field $\mathbb{F}$ of characteristic p, q is a power of p, and $u_1, \ldots, u_n \in V$ and $z_1, \ldots, z_n \in V^*$ are dual bases, then the operators $\mathbf{P} : \mathbb{F}[V] \longrightarrow \mathbb{F}[V]$ and $\mathbf{P} : \Gamma(V) \longrightarrow \Gamma(V)$*

defined by

$$\mathbf{P}(f) = (z_1 \cdots z_n)^{q-1} f^q \qquad f \in \mathbb{F}[V]$$
$$\mathbf{P}(\theta) = \gamma_{q-1}(u_1) \cdots \gamma_{q-1}(u_n) \Upsilon_q(\theta) \qquad \theta \in \Gamma(V)$$

are called **periodicity operators.**

REMARK: The operators **P** just defined do not depend on the choice of bases for V or V^*.

A result of W. Vasconcelos [102] implies that an $\overline{\mathbb{F}[x, y]}$-primary irreducible ideal is generated by a regular sequence of length two. The generalization to more variables may fail: an irreducible ideal in $\mathbb{F}[x, y, z]$ that is $\overline{\mathbb{F}[x, y, z]}$-primary need not be generated by a regular sequence. An example is provided by Macaulay [49] Section 71 which we described in Section I.5 Example 1. We use it next to illustrate our results on Frobenius powers.

EXAMPLE 1: Consider the ideal

$$L = (xy, xz, yz, z^2 - x^2, z^2 - y^2) \subset \mathbb{F}[x, y, z].$$

The corresponding Poincaré duality quotient algebra is the connected sum of three copies of the algebra $\mathbb{F}[u]/(u^3)$. The number of generators of an $\overline{\mathbb{F}[x, y, z]}$-primary irreducible ideal must be odd (see [103]). So the minimal number of generators of an $\overline{\mathbb{F}[x, y, z]}$-primary irreducible ideal that is not generated by a regular sequence is five. Therefore L has the minimal number of generators for an $\overline{\mathbb{F}[x, y, z]}$-primary irreducible ideal not generated by a regular sequence.

For $\mathbb{F} = \mathbb{F}_2$ the fundamental class of the Poincaré duality quotient

$$H = \mathbb{F}_2[x, y, z]/(xy, xz, yz, z^2 - x^2, z^2 - y^2)$$

is represented by $z^2 \in \mathbb{F}_2[x, y, z]$. By Theorem II.6.5 the first Frobenius power $L^{[2]}$ of the ideal L is primary for the maximal ideal, irreducible, and has Poincaré duality quotient

$$\mathbb{F}_2[x, y, z]/(x^2y^2, x^2z^2, y^2z^2, z^4 - x^4, z^4 - y^4),$$

whose fundamental class is represented by

$$\mathbf{P}(z^2) = xyz(z^2)^2 = xyz^5$$

by Theorem II.6.6. A generator for the dual principal system of L is

$$\theta = \gamma_2(u) + \gamma_2(v) + \gamma_2(w) \in \Gamma(u, v, w)_2,$$

where $u, v, w \in V$ is the basis dual to $x, y, z \in V^*$. Therefore by Theorem II.6.6 the dual principal system of $L^{[2]}$ is generated by

$$\gamma_1(u)\gamma_1(v)\gamma_1(w)\Upsilon_2\big(\gamma_2(u) + \gamma_2(v) + \gamma_2(w)\big) \in \Gamma(u, v, w)_7.$$

Using the formulae in Table II.1.1 we find

$$\gamma_2\big(\gamma_2(u) + \gamma_2(v) + \gamma_2(w)\big)$$
$$= \gamma_4(u) + \gamma_4(v) + \gamma_4(w) + \gamma_2(u)\gamma_2(v) + \gamma_2(v)\gamma_2(w) + \gamma_2(w)\gamma_2(u),$$

so the dual principal generator $\theta^{[2]}$ for $L^{[2]}$ turns out to be

$$\theta^{[2]} = \mathbf{P}(\theta) = \gamma_5(u)\gamma_1(v)\gamma_1(w) + \gamma_1(u)\gamma_5(v)\gamma_1(w) + \gamma_1(u)\gamma_1(v)\gamma_5(w)$$
$$+ \gamma_1(u)\gamma_3(v)\gamma_3(w) + \gamma_3(u)\gamma_1(v)\gamma_3(w) + \gamma_3(u)\gamma_3(v)\gamma_1(w).$$

From this it follows that

$$\operatorname{supp}(\theta^{[2]}) = \left\{x^5yz,\ xy^5z,\ xyz^5,\ xy^3z^3,\ x^3yz^3,\ x^3y^3z\right\},$$

which is in accord with the fact that xyz^5 represents a fundamental class for $\mathbb{F}_2[x, y, z]/L^{[2]}$.

The following proposition provides a direct description of the algebra $\mathbb{F}[V]/I^{[p]}$ for an irreducible ideal $I \subset \mathbb{F}[V]$. It is the result of discussions with N. Nossem. We use the same notation as in the proof of Proposition II.6.3. In addition recall that for any graded vector space, algebra, or module – over a field of characteristic p we denote by $\Phi(-)$ the *same* object regraded by

$$\Phi(-)_k = \begin{cases} -_{k/p} & \text{if } p \text{ divides } k \\ 0 & \text{otherwise.} \end{cases}$$

A sequence of graded connected commutative algebras $A' \xrightarrow{f'} A \xrightarrow{f''} A''$ is said to be **coexact** if $\operatorname{Im}(f'') \cong \mathbb{F} \otimes_{A'} A$ (see [62] and the references there for more about coexact sequences).

PROPOSITION II.6.7: *Let $\mathbb{F}$ be a field of characteristic $p \neq 0$, $\mathbb{F}[V] = \mathbb{F}[z_1, \ldots, z_n]$, and I an $\overline{\mathbb{F}[V]}$-primary irreducible ideal in $\mathbb{F}[V]$. Then there is a coexact sequence of Poincaré duality algebras*

$$\mathbb{F} \longrightarrow \Phi(\mathbb{F}[V]/I) \longrightarrow \mathbb{F}[V]/I^{[p]} \longrightarrow \mathbb{F}[z_1, \ldots, z_n]/(z_1^p, \ldots, z_n^p) \longrightarrow \mathbb{F}.$$

Moreover, $\mathbb{F}[V]/I^{[p]}$ is a free $\Phi(\mathbb{F}[V]/I)$-module and the sequence splits as a sequence of $\Phi(\mathbb{F}[V]/I)$-modules. Finally

$$\text{f-dim}(\mathbb{F}[V]/I^{[p]}) = n(p-1) + p \cdot \text{f-dim}(\mathbb{F}[V]/I)$$
$$= \text{f-dim}(\mathbb{F}[z_1, \ldots, z_n]/(z_1^p, \ldots, z_n^p)) + \text{f-dim}(\Phi(\mathbb{F}[V]/I)).$$

PROOF: As before let

$$\lambda : \Phi(\mathbb{F}[z_1, \ldots, z_n]) \longrightarrow \mathbb{F}[z_1, \ldots, z_n]$$

be the map defined by $\lambda(\Phi(f)) = f^p$ for any $f \in \mathbb{F}[z_1, \ldots, z_n]$. The grading on $\Phi(\mathbb{F}[z_1, \ldots, z_n])$ has been arranged to make this a map of algebras.[4] Since there is an inclusion $\lambda(\Phi(I)) \subseteq I^{[p]}$, there is also an induced map of algebras $\Phi(\mathbb{F}[V]/I) \longrightarrow \mathbb{F}[V]/I^{[p]}$ given by $\Phi(f + I) \longmapsto f^p + I^{[p]}$, where $f \in \mathbb{F}[V]$. Denote this map by ζ. The map ζ is monic: to see this note that an element $\Phi(f + I) \in \Phi(\mathbb{F}[V]/I)$ belongs to $\ker(\zeta)$ if and only if $f^p \in I^{[p]}$. By Proposition II.6.3 this means that $1 \in (I^{[p]} : f^p) = (I : f)^{[p]} \subseteq \mathbb{F}[V]$, which says $f \in I$ so $\Phi(f) \in \Phi(I)$, and hence $\ker(\zeta) = 0$ as claimed.

Next, observe that as $I^{[p]} \subseteq (z_1^p, \ldots, z_n^p)$, the cokernel of ζ is

$$\begin{aligned}\mathbb{F} \otimes_{\Phi(\mathbb{F}[V]/I)} \mathbb{F}[V]/I^{[p]} &= \mathbb{F}[z_1, \ldots, z_n]/(I^{[p]} + (z_1^p, \ldots, z_n^p)) \\ &= \mathbb{F}[z_1, \ldots, z_n]/(z_1^p, \ldots, z_n^p),\end{aligned}$$

since the ideal generated by the image of $\Phi(I) \subset \Phi(\mathbb{F}[V])$ under the map λ is $I^{[p]}$. This proves the coexactness of the sequence.

The formula for the formal dimensions follows from Theorem II.6.6. This in turn implies by Lemma VI.4.11 that $\mathbb{F}[V]/I^{[p]}$ is a free $\Phi(\mathbb{F}[V]/I)$-module. Although Lemma VI.4.11 is proved further along in this manuscript, in Part VI, its proof is independent of Proposition II.6.7. □

Note that for an irreducible $\mathfrak{m}$-primary ideal I of $\mathbb{F}[z_1, \ldots, z_n]$ the split coexact sequence given in Proposition II.6.7 representing $\mathbb{F}[z_1, \ldots, z_n]/I^{[p]}$ is not the trivial split coexact sequence, viz.,

$$\mathbb{F} \longrightarrow \Phi(\mathbb{F}[V]/I) \longrightarrow \Phi(\mathbb{F}[V]/I) \otimes \big(\mathbb{F}[z_1, \ldots, z_n]/(z_1^p, \ldots, z_n^p)\big) \longrightarrow \big(\mathbb{F}[z_1, \ldots, z_n]/(z_1^p, \ldots, z_n^p)\big) \longrightarrow \mathbb{F}.$$

The coexact sequence of Proposition II.6.7 does not split except in very special degenerate cases. The following property of this sequence explains why. The elements $\Phi(z_i)$ of degree p in the tensor product algebra $\Phi(\mathbb{F}[V]/I) \otimes \big(\mathbb{F}[z_1, \ldots, z_n]/(z_1^p, \ldots, z_n^p)\big)$ are indecomposable, whereas $\mathbb{F}[V]/I^{[p]}$ is generated by forms of degree 1. In fact, one sees that the multiplication on $\mathbb{F}[V]/I^{[p]}$ satisfies $z_i^p = \lambda(\Phi(z_i))$ for $i = 1, \ldots, n$. So this extension is in some sense universal for converting generators truncated at height p into generators truncated at height p^2.

Implicit in the proof of Proposition II.6.7 is a definition of the Frobenius functor (see [73]) in the graded case. We intend to return to this functor in a further joint manuscript with N. Nossem.

As an illustration of how the tools developed to this point interact we close this part with a result that makes use of several of them in its short proof.

---

[4] Apart from perhaps a Frobenius automorphism of the ground field if it has more than p elements.

PROPOSITION II.6.8: *Let $\mathbb{F}$ be a field of characteristic p, $V' = \mathbb{F}^{n'}$, and $V'' = \mathbb{F}^{n''}$. Suppose that $H' = \mathbb{F}[V']/I'$ and $H'' = \mathbb{F}[V'']/I''$ are Poincaré duality quotients of $\mathbb{F}[V']$, $\mathbb{F}[V'']$ respectively, and set*

$$H = H' \# H'' = \mathbb{F}[V' \oplus V'']/I.$$

If $\gamma' \in \Gamma(V')$ and $\gamma'' \in \Gamma(V'')$ are Macaulay duals for I' respectively I'', then a Macaulay dual for $I^{[p]}$ is

$$\gamma_{p-1}(u_1') \cdots \gamma_{p-1}(u_{n'}') \cdot \gamma_{p-1}(u_1'') \cdots \gamma_{p-1}(u_{n''}'') \cdot \gamma_p(\gamma' + \gamma'') \in \Gamma(V' \oplus V''),$$

where $u_1', \ldots, u_{n'}'$ is a basis for V' and $u_1'', \ldots, u_{n''}''$ is a basis for V''.

PROOF: By Proposition II.2.4 $\gamma' + \gamma'' \in \Gamma(V' \oplus V'')$ is a Macaulay dual for I. The result then follows from Theorem II.6.6. □

Part III
Poincaré duality and the Steenrod algebra

POINCARÉ duality quotients of $\mathbb{F}_q[V]$, where $\mathbb{F}_q$ is a Galois field, may support an additional structure derived from the Frobenius homomorphism.[1] Specifically the operation $\mathbf{\Phi}$ of raising linear forms in $\mathbb{F}_q[V]$ to the q-th power defines an algebra homomorphism $\mathbf{\Phi} : \mathbb{F}_q[V] \longrightarrow \mathbb{F}_q[V]$ that can be used to define the Steenrod algebra $\mathscr{P}^*$ of a Galois field and $\mathbb{F}_q[V]$ becomes a $\mathscr{P}^*$-module (see e.g. [87] Chapter 10). To demand that an ideal I in $\mathbb{F}_q[V]$ be stable under the $\mathscr{P}^*$-action imposes severe restrictions, e.g., the prime ideals in $\mathbb{F}_q[V]$ stable under $\mathscr{P}^*$ are generated by the linear forms they contain (see [78] Section 2 Proposition 1, or [68] Theorem 9.2.1). If an irreducible $\overline{\mathbb{F}_q[V]}$-primary ideal is stable under $\mathscr{P}^*$ then the Steenrod operations pass to the Poincaré duality quotient defined by it, which becomes an unstable algebra over the Steenrod algebra. This allows us to define *characteristic classes* for such quotients.

Although Steenrod operations were originally conceived in the 1940s as a refinement of the cup-product in cohomology, we prefer to view them as a mechanism for bringing to the fore information otherwise left hidden behind the Frobenius homomorphism $\mathbf{\Phi}$. For an introduction to the Steenrod algebra $\mathscr{P}^*$ of a Galois field from this viewpoint see for example [68] Chapter 8, [87] Chapters 9 and 10, or [90]. The latter provides a complete account of the basic structure of $\mathscr{P}^*$ as a Hopf algebra, and presents a number of examples of the use of Steenrod operations in modular invariant theory. *We assume the reader is familiar with one of these, or at a minimum the basic algebraic material on the Steenrod algebra in* [97] *Chapters* 1, 2, and 4.

---

[1] The term *Frobenius homomorphism* is not unambiguously defined, not even in this manuscript! We hope we have made clear what we mean by it in each context where it appears.

Many other points of departure to introduce the Steenrod algebra are possible. See for example, [108] for an interpretation of Steenrod operations as differential operators (in this connection see also [26]). For a categorical approach see [7]. Steenrod-like operations also occur in connection with Hasse–Schmidt differentials (see e.g. [29] or [101]). For a completely different approach to deriving information from the Frobenius homorphism see [73] and [71].

After introducing some basic concepts related to Steenrod operations and the Steenrod algebra we extend Macaulay's theory of $\mathfrak{m}$-primary irreducible ideals to this new context. We then define Wu classes for Poincaré duality quotients of $\mathbb{F}_q[V]$ supporting Steenrod operations. These provide yet another perspective in our study, derived from one which was pioneered by Wu Wen-Tsün [111]and [112] for smooth manifolds. Those Poincaré duality quotients of $\mathbb{F}_q[V]$ supporting Steenrod operations which have trivial Wu classes are closely related to the *Hit Problem* of finding a minimal generating set for $\mathbb{F}_q[V]$ as a module over $\mathscr{P}^*$. We make use of the Frobenius powers of ideals (see Section II.6) and the $K \subset L$ paradigm (see Section II.5) to study families of Poincaré duality quotients with trivial Wu classes at the end of this Part as well as in Parts IV, V and VI.

III.1 $\mathscr{P}^*$-Unstable Poincaré duality quotients of $\mathbb{F}_q[V]$

In the case of a finite field $\mathbb{F}_q$ with $q = p^\nu$ elements, p a prime, there are a number of additional features that arise in the study of Poincaré duality quotients of $\mathbb{F}_q[V]$ that depend on the action of the Steenrod operations on $\mathbb{F}_q[V]$ (see e.g. [68] Chapter 8, [87] Chapters 10 and 11, or [90]).

NOTATION: *We write $\mathscr{P}^*$ for the Steenrod algebra of a Galois field $\mathbb{F}_q$. In the special case $q = 2$ we also use the notation $\mathscr{A}^*$.*

By an **unstable Poincaré duality algebra** over $\mathbb{F}_q$ we mean a Poincaré duality algebra over $\mathbb{F}_q$ that is also an unstable algebra over the Steenrod algebra $\mathscr{P}^*$ of the given Galois field $\mathbb{F}_q$. In the same spirit we use the terminology unstable Noetherian algebra, unstable Cohen–Macaulay algebra, etc. If A is an unstable algebra over the Steenrod algebra $\mathscr{P}^*$ recall that an ideal $I \subset A$ is said to be $\mathscr{P}^*$**-invariant** if $\Theta(I) \subseteq I$ for all $\Theta \in \mathscr{P}^*$. We begin by explaining how the results of the previous Parts extend to encompass an unstable Steenrod algebra action. As a first step we reformulate Proposition I.1.5 as follows.

PROPOSITION III.1.1: *Let A be a Noetherian commutative graded connected algebra over the Galois field $\mathbb{F}_q$ which is an unstable algebra over* *the Steenrod algebra $\mathscr{P}^*$ of the given Galois field. Then there is a bijective correspondence between*

$\mathscr{P}^*$-unstable Poincaré duality algebras that are quotients of A,

on the one hand, and

$\mathscr{P}^*$-invariant $\overline{A}$-primary irreducible ideals $I \subset A$,

on the other. The correspondence associates to a $\mathscr{P}^*$-unstable Poincaré duality quotient algebra of A with quotient map $\pi : A \longrightarrow H$ the ideal $\ker(\pi)$, *and to a $\mathscr{P}^*$-invariant irreducible $\overline{A}$-primary ideal I the quotient algebra A/I.* □

If A is an unstable Gorenstein algebra, then the characterization of the $\overline{A}$-primary irreducible ideals in A as those ideals that are principal over a parameter ideal contained in them (Theorem I.2.1 or [114] Chapter IV Theorem 34) also extends to the new context. To prove this we require two lemmas.

LEMMA III.1.2: *Let A be a commutative graded connected $\mathscr{P}^*$-unstable Noetherian algebra over the Galois field $\mathbb{F}_q$. If $I \subset A$ is an $\overline{A}$-primary ideal then I contains a $\mathscr{P}^*$-invariant parameter ideal.*

PROOF: By [87] Proposition 11.2.3, the minimal prime ideals of A are $\mathscr{P}^*$-invariant, so passing from A to the quotient by a minimal prime allows us to assume that A is an integral domain. By [66] Theorem 7.4.3 or [53] Theorem 5.2.1, A contains a $\mathscr{P}^*$-invariant parameter ideal, say Q. Let $Q = (a_1, \ldots, a_n)$ and set $d_i = \deg(a_i)$ for $i = 1, \ldots, n$. Since I is $\overline{A}$-primary the algebra A/I is zero in all sufficiently large degrees, say $(A/I)_j = 0$ if $j > d$. Choose $\lambda \in \mathbb{N}$ such that $q^\lambda \cdot d_i > d$ for $i = 1, \ldots, n$. Then $a_1^{q^\lambda}, \ldots, a_n^{q^\lambda}$ belong to I and the ideal they generate, say Q', is a parameter ideal. A short computation with the formula

$$\mathscr{P}^k(u^q) = \begin{cases} (\mathscr{P}^{k/q}(u))^q & \text{if } k \text{ is divisible by } q \\ 0 & \text{otherwise} \end{cases}$$

shows that Q' is $\mathscr{P}^*$-invariant. □

LEMMA III.1.3: *Let A be a commutative graded connected unstable algebra over the Galois field $\mathbb{F}_q$ and let $I, J \subset A$ be $\mathscr{P}^*$-invariant ideals. Then the ideal $(I : J)$ is also $\mathscr{P}^*$-invariant.*

PROOF: Suppose that $a \in (I : J)$ and that inductively we have shown that $a, \ldots, \mathscr{P}^{k-1}(a)$ belong to $(I : J)$ for some integer $k > 0$. Let $b \in J$. Then $ab \in I$, so $\mathscr{P}^k(ab) \in I$ since I is $\mathscr{P}^*$-invariant. If we apply the Cartan formula to $\mathscr{P}^k(ab)$ we get

$$\mathscr{P}^k(ab) = \big(\mathscr{P}^k(a)\big)b + \big(\mathscr{P}^{k-1}(a)\mathscr{P}^1(b)\big) + \cdots + a\big(\mathscr{P}^k(b)\big).$$

The elements $\mathcal{P}^{k-1}(a), \ldots, \mathcal{P}^1(a)$, a belong to $(I:J)$ by the induction assumption, and the elements $\mathcal{P}^1(b), \ldots, \mathcal{P}^k(b)$, b belong to J because J is $\mathscr{P}^*$-invariant. Rearranging the previous equation gives

$$\big(\mathcal{P}^k(a)\big)b = \mathcal{P}^k(ab) - \big(\mathcal{P}^{k-1}(a)\mathcal{P}^1(b) + \cdots + a\mathcal{P}^k(b)\big),$$

and since the right hand side is in I it follows that $\mathcal{P}^k(a)b$ belongs to I. So $\mathcal{P}^k(a)$ belongs to $(I:J)$, and hence $(I:J)$ is $\mathscr{P}^*$-invariant as claimed. □

ALTERNATIVE PROOF OF LEMMA III.1.3: Introduce the **total Steenrod operation**

$$\mathcal{P} = 1 + \mathcal{P}^1 + \cdots + \mathcal{P}^k + \cdots \in \mathscr{P}^{**} = \prod_{i=0}^{\infty} \mathscr{P}^i.$$

It is an algebra automorphism[2] of the ungraded algebra $\mathrm{Tot}(A)$ with inverse $\chi(\mathcal{P})$ where χ is the canonical anti-automorphism of the Steenrod algebra regarded as a Hopf algebra (see e.g. [59] or [58]). For a $\mathscr{P}^*$-invariant ideal K one has $\mathcal{P}(K) = K$. Since $a \in (I:J)$ if and only if $a \cdot J \subseteq I$ we find if we apply $\mathcal{P}$ to this equation that $\mathcal{P}(a) \cdot J = \mathcal{P}(a) \cdot \mathcal{P}(J) = \mathcal{P}(a \cdot J) \subseteq \mathcal{P}(I) = I$ and the result follows. □

We say that an element a in an unstable $\mathscr{P}^*$-algebra is a **Thom class** if the principal ideal $(a) \subset A$ is closed under the action of the Steenrod algebra. This differs a bit from the usage in [66] Section 4.3 where in addition it is required that the ideal (a) have height one, or in [10] where it is required that (a) be a free A-module.

THEOREM III.1.4: *Let A be a commutative graded connected $\mathscr{P}^*$-unstable Noetherian algebra over the Galois field $\mathbb{F}_q$. If $K \subset L$ are $\mathscr{P}^*$-invariant $\overline{A}$-primary irreducible ideals then $(K:L) = (a) + K$ and $L = (K:a)$ for some $a \in A$ which becomes a Thom class in A/K. Conversely, if $K \subset A$ is a $\mathscr{P}^*$-invariant $\overline{A}$-primary irreducible ideal and $a \in A$ becomes a Thom class in A/K, then $L = (K:a)$ is also a $\mathscr{P}^*$-invariant $\overline{A}$-primary irreducible ideal.*

PROOF: By Theorem I.2.1 $(K:L) = (a) + K$ and $L = (K:a)$ for some element $a \in A$. By Lemma III.1.3 the ideal $(K:L)$ is $\mathscr{P}^*$-invariant as both K and L are. The principal ideal generated by u, the image of a in A/K, is $(K:L)/K$ and is therefore a $\mathscr{P}^*$-invariant ideal. Hence $u \in A/K$ is a Thom class.

Conversely, if $K \subset A$ is a $\mathscr{P}^*$-invariant $\overline{A}$-primary irreducible ideal and $a \in A$ becomes a Thom class in A/K, then the principal ideal (u) generated by

---

[2] The group of units in the algebra $\mathscr{P}^{**}$ turns out to be the Nottingham group (see e.g. [89], and [38] and [113] for the group theoretic background).

the image of a in A/K is $\mathscr{P}^*$-invariant. Since the quotient map $A \longrightarrow A/K$ is a map of $\mathscr{P}^*$-algebras, the preimage $(a) + K$ of the ideal (u) is $\mathscr{P}^*$-invariant and hence so is $L = (K : a)$. By Theorem I.2.1 it is also $\overline{A}$-primary and irreducible. □

COROLLARY III.1.5: *Let A be a commutative graded connected $\mathscr{P}^*$-unstable Gorenstein algebra over the Galois field $\mathbb{F}_q$ and I a $\mathscr{P}^*$-invariant $\overline{A}$-primary ideal of A. Then I contains $\mathscr{P}^*$-invariant parameter ideals. It is irreducible if and only if for every $\mathscr{P}^*$-invariant parameter ideal $Q \subset I$ there is an element $a \in A$ with $(Q : I) = (a) + Q$ such that a becomes a Thom class in A/Q.*

PROOF: Combine Lemma III.1.2 and Theorem III.1.4. □

We can apply these results to the study of $\mathscr{P}^*$-Poincaré duality quotient algebras of $\mathbb{F}_q[V]$ as follows: let $\mathbb{F}_q[V] \xrightarrow{\pi} H$ be a $\mathscr{P}^*$-Poincaré duality quotient of $\mathbb{F}_q[V]$ with $I = \ker(\pi) \subset \mathbb{F}_q[V]$. Choose a basis $z_1, \ldots, z_n \in V^*$ for the linear forms of $\mathbb{F}_q[V]$ and an integer $k \in \mathbb{N}$ such that the $\mathscr{P}^*$-invariant parameter ideal $Q^{[q^k]} = (z_1^{q^k}, \ldots, z_n^{q^k})$ is contained in I. Then $I = (Q^{[q^k]}) : f)$ for some form $f \in \mathbb{F}_q[V]$ which becomes a Thom class in $\mathbb{F}_q[V]/Q^{[q^k]}$ and determines H. From this point of view the problem of classifying the $\mathscr{P}^*$-invariant Poincaré duality quotients of $\mathbb{F}_q[V]$ translates into the problem of determining the Thom classes in the quotient algebras $\mathbb{F}_q[z_1, \ldots, z_n]/(z_1^{q^k}, \ldots, z_n^{q^k})$. The following result and example show that this is quite a bit different from finding Thom classes in $\mathbb{F}_q[V]$.

PROPOSITION III.1.6: *Suppose $f \in \mathbb{F}_q[V]$ is a Thom class, then f is a product of linear forms.*

PROOF: If f has degree one there is nothing to prove, so we may proceed by induction on the degree of f and assume the result holds for all Thom classes whose degree is strictly smaller than that of f. Let $\mathfrak{p} \supseteq (f)$ be an associated minimal prime ideal of the principal ideal (f). By [67] (or see [87] Corollary 11.2.4) $\mathfrak{p}$ is a $\mathscr{P}^*$-invariant ideal, and by Krull's Principal Ideal Theorem ([5] Theorem 1.2.10) $ht(\mathfrak{p}) = 1$. Hence by Serre's theorem ([87] Theorem 11.3.1) $\mathfrak{p} = (\ell)$ for some nonzero linear form ℓ. Since $(\ell) \supseteq (f)$ it follows that ℓ divides f. Let $h = f/\ell$. We claim that h is a Thom class. To see this apply the total Steenrod operation

$$\mathscr{P} = 1 + \mathscr{P}^1 + \cdots + \mathscr{P}^k + \cdots,$$

to f. Since f is a Thom class the principal ideal (f) is $\mathscr{P}^*$-invariant and we may write $\mathcal{P}(f) = f \cdot F$ for some inhomogeneous $F = 1 + f_1 + \cdots$. Then

$$\mathcal{P}(h) = \frac{\mathcal{P}(f)}{\mathcal{P}(\ell)} = \frac{f \cdot F}{\ell + \ell^q} = \frac{f}{\ell} \cdot \frac{F}{1 + \ell^{q-1}} = hF\Big(\sum_{i=0}^{\infty}(-\ell^{q-1})^i\Big).$$

Taking homogeneous components then shows that h divides $\mathcal{P}^k(h)$ for all $k \in \mathbb{N}_0$. So h is indeed a Thom class and we can apply the inductive hypothesis to complete the proof. □

Thus, there are two obvious sources of Thom classes in $\mathbb{F}_q[V]/Q$ when $Q \subset \mathbb{F}_q[V]$ is a $\mathscr{P}^*$-invariant parameter ideal: elements whose degree is strictly larger than f-dim$(\mathbb{F}_q[V]/Q) - (q-1)$ and elements that are the image of a product of linear forms in $\mathbb{F}_q[V]$. Here is an example of a less obvious sort.

EXAMPLE 1: Consider the ideal $K = (x^4, y^4, z^4) \subset \mathbb{F}_2[x, y, z]$. It is $\mathscr{A}^*$-invariant so the corresponding Poincaré duality quotient $H = \mathbb{F}_2[x, y, z]/K$ has formal dimension 9 and is an unstable $\mathscr{A}^*$-algebra. The cubic form $f = x^3 + y^3 + z^3 \in \mathbb{F}_2[x, y, z]$ is not a product of linear forms, so by Proposition III.1.6 is not a Thom class. However since

$$\mathrm{Sq}^i(u^k) = \binom{k}{i} u^{i+k}$$

for any linear form u we see that f becomes a Thom class in the quotient algebra H. In fact $\mathrm{Sq}^i(f) = 0$ for all $i > 0$ in H. From this it follows that $\mathrm{Ann}_H(f) \subset H$ is an $\mathscr{A}^*$-invariant irreducible ideal and $H/\mathrm{Ann}_H(f)$ is an unstable Poincaré duality algebra over the Steenrod algebra $\mathscr{A}^*$.

III.2 The $\mathscr{P}^*$-Double Annihilator Theorem

We next reformulate Macaulay's Double Annihilator Theorem, Theorem II.2.2, in the extended context of $\mathscr{P}^*$-algebras. To this end let V be a finite dimensional $\mathbb{F}_q$-vector space. Introduce a $\mathscr{P}^*$-action on $\Gamma(V)$ by the requirement

$$<\Theta(\gamma) \mid f> = <\gamma \mid \chi(\Theta)(f)> \qquad \forall\, \Theta \in \mathscr{P}^*,\ \gamma \in \Gamma(V),\ f \in \mathbb{F}_q[V]_{|\gamma| - |\Theta|},$$

where $\chi : \mathscr{P}^* \longrightarrow \mathscr{P}^*$ is the canonical anti-automorphism of the Steenrod algebra. As with the $\mathbb{F}_q[V]$-module structure on $\Gamma(V)$ this action of $\mathscr{P}^*$ on $\Gamma(V)$ *lowers* degrees, i.e., $\deg(\Theta(\gamma)) = \deg(\gamma) - \deg(\Theta)$.

To make computations with the action of $\mathscr{P}^*$ on $\Gamma(V)$ we note that for any linear form z

$$z = \chi(\mathcal{P})(\mathcal{P}(z)) = \chi(\mathcal{P})(z + z^q) = \chi(\mathcal{P})(z) + \big(\chi(\mathcal{P})(z)\big)^q,$$

since $\chi(\mathcal{P}) \cdot \mathcal{P} = 1 \in \mathscr{P}^{**}$. So comparison of coefficients gives

$$\chi(\mathcal{P})(z) = z - z^q + z^{q^2} - \cdots$$

and taking homogeneous components yields

$$\chi(\mathcal{P}^i)(z) = \begin{cases} (-1)^r z^{q^r} & \text{if } i = \frac{q^r-1}{q-1} \\ 0 & \text{otherwise.} \end{cases}$$

From this one may derive formulae for the action of $\mathscr{P}^*$ on the generators $\gamma_{q^r}(u_1), \ldots, \gamma_{q^r}(u_n)$ of $\Gamma(V)$. For example, the element $z_i^{q^r} \in \mathbb{F}_q[V]$ is dual to $\gamma_{q^r}(u_i) \in \Gamma(V)$, so we obtain

$$\mathcal{P}^{\frac{q^r-1}{q-1}}\left(\gamma_{q^r}(u_i)\right) = (-1)^r \gamma_1(u_i) = (-1)^r u_i \in \Gamma(V)$$

since

$$(-1)^r = <\gamma_{q^r}(u_i) \mid \chi(\mathcal{P}^{\frac{q^r-1}{q-1}})(z_i)> = <\mathcal{P}^{\frac{q^r-1}{q-1}}\left(\gamma_{q^r}(u_i)\right) \mid z_i>.$$

See e.g., [17] for other such formulae.

Note that the action of $\chi(\mathcal{P}) = 1 + \chi(\mathcal{P}^1) + \chi(\mathcal{P}^2) + \cdots$ on $\mathbb{F}_q[V]$ commutes with the coproduct ∇ of $\mathbb{F}_q[V]$. This may be seen by reducing to the case where $\dim_{\mathbb{F}_q}(V) = 1$, using that $\mathbb{F}_q[z_1, \ldots, z_n] \cong \mathbb{F}_q[z_1] \otimes \cdots \otimes \mathbb{F}_q[z_n]$ as Hopf algebras, and direct computation. Hence the action of $\mathcal{P}$ on $\Gamma(V)$ is an algebra automorphism. This means that the action of $\mathscr{P}^*$ on $\Gamma(V)$ satisfies the Cartan formula, viz.,

$$\mathcal{P}^k(\gamma' \cdot \gamma'') = \sum_{k'+k''=k} \mathcal{P}^{k'}(\gamma') \cdot \mathcal{P}^{k''}(\gamma'').$$

The $\mathbb{F}_q[V]$- and $\mathscr{P}^*$-module structures on $\Gamma(V)$ are compatible in the sense that they also satisfy a Cartan like formula, viz.,

$$\mathcal{P}^k(f \cap \gamma) = \sum_{i+j=k} \mathcal{P}^i(f) \cap \mathcal{P}^j(\gamma).$$

This makes $\Gamma(V)$ into a module over the algebra $\mathbb{F}_q[V] \odot \mathscr{P}^*$, the semitensor product of $\mathbb{F}_q[V]$ with $\mathscr{P}^*$ (see [51], in fact it is an algebra over the Hopf algebra $\mathbb{F}_q[V] \odot \mathscr{P}^*$ [85]). We do not need the construction of the semitensor product but find the notation a convenient shorthand for this type of structure.

Given this, the proof of Theorem II.2.2 shows the following.

THEOREM III.2.1 (F. S. Macaulay): *There is a bijective correspondence between nonzero $\overline{\mathbb{F}_q[V] \odot \mathscr{P}^*}$-submodules of $\Gamma(V)$ that are cyclic as $\mathbb{F}_q[V]$-modules, and proper $\overline{\mathbb{F}_q[V]}$-primary irreducible $\mathscr{P}^*$-invariant ideals in $\mathbb{F}_q[V]$. The correspondence is given by associating to a nonzero cyclic $\mathbb{F}_q[V]$-submodule $M(\gamma) \subset \Gamma(V)$ that is closed under the action of the*

Steenrod algebra on $\Gamma(V)$ its annihilator ideal $I(\gamma) = \mathrm{Ann}_{\mathbb{F}_q[V]}(M(\gamma)) \subset \mathbb{F}_q[V]$, and to a proper $\overline{\mathbb{F}_q[V]}$-primary irreducible $\mathscr{P}^*$-invariant ideal $I \subset \mathbb{F}_q[V]$ the submodule $\mathrm{Ann}_{\Gamma(V)}(I)$ of elements annihilated by I. □

Examples of cyclic $\mathbb{F}_q[V] \odot \mathscr{P}^*$-submodules of $\Gamma(V)$ and the application of this $\mathscr{P}^*$ version of Macaulay's Double Annihilator Theorem appear in particular in Sections III.4 and III.7.

III.3 Wu classes

$\mathscr{P}^*$-unstable Poincaré duality algebras support an additional structure which is an analog of the characteristic classes of algebraic topology. To describe this we introduce some additional notation and terminology. If H is an unstable Poincaré duality algebra with fundamental class $[H]$ then we define the **Wu classes** of H, denoted by $\mathrm{Wu}_k(H) \in H_{k(q-1)}$ for $k \in \mathbb{N}_0$, by requiring (see e.g. [1])

$$<\mathcal{P}^k(f) \mid [H]> = <f \cdot \mathrm{Wu}_k(H) \mid [H]>,$$

where $<- \mid [H]>$ is the map of H to $\mathbb{F}_q$ defined as follows:

$$<a \mid [H]> = \begin{cases} \lambda & \text{if } \deg(a) = \text{f-dim}(H) \text{ and } a = \lambda[H] \\ 0 & \text{otherwise.} \end{cases}$$

In the special case where $H = \mathbb{F}_q[V]/I$ with $I \subset \mathbb{F}_q[V]$ a $\mathscr{P}^*$-invariant $\overline{\mathbb{F}[V]}$-primary irreducible ideal, the contraction pairing $\Gamma(V) \times \mathbb{F}_q[V] \longrightarrow \Gamma(V)$ induces a pairing

$$<- \mid -> : I^{\perp} \times \mathbb{F}_q[V]/I \longrightarrow \mathbb{F}_q.$$

One readily sees that the Wu classes of H are then uniquely defined by the requirement

$$(\because) \qquad <\theta_I \mid \mathcal{P}^k(f)> = <\theta_I \mid f \cdot \mathrm{Wu}_k(H)> \qquad \forall\, f \in H$$

where $\theta_I \in \Gamma(V)$ is a generator of the dual system $I^{\perp}$ of I.

The unstability condition for the Steenrod algebra action implies the following vanishing result for Wu classes.

PROPOSITION III.3.1: *Suppose that H is an unstable Poincaré duality algebra with $\text{f-dim}(H) = d$. Then $\mathrm{Wu}_k(H) = 0$ if $k > \left[\frac{d}{q}\right]$ where q is the number of elements in the ground field.*

PROOF: If $\mathrm{Wu}_k(H) \neq 0$ and $m = d - k(q-1)$ there must be an element $h \in H_m$ with $\mathcal{P}^k(h) \neq 0$. Therefore by unstability $k \leq m = d - k(q-1)$, i.e., $\mathrm{Wu}_k(H) \neq 0$ requires that $kq = k(q-1) + k \leq d$. □

We say that H has **trivial Wu classes** if $\mathrm{Wu}_k(H) = 0$ for all $k > 0$.

PROPOSITION III.3.2: *Suppose that H is a $\mathcal{P}^*$-unstable Poincaré duality algebra with trivial Wu classes of formal dimension d over the Galois field $\mathbb{F}_q$. If $a \in H_i$, $b \in H_j$ and $k \in \mathbb{N}_0$ are such that $i + k(q-1) + j = d$ then*

$$\mathcal{P}^k(a) \cdot b = a \cdot \chi(\mathcal{P}^k)b.$$

PROOF: Let $[H] \in H_d$ be a fundamental class. For $h = h_0 + h_1 + \cdots \in \mathrm{Tot}(H)$ define $<h \mid [H]> \in \mathbb{F}_q$ by demanding that $h_d = <h \mid [H]> \cdot [H]$. Using this notation we need to prove the identity

$$<\mathcal{P}^k(a) \cdot b \mid [H]> = <a \cdot \chi(\mathcal{P}^k)(b) \mid [H]>.$$

Since $\mathcal{P}^{-1} = \chi(\mathcal{P})$ we have for degree reasons

$$\begin{aligned} &<\mathcal{P}^k(a) \cdot b \mid [H]> \\ &\quad = <\mathcal{P}(a) \cdot b \mid [H]> = <\mathcal{P}\big(a \cdot \chi(\mathcal{P})(b)\big) \mid [H]> \\ &\quad = <\mathrm{Wu}(H) \cdot a \cdot \chi(\mathcal{P})(b) \mid [H]> = <a \cdot \chi(\mathcal{P})(b) \mid [H]> \\ &\quad = <a \cdot \chi(\mathcal{P}^k)(b) \mid [H]> \end{aligned}$$

as $\mathrm{Wu}(H) = 1$. □

If $I \subset \mathbb{F}_q[V]$ is a $\mathcal{P}^*$-invariant $\overline{\mathbb{F}_q[V]}$-primary irreducible ideal we take the freedom to write $\mathrm{Wu}_k(I)$ for $\mathrm{Wu}_k(\mathbb{F}_q[V]/I)$ in the belief that this simplified notation is unlikely to cause confusion. Likewise we say that I has **trivial Wu classes** if $\mathrm{Wu}_k(I) = 0$ for $k > 0$.

The Wu classes can be used to relate the actions of the Steenrod algebra on $\mathbb{F}_q[V]/I$ and on $I^\perp$. To this end we introduce the **conjugate Wu classes**, denoted by $\chi\mathrm{Wu}(H) \in H_{k(q-1)}$ for $k \in \mathbb{N}_0$, which are defined for any unstable Poincaré duality algebra by the requirement that

$$<\chi(\mathcal{P}^k)(f) \mid [H]> = <f \cdot \chi\mathrm{Wu}_k(H) \mid [H]>.$$

Analogous to formula (·:·) the conjugate Wu classes of a $\mathcal{P}^*$-Poincaré duality quotient algebra $H = \mathbb{F}_q[V]/I$ are characterized by the condition

$$(✠) \qquad <\theta_I \mid \chi(\mathcal{P}^k)(f)> = <\theta_I \mid f \cdot \chi\mathrm{Wu}_k(H)>.$$

We say H has **trivial conjugate Wu classes** if $\chi\mathrm{Wu}(H) = 1$. The lemma that follows shows it is unnecessary to make a distinction between trivial Wu classes and trivial conjugate Wu classes.

LEMMA III.3.3: *If H is an unstable Poincaré duality algebra of formal dimension d with total Wu class $\mathrm{Wu}(H) = 1 + \mathrm{Wu}_1(H) + \cdots$ and total conjugate Wu class $\chi\mathrm{Wu}(H) = 1 + \chi\mathrm{Wu}_1(H) + \cdots$, then $\mathrm{Wu}(H)$ and $\chi\mathrm{Wu}(H)$*

are related to each other in $\mathrm{Tot}(H)$ *by the formulae*

$$\mathcal{P}(\mathrm{Wu}(H)) \cdot \chi\mathrm{Wu}(H) = 1 = \mathrm{Wu}(H) \cdot \chi(\mathcal{P})(\chi\mathrm{Wu}(H)).$$

In particular, H *has trivial Wu classes if and only if it has trivial conjugate Wu classes.*

PROOF: It follows from the perfectness of the pairing $<-\mid->$ that $\mathcal{P}(\mathrm{Wu}(H)) \cdot \chi\mathrm{Wu}(H) = 1 \in \mathrm{Tot}(H)$ if and only if for every $h \in H$

$$<h \mid [H]> = <\big(\mathcal{P}\big(\mathrm{Wu}(H)\big) \cdot \chi\mathrm{Wu}(H)\big) \cdot h \mid [H]>.$$

For any $h \in H$ we have

$$\begin{aligned} &<\big(\mathcal{P}\big(\mathrm{Wu}(H)\big) \cdot \chi\mathrm{Wu}(H)\big) \cdot h \mid [H]> = <\chi\mathrm{Wu}(H) \cdot \mathcal{P}\big(\mathrm{Wu}(H)\big) \cdot h \mid [H]> \\ &\quad = <\chi(\mathcal{P})\big(\mathcal{P}\big(\mathrm{Wu}(H)\big) \cdot h\big) \mid [H]> = <\mathrm{Wu}(H) \cdot \chi(\mathcal{P})(h) \mid [H]> \\ &\quad = <\mathcal{P}\big(\chi(\mathcal{P})(h)\big) \mid [H]> = <h \mid [H]>. \end{aligned}$$

Therefore $\mathcal{P}(\mathrm{Wu}(H)) \cdot \chi\mathrm{Wu}(H) = 1$. The last assertion follows by applying $\chi(\mathcal{P})$ to this equation since $\mathcal{P}$ is an automorphism with $\mathcal{P}^{-1} = \chi(\mathcal{P})$. □

REMARK: Topologists familiar with the work of Thom and Wu should recognize from this lemma that the conjugate Wu classes are an analog of Stiefel–Whitney classes when $q = 2$. Note that in contrast to $\mathcal{P}$ and $\chi(\mathcal{P})$ which are inverse to each other in $\mathscr{P}^{**}$, $\mathrm{Wu}(H)$ and $\chi\mathrm{Wu}(H)$ are *not* inverse to each other in $\mathrm{Tot}(H)$.

If $\mathbb{F}_q[V] \xrightarrow{\pi} H = \mathbb{F}_q[V]/I$ is an unstable Poincaré duality quotient algebra of $\mathbb{F}_q[V]$ then Lemma III.1.2 tells us that there is a $\mathscr{P}^*$-invariant parameter ideal $Q \subset \mathbb{F}_q[V]$ contained in I, and Theorem III.1.4 that there is an element $h \in \mathbb{F}_q[V]$ such that $(Q : I)/Q = (h)$ in $\mathbb{F}_q[V]/Q$. Moreover the principal ideal $(h) \subset \mathbb{F}_q[V]/Q$ is $\mathscr{P}^*$-invariant. Given such a Q and h for an $\overline{\mathbb{F}_q[V]}$-primary irreducible $\mathscr{P}^*$-invariant ideal $I \subset \mathbb{F}_q[V]$ we would like to be able to compute the Wu classes of $\mathbb{F}_q[V]/I$ from those of $\mathbb{F}_q[V]/Q$. The following results explain how to do this making use of Theorem II.5.1 in a crucial way.

THEOREM III.3.4: *Let* V *be a finite dimensional vector space over the Galois field* $\mathbb{F}_q$. *If* $I \subset \mathbb{F}_q[V]$ *is a* $\mathscr{P}^*$*-invariant* $\overline{\mathbb{F}_q[V]}$*-primary irreducible ideal and* $\theta_I \in \Gamma(V)$ *is a generator of Macaulay's dual principal system, then*

$$\chi(\mathcal{P})(\theta_I) = \mathrm{Wu}(I) \cap \theta_I$$

and

$$\mathcal{P}(\theta_I) = \chi\mathrm{Wu}(I) \cap \theta_I.$$

Proof: It suffices to prove one of the formulae as the other formula is proven analogously. For any $f \in \mathbb{F}_q[V]/I$ we have by definition of the $\cap$-product

$$\begin{aligned}<\mathrm{Wu}(I) \cap \theta_I \mid f> &= <\theta_I \mid \mathrm{Wu}(I) \cdot f> \\ &= <\theta_I \mid \mathcal{P}(f)> = <\chi(\mathcal{P})(\theta_I) \mid f>.\end{aligned}$$

The module $I^{\perp} \subset \Gamma(V)$ is cyclic with generator θ_I and closed under the action of the Steenrod algebra on $\Gamma(V)$. Therefore

$$\chi(\mathcal{P})(\theta_I) = h \cap \theta_I$$

for some element $h \in \mathbb{F}_q[V]$. Choose such an h. Then

$$<\mathrm{Wu}(I) \cap \theta_I \mid f> = <h \cap \theta_I \mid f>$$

for all $f \in \mathbb{F}[V]/I$ and hence the difference $h - \mathrm{Wu}(I)$ annihilates θ_I establishing the first formula. □

Theorem III.3.5: *Let $K \subsetneq L \subset \mathbb{F}_q[V]$ be $\mathcal{P}^*$-invariant $\overline{\mathbb{F}_q[V]}$-primary irreducible ideals. Choose $h \in \mathbb{F}_q[V]$ so that $(K : L) = (h) + K$. Then h becomes a Thom class in $\mathbb{F}_q[V]/K$. If $\chi(\mathcal{P})(h) \equiv w \cdot h \bmod K$ and $\mathcal{P}(h) \equiv \overline{w} \cdot h \bmod K$ then*

$$\mathrm{Wu}(L) = w \cdot \mathrm{Wu}(K) \in \mathbb{F}_q[V]/L$$

and

$$\chi\mathrm{Wu}(L) = \overline{w} \cdot \chi\mathrm{Wu}(K) \in \mathbb{F}_q[V]/L.$$

In particular if $\mathrm{Wu}(K) = 1$ then $\mathrm{Wu}(L) = w$ and $\chi\mathrm{Wu}(L) = \overline{w}$. So, if K has trivial Wu classes, then L has trivial Wu classes if and only if $w \in L$ or equivalently $\overline{w} \in L$.

Proof: Let θ_K be a Macaulay dual for K. By Theorem II.5.1 we may choose $\theta_L = h \cap \theta_K$ as a Macaulay dual for L. If we apply $\chi(\mathcal{P})$ to this equation we obtain from Theorem III.3.4 that

$$\begin{aligned}\mathrm{Wu}(L) \cap \theta_L &= \chi(\mathcal{P})(\theta_L) = \chi(\mathcal{P})(h \cap \theta_K) \\ &= \chi(\mathcal{P})(h) \cap \chi(\mathcal{P})(\theta_K) = (w \cdot h) \cap (\mathrm{Wu}(K) \cap \theta_K) \\ &= \big(w \cdot \mathrm{Wu}(K)\big) \cap (h \cap \theta_K) = \big(w \cdot \mathrm{Wu}(K)\big) \cap \theta_L.\end{aligned}$$

This says that $\mathrm{Wu}(L) - w \cdot \mathrm{Wu}(K)$ annihilates θ_L so by Theorem II.2.2 $\mathrm{Wu}(L) - w \cdot \mathrm{Wu}(K) \in L$. This establishes the first formula. The rest follows from the definitions or by applying χ. □

The following example illustrates the use of these results.

Example 1: Let $\mathbf{d}_{2,0} = x^2y + xy^2$, $\mathbf{d}_{2,1} = x^2 + xy + y^2 \in \mathbb{F}_2[x, y]$ be the two Dickson polynomials. Consider the ideal $(\mathbf{d}_{2,0}, \mathbf{d}_{2,1}) \subset \mathbb{F}_2[x, y]$. This ideal is $\mathscr{A}^*$-invariant. Either of the monomials x^2y, xy^2 represents a fundamental class of the Poincaré duality quotient algebra $\mathbb{F}_2[x, y]/(\mathbf{d}_{2,0}, \mathbf{d}_{2,1})$. Direct computation[3] shows that this algebra has trivial Wu classes. The ideal $(\mathbf{d}_{2,0}, \mathbf{d}_{2,1})$ contains the parameter ideal (x^3, y^3) as was shown in Example 1 of Section II.5 and hence also the parameter ideal (x^4, y^4). Reasoning as in that example we see that

$$\big((x^4, y^4) : (\mathbf{d}_{2,0}, \mathbf{d}_{2,1})\big) = (x^2y + xy^2) + (x^4, y^4) = (\mathbf{d}_{2,0}) + (x^4, y^4).$$

By Corollary II.6.4 taking Frobenius powers gives $\big((x^8, y^8) : (\mathbf{d}^2_{2,0}, \mathbf{d}^2_{2,1})\big) = (\mathbf{d}^2_{2,0}) + (x^8, y^8)$ so $\mathbf{d}^2_{2,0}$ is a transition element for $(\mathbf{d}^2_{2,0}, \mathbf{d}^2_{2,1})$ over (x^8, y^8). Introduce the **total Steenrod squaring operation**

$$\mathbf{Sq} = 1 + \mathrm{Sq}^2 + \mathrm{Sq}^2 + \cdots.$$

One has by direct computation that

$$\mathbf{Sq}(\mathbf{d}_{2,1}) = \mathbf{d}_{2,1} + \mathbf{d}_{2,0} + \mathbf{d}^2_{2,1}$$
$$\mathbf{Sq}(\mathbf{d}_{2,0}) = \mathbf{d}_{2,0}(1 + \mathbf{d}_{2,1} + \mathbf{d}_{2,0}),$$

so $\mathbf{Sq}(\mathbf{d}^2_{2,0}) = \mathbf{d}^2_{2,0}(1 + \mathbf{d}^2_{2,1} + \mathbf{d}^2_{2,0})$. By Theorem III.3.5 we see that the algebra $\mathbb{F}_2[x, y]/(\mathbf{d}^2_{2,0}, \mathbf{d}^2_{2,1})$ also has trivial Wu classes.[4]

More interesting is to consider the ideal $(\mathbf{d}_{2,0}, \mathbf{d}^2_{2,1}) \subset \mathbb{F}_2[x, y]$ which lies between $(\mathbf{d}_{2,0}, \mathbf{d}_{2,1})$ and its first Frobenius power $(\mathbf{d}^2_{2,0}, \mathbf{d}^2_{2,1})$. It too is $\mathscr{A}^*$-invariant, as the formulae for $\mathbf{Sq}(\mathbf{d}_{2,0})$ and $\mathbf{Sq}(\mathbf{d}_{2,1})$ above show. We have

$$(\mathbf{d}_{2,0}, \mathbf{d}^2_{2,1}) = \big((\mathbf{d}^2_{2,0}, \mathbf{d}^2_{2,1}) : \mathbf{d}_{2,0}\big).$$

Since $\mathbf{d}_{2,0}$ is a Thom class in $\mathbb{F}_2[x, y]$ the principal ideal it generates is $\mathscr{A}^*$-invariant. We apply Theorem III.3.5 and use the formula for $\mathbf{Sq}(\mathbf{d}_{2,0})$ above. The result is

$$\chi\mathrm{Wu}(\mathbf{d}_{2,0}, \mathbf{d}^2_{2,1}) = 1 + \mathbf{d}_{2,1}.$$

After a bit of computation using Lemma III.3.3 we conclude that

$$\mathrm{Wu}(\mathbf{d}_{2,0}, \mathbf{d}^2_{2,1}) = 1 + \mathbf{d}_{2,1} \in \mathbb{F}_2[x, y]/(\mathbf{d}_{2,0}, \mathbf{d}^2_{2,1}).$$

So the ideal $(\mathbf{d}_{2,0}, \mathbf{d}^2_{2,1})$ does not have trivial Wu classes. See Section V.1 for the complete list of ideals generated by powers of the Dickson polynomials $\mathbf{d}_{2,0}, \mathbf{d}_{2,1} \in \mathbb{F}_2[x, y]$ that are invariant under the Steenrod algebra and a sublist of those with trivial Wu classes.

---

[3] This is a very special case of a result of S. A. Mitchell [60] Appendix B that we will reprove as Corollary IV.2.3.

[4] These computations could be avoided by using results proved further on in Sections III.6 (Corollary III.6.5) and IV.1 (Proposition IV.1.3).

III.4 $\mathscr{P}^*$-Indecomposables: Hit Problems

An element $f \in \mathbb{F}_q[V]$ is called **$\mathscr{P}^*$-decomposable** if for some $k \in \mathbb{N}$ there are elements $\Theta_i \in \mathscr{P}^*$ of strictly positive degree, and elements $f_i \in \mathbb{F}_q[V]$, such that $f = \sum_{i=1}^{k} \Theta_i(f_i)$. Two elements $f', f'' \in \mathbb{F}_q[V]$ are called **$\mathscr{P}^*$-equivalent** if their difference is $\mathscr{P}^*$-decomposable. If a form f is not $\mathscr{P}^*$-decomposable it is called **$\mathscr{P}^*$-indecomposable**. The **module of $\mathscr{P}^*$-indecomposable elements** is per definition $\mathbb{F}_q \otimes_{\mathscr{P}^*} \mathbb{F}_q[V]$ and is denoted by either $Q_{\mathscr{P}^*}(\mathbb{F}_q[V])$ or $\mathbb{F}_q[V]_{\mathscr{P}^*}$. A problem of fundamental importance for algebraic topologists is to determine this module $\mathbb{F}_q[V]_{\mathscr{P}^*}$ of $\mathscr{P}^*$-indecomposable elements. This is often referred to as the *Hit Problem* for $\mathbb{F}_q[V]$. See e.g. [107], [108], [109] and the references there. A basis for the vector space $Q_{\mathscr{P}^*}(\mathbb{F}_q[V])$ is known in the so called *generic degrees* (see e.g. [65] and [110] II.5). We find it convenient to reformulate things for $\Gamma(V)$ using Hopf algebra duality as in [16] to bring out the hidden connection with the unstable Poincaré duality quotients of $\mathbb{F}_q[V]$.

An element $\gamma \in \Gamma(V)$ is called **$\mathscr{P}^*$-invariant** if $\mathcal{P}^k(\gamma) = 0$ for all $k > 0$. This is consistent with the terminology of R. M. Fossum [22] for invariants of cocommutative Hopf algebras acting on rings, as well as with the fact that the total Steenrod operation $\mathcal{P} = 1 + \mathcal{P}^1 + \cdots + \mathcal{P}^k + \cdots$ induces an action of $\mathbb{Z}$ on $\mathbb{F}_q[V]$ as well as $\Gamma(V)$ whose invariants in the classical sense are those just defined (see e.g. [89]). The subalgebra of $\mathscr{P}^*$-invariant elements in $\Gamma(V)$ will be denoted by $\Gamma(V)^{\mathscr{P}^*}$: as a graded vector space over $\mathbb{F}_q$ it is dual to $\mathbb{F}_q \otimes_{\mathscr{P}^*} \mathbb{F}_q[V] = \mathbb{F}_q[V]_{\mathscr{P}^*}$, the module of $\mathscr{P}^*$-indecomposable elements of $\mathbb{F}_q[V]$.

After all these preliminaries we come to one of the most surprising results so far in this study. It exhibits an unexpected connection between two seemingly unrelated things, viz., the *Hit Problem* and $\mathscr{P}^*$-Poincaré duality quotients of $\mathbb{F}_q[V]$. It should come as no surprise however that it is easy to prove this result after these massive amounts of background material.

THEOREM III.4.1: *Macaulay's double annihilator theorem establishes a correspondence between nonzero elements of $\Gamma(V)^{\mathscr{P}^*}$ and the $\mathscr{P}^*$-Poincaré duality quotients of $\mathbb{F}_q[V]$ with trivial Wu classes. Two nonzero elements $\gamma', \gamma'' \in \Gamma(V)^{\mathscr{P}^*}$ define the same $\mathscr{P}^*$-Poincaré duality quotient if and only if they are nonzero scalar multiples of each other.*

PROOF: Let $0 \neq \gamma \in \Gamma(V)^{\mathscr{P}^*}$ generate the $\mathbb{F}_q[V]$-submodule $M(\gamma)$ of $\Gamma(V)$. Then $M(\gamma)$ is a $\mathscr{P}^*$-submodule. Let $I(\gamma) = \mathrm{Ann}_{\mathbb{F}_q[V]}(M(\gamma)) \subset \mathbb{F}_q[V]$ be the corresponding ideal whose Poincaré duality quotient H is an unstable $\mathscr{P}^*$-algebra by Theorem III.2.1. By Theorem III.3.5 the action map

$$\alpha : \mathbb{F}_q[V] \longrightarrow \Gamma(V) \quad f \longmapsto f \cap \gamma$$

defines a bijection[5] of $\mathbb{F}_q[V]/I(\gamma)$ onto $M(\gamma)$ which sends the element $\mathrm{Wu}_k(H)$ to $\chi(\mathcal{P}^k)(\gamma)$. If γ is $\mathscr{P}^*$-invariant then $\chi(\mathcal{P}^k)(\gamma) = 0$ for all $k > 0$, so the Wu classes of H are trivial, and conversely. By Proposition I.4.2 two nonzero elements γ', $\gamma'' \in \Gamma(V)$ determine the same Poincaré duality quotient if and only if they are nonzero scalar multiples of each other, so the result is established. □

If a Poincaré duality quotient $\mathbb{F}_q[V]/I$ has trivial Wu classes and $z_1, \ldots, z_n$ is a basis for V^*, then clearly no monomial z^D representing a fundamental class can be $\mathscr{P}^*$-decomposable. For, if one were, say z^D, then there would be a monomial z^E and a Steenrod operation $\mathcal{P}^k$ with $k > 0$ and $\mathcal{P}^k(z^E) = z^D$ in $\mathbb{F}_q[V]/I$. This would imply that $\mathrm{Wu}_k(I) \neq 0$. The following example shows that the naive converse fails: a representing monomial for the fundamental class of $\mathbb{F}_q[V]/I$ may well be $\mathscr{P}^*$-indecomposable without the Wu classes of $\mathbb{F}_q[V]/I$ being trivial. See Proposition III.5.3 for the correct converse, and Section III.7 for more examples of this type.

EXAMPLE 1: We consider in $\Gamma(u, v, w)$ the quadratic divided polynomial $\gamma = \gamma_2(w) + \gamma_1(u)\gamma_1(v)$. The ideal $I(\gamma) \subset \mathbb{F}_q[x, y, z]$ corresponding to γ under the double annihilator theorem is [6]

$$I(\gamma) = (x^2, y^2, xz, yz, z^2 - xy) \subset \mathbb{F}_q[x, y, z].$$

This is most easily seen using the catalecticant matrices to make the computation: see Section VI.2. Again we see that in more than two variables $\mathfrak{m}$-primary irreducible ideals in $\mathbb{F}_q[z_1, \ldots, z_n]$ need not be generated by a regular sequence. The monomials xy and z^2 are representatives for a fundamental class of the Poincaré duality quotient $\mathbb{F}_q[x, y, z]/I(\gamma)$. If $\mathbb{F}_q = \mathbb{F}_2$, then since four of the five generators are Thom classes and

$$\mathrm{Sq}^1(z^2 + xy) = 0 + x^2y + xy^2 \in I(\gamma)$$

the ideal $I(\gamma) \subset \mathbb{F}_2[x, y, z]$ is closed under the action of the Steenrod algebra $\mathscr{A}^*$. Of the elements representing a fundamental class, xy is $\mathscr{A}^*$-indecomposable, but $z^2 = \mathrm{Sq}^1(z)$ is $\mathscr{A}^*$-decomposable, and therefore $\mathrm{Wu}_1(I(\gamma)) \neq 0$.

This is by no means an isolated example. $\mathbb{F}_2[x, y, z]/I(\gamma)$ is isomorphic to $\mathbb{F}_2[x, y]/(x^2, y^2)\#\mathbb{F}_2[z]/(z^3)$, where $\mathbb{F}_2[x, y]/(x^2, y^2)$ has trivial Wu

---

[5] This bijection sends elements of degree k to elements of degree $\deg(\gamma) - k$.

[6] Topologists should recognize this example as $H^*((S^2 \times S^2)\#\mathbb{CP}(2); \mathbb{F}_q)$ with the grading degrees halved.

classes and $\mathrm{Wu}_1(\mathbb{F}_2[z]/(z^3)) = z$. More examples of this type arise by taking connected sums $H' \# H''$ where H' has trivial Wu classes but H'' does not.

We round out this part of the discussion with a property of $\mathscr{P}^*$-indecomposables that does not seem to be as well known as it might. We therefore offer two distinct proofs of this result. Here is the first one, the second one appears in Section III.5.

PROPOSITION III.4.2: *Let $\mathbb{F}_q$ be the Galois field with $q = p^\nu$ elements, p a prime. If V' and V'' are finite dimensional vector spaces over $\mathbb{F}_q$ and $f' \in \mathbb{F}_q[V']$ and $f'' \in \mathbb{F}_q[V'']$ are $\mathscr{P}^*$-indecomposable elements, then $f' \otimes f'' \in \mathbb{F}_q[V' \oplus V'']$ is also $\mathscr{P}^*$-indecomposable.*

PROOF: Consider the natural map

$$(\mathbb{F}_q[V'] \otimes \mathbb{F}_q[V'']) \otimes_{\mathscr{P}^*} \mathbb{F}_q \longrightarrow (\mathbb{F}_q[V'] \otimes \mathbb{F}_q[V'']) \otimes_{\mathscr{P}^* \otimes \mathscr{P}^*} \mathbb{F}_q$$

induced by change of rings along the diagonal map $\nabla : \mathscr{P}^* \longrightarrow \mathscr{P}^* \otimes \mathscr{P}^*$ of the Steenrod algebra. This map is an epimorphism. Since

$$(\mathbb{F}_q[V'] \otimes \mathbb{F}_q[V'']) \otimes_{\mathscr{P}^* \otimes \mathscr{P}^*} \mathbb{F}_q \cong (\mathbb{F}_q[V'] \otimes_{\mathscr{P}^*} \mathbb{F}_q) \otimes (\mathbb{F}[V''] \otimes_{\mathscr{P}^*} \mathbb{F}_q)$$

we therefore obtain an epimorphism

$$(\mathbb{F}_q[V'] \otimes \mathbb{F}_q[V''])_{\mathscr{P}^*} \longrightarrow \mathbb{F}_q[V']_{\mathscr{P}^*} \otimes \mathbb{F}_q[V'']_{\mathscr{P}^*}.$$

For the element f of $\mathbb{F}_q[V' \oplus V'']_{\mathscr{P}^*}$ represented by $f' \otimes f''$ the image of f in $\mathbb{F}_q[V']_{\mathscr{P}^*} \otimes \mathbb{F}_q[V'']_{\mathscr{P}^*}$ under this map is the tensor product of the elements represented by f' and f'' respectively, which are nonzero yielding the desired conclusion. □

III.5 A dual interpretation of the $\mathscr{P}^*$-Double Annihilator Theorem

There is a dual interpretation for the results of Section III.4 that arises from the vector space duality between $\Gamma(V)^{\mathscr{P}^*}$ and $\mathbb{F}_q[V]_{\mathscr{P}^*}$ given by the contraction pairing between $\Gamma(V)$ and $\mathbb{F}_q[V]$. Hence we have the following.

PROPOSITION III.5.1: *An element $\theta \in \Gamma(V)_d$ is $\mathscr{P}^*$-invariant if and only if when regarded as a linear form $\mathbb{F}_q[V]_d \longrightarrow \mathbb{F}_q$ there is a factorization*

$$\begin{array}{ccc} \mathbb{F}_q[V]_d & \xrightarrow{\ \theta\ } & \mathbb{F}_q \\ \pi \big\downarrow & \nearrow_{\bar{\theta}} & \\ (\mathbb{F}_q[V]_{\mathscr{P}^*})_d & & \end{array}$$

where π is the canonical quotient map. □

A subspace $W \subset \mathbb{F}_q[V]_d$ is called $\mathscr{P}^*$**-absorbing** if whenever $f \in \mathbb{F}_q[V]_{d-k}$ and $\Theta \in \mathscr{P}^*$ has degree k then $\Theta(f) \in W$. A $\mathscr{P}^*$-absorbing subspace W of codimension one in $\mathbb{F}_q[V]_d$ defines a linear form $\gamma : \mathbb{F}_q[V]_d \longrightarrow \mathbb{F}_q = \mathbb{F}_q[V]_d/W$ which may be regarded as an element of $\Gamma(V)$. A short exercise in vector space duality with the $\mathscr{P}^*$-module structure shows that such a form γ lies in $\Gamma(V)^{\mathscr{P}^*}$. Conversely, if $\gamma \in \Gamma(V)^{\mathscr{P}^*}$ has degree d and is regarded as a linear form $\mathbb{F}_q[V]_d \longrightarrow \mathbb{F}_q$, then $\ker(\gamma) \subset \mathbb{F}_q[V]_d$ is a $\mathscr{P}^*$-absorbing subspace.

LEMMA III.5.2: *If $f \in \mathbb{F}_q[V]$ is $\mathscr{P}^*$-indecomposable, then there exists a $\mathscr{P}^*$-Poincaré duality quotient algebra with trivial Wu classes $H = \mathbb{F}_q[V]/I$ such that f represents a fundamental class for H.*

PROOF: The image of the element f in $\mathbb{F}_q[V]_{\mathscr{P}^*}$ is nonzero, so may be extended to an $\mathbb{F}_q$-vector space basis $f = f_1, \ldots, f_m$ for $(\mathbb{F}_q[V]_{\mathscr{P}^*})_d$, where $d = \deg(f)$. Define $\overline{\delta} : (\mathbb{F}_q[V]_{\mathscr{P}^*})_d \longrightarrow \mathbb{F}_q$ by requiring $\overline{\delta}(f_i) = \delta_{1,i}$. If $\pi : \mathbb{F}_q[V] \longrightarrow \mathbb{F}_q[V]_{\mathscr{P}^*}$ is the canonical projection and $W \subset \mathbb{F}_q[V]_d$ is the kernel of the composition $\delta = \overline{\delta} \circ \pi$, then W is an absorbing subspace. Hence $I = \mathfrak{A}(W) \subset \mathbb{F}_q[V]$ is $\mathscr{P}^*$-invariant, $\overline{\mathbb{F}_q[V]}$-primary, and irreducible. By construction f represents a fundamental class of $H = \mathbb{F}_q[V]/I$ and the result follows. □

Using this we offer an alternative proof of Proposition III.4.2.

PROOF OF PROPOSITION III.4.2: Using the construction from the proof of Lemma III.5.2 one sees that the $\mathscr{P}^*$-indecomposable elements f', f'' correspond to Poincaré duality quotients H' and H'' of $\mathbb{F}_q[V']$ and $\mathbb{F}_q[V'']$ with trivial Wu classes. Then $H' \otimes H''$ also has trivial Wu classes: it corresponds via Theorem III.4.1 to $f' \otimes f'' \in \mathbb{F}_q[V'] \otimes \mathbb{F}_q[V''] = \mathbb{F}_q[V' \oplus V'']$ whence the result. □

Suppose $\theta \in \Gamma(V)$ defines a $\mathscr{P}^*$-invariant ideal whose associated Poincaré duality quotient has trivial Wu classes, i.e., by Theorem III.4.1 $\theta \in \Gamma(V)^{\mathscr{P}^*}$. Starting with the support of $\theta \in \Gamma(V)_d$ regarded as a linear form on $\mathbb{F}_q[V]_d$ it is very subtle to determine whether θ belongs to $\Gamma(V)^{\mathscr{P}^*}$, and if it does not, what the Wu classes of the corresponding Poincaré duality quotient might be. Several examples that illustrate this appear in Section III.7. However the support cannot be arbitrary: if $\theta \in \Gamma(V)^{\mathscr{P}^*}$, then $\mathrm{supp}(\theta)$ turns out to be a union of $\mathscr{P}^*$-equivalence classes of $\mathscr{P}^*$-indecomposable monomials. Specifically, we have the following.

PROPOSITION III.5.3: *Let $\mathbb{F}_q[V] \xrightarrow{\pi} \mathbb{F}_q[V]/I$ be a $\mathscr{P}^*$-Poincaré duality quotient with trivial Wu classes defined by $\theta \in \Gamma(V)_d^{\mathscr{P}^*}$. If a monomial z^D represents a fundamental class of $\mathbb{F}_q[V]/I$ then z^D is $\mathscr{P}^*$-indecomposable.*

Furthermore, if z^E is $\mathscr{P}^*$-equivalent to z^D then z^E also represents a fundamental class. Therefore supp(θ) *is a union of $\mathscr{P}^*$-equivalence classes of $\mathscr{P}^*$-indecomposable monomials.*

Proof: Consider the diagram

$$\begin{array}{ccc} \mathbb{F}_q[V]_d & \xrightarrow{\theta} & \mathbb{F}_q \\ \downarrow \pi & \nearrow \bar{\theta} & \\ (\mathbb{F}_q[V]_{\mathscr{P}^*})_d & & \end{array}$$

where π is the natural quotient map. Since $\mathbb{F}_q[V]/I$ has trivial Wu classes there is a factorization $\bar{\theta}$ as indicated (see Proposition III.5.1 and the discussion preceding Theorem III.4.1). If z^D represents a fundamental class for $\mathbb{F}_q[V]/I$ then $0 \neq \theta(z^D) = \bar{\theta}\pi(z^D)$ implies that $\pi(z^D) \neq 0$, so z^D is $\mathscr{P}^*$-indecomposable. By the same token, if z^E is $\mathscr{P}^*$-equivalent to z^D then $\theta(z^E) = \bar{\theta}\pi(z^E) = \bar{\theta}\pi(z^D) = \theta(z^D) \neq 0$ so z^E also represents a fundamental class of $\mathbb{F}_q[V]/I$. □

Example 1: Recall Example 1 of Section II.6, viz.,

$$L = (xy,\ xz,\ yz,\ z^2 - x^2,\ z^2 - y^2) \subset \mathbb{F}_2[x,\ y,\ z].$$

The corresponding Poincaré duality quotient algebra is the connected sum of three copies of the algebra $\mathbb{F}_2[u]/(u^3)$. We found that the element xyz^5 represents a fundamental class of $\mathbb{F}_2[x,\ y,\ z]/L^{[2]}$. This element is $\mathscr{A}^*$-indecomposable. The dual principal generator $\theta^{[2]}$ for $L^{[2]}$ has support

$$\operatorname{supp}(\theta^{[2]}) = \left\{x^5yz,\ xy^5z,\ xyz^5,\ xy^3z^3,\ x^3yz^3,\ x^3y^3z\right\}.$$

This is not the union of $\mathscr{A}^*$-equivalence classes of monomials, since $x^2y^2z^3$ is $\mathscr{A}^*$-equivalent to xyz^5 but $x^2y^2z^3$ does not belong to the support of $\theta^{[2]}$. From Proposition III.5.3 it follows that the corresponding Poincaré duality quotient algebra $\mathbb{F}_2[x,\ y,\ z]/L^{[2]}$ has a nontrivial Wu class. Indeed, since

$$\mathrm{Sq}^2(xyz^3) = x^2y^2z^3 + x^2yz^4 + xy^2z^4 + xyz^5,$$

and in the quotient algebra $\mathbb{F}_2[x,\ y,\ z]/L^{[2]}$ we have

$$x^2y^2z^3 = 0, \quad x^2yz^4 = 0, \quad \text{and } xy^2z^4 = 0,$$

we see that

$$\mathrm{Sq}^2(xyz^3) = xyz^5 \in \mathbb{F}_2[x,\ y,\ z]/L^{[2]}.$$

Hence $\mathrm{Wu}_2(\mathbb{F}_2[x,\ y,\ z]/L^{[2]}) \neq 0$.

These computations also show that it is not enough for a fundamental class of a $\mathscr{P}^*$-Poincaré duality quotient of $\mathbb{F}_q[V]$ to be represented by a $\mathscr{P}^*$-indecomposable monomial to conclude that the Wu classes of the $\mathscr{P}^*$-Poincaré duality algebra are trivial: in the example, $xyz^5 \in \mathbb{F}_2[x, y, z]$ is $\mathscr{A}^*$-indecomposable but represents a fundamental class of the unstable Poincaré duality algebra $\mathbb{F}_2[x, y, z]/L^{[2]}$ which has nontrivial Wu classes.

The action of the Steenrod algebra on $\mathbb{F}_q[V]$ is very far from being a monomial action, i.e., one that sends monomials to monomials. Proposition III.5.3 says that the set of monomials that represent a fundamental class for an unstable $\mathscr{P}^*$-Poincaré duality quotient $\mathbb{F}_q[V]/I$ with trivial Wu classes is a union of $\mathscr{P}^*$-equivalence classes of $\mathscr{P}^*$-indecomposable monomials. One could therefore wonder about the converse, i.e., if a linear form $\theta \in \Gamma(V)$ generates a $\mathscr{P}^*$-submodule and is supported on a union of $\mathscr{P}^*$-equivalence classes of $\mathscr{P}^*$-indecomposable monomials, does it define a $\mathscr{P}^*$-Poincaré duality quotient with trivial Wu classes? The following example shows that this need not be the case, even when the support consists of a single monomial.

EXAMPLE 2: In this example the ground field is $\mathbb{F}_2$. We begin by observing

$$\mathrm{Sq}^1(xyz) = x^2yz + xy^2z + xyz^2 \in \mathbb{F}_2[x, y, z].$$

Each of the monomials x^2yz, xy^2z, and xyz^2 is an $\mathscr{A}^*$-equivalence class all by itself. The module of $\mathscr{A}^*$-indecomposables $\mathbb{F}_2[x, y, z]_{\mathscr{A}^*}$ is generated in homogeneous degree 4 by the images of the 9 monomials

$$x^3y,\ xy^3,\ x^3z,\ xz^3,\ y^3z,\ yz^3,\ x^2yz,\ xy^2z,\ zyz^3.$$

However, these monomials satisfy a linear relation, to wit,

$$x^2yz + xy^2z + xyz^2 = 0 \in (\mathbb{F}_2[x, y, z]_{\mathscr{A}^*})_4 .$$

Let us define $\theta \in \Gamma(u, v, w)_4$ to be the linear form supported on x^2yz. Then the support of θ is a single $\mathscr{A}^*$-equivalence class. One easily sees that the $\mathbb{F}_q[x, y, z]$-submodule of $\Gamma(u, v, w)$ generated by θ is stable under $\mathscr{P}^*$. However, there can be no factorization of $\theta : \mathbb{F}_2[x, y, z]_4 \longrightarrow \mathbb{F}_2$ as in the diagram

$$\begin{array}{ccc} \mathbb{F}_2[x, y, z]_4 & \xrightarrow{\ \theta\ } & \mathbb{F}_2 \\ \downarrow{\scriptstyle \pi} & \nearrow{\scriptstyle \bar{\theta}} & \\ (\mathbb{F}_2[x, y, z]_{\mathscr{A}^*})_4 & & \end{array}$$

since, if there were, we would have the contradiction

$$\begin{aligned} 0 &= \theta(x^2yz + xy^2z + xyz^2) \\ &= \theta(x^2yz) + \theta(xy^2z) + \theta(xyz^2) = 1 + 0 + 0 \neq 0 \in \mathbb{F}_2. \end{aligned}$$

Thus the Poincaré duality quotient defined by θ cannot have trivial Wu classes. In fact $\mathrm{Wu}_1 \neq 0$. Since this quotient algebra may be identified with

$H^*(\mathbb{RP}(2) \times S^1 \times S^1; \mathbb{F}_2)$ as an algebra over the Steenrod algebra this is clear from topological considerations also.

III.6 Steenrod operations and Frobenius powers

The proof of the following proposition is completely elementary; it is a consequence of the formula

$$\mathcal{P}^k(f^q) = \begin{cases} (\mathcal{P}^{k/q}(f))^q & \text{if } k \equiv 0 \bmod q \\ 0 & \text{otherwise.} \end{cases}$$

PROPOSITION III.6.1: *Let $\mathbb{F}_q$ be the Galois field with $q = p^\nu$ elements, p a prime, and let V be a finite dimensional vector space over $\mathbb{F}_q$. If $I \subset \mathbb{F}_q[V]$ is a $\mathcal{P}^*$-invariant ideal then so are the Frobenius powers $I^{[q^k]}$ of I.* □

We next examine how the periodicity operators $\mathbf{P}$ introduced in Section II.6 and the γ-operations interact with respect to Steenrod operations and Wu classes. For this we need a pair of lemmas.

LEMMA III.6.2: *Let $\mathbb{F}_q$ be the Galois field with $q = p^\nu$ elements, p a prime. Let V be a finite dimensional vector space over $\mathbb{F}_q$. The total Steenrod operation*

$$\mathcal{P} = 1 + \mathcal{P}^1 + \cdots + \mathcal{P}^k + \cdots \in \mathcal{P}^{**}$$

acts on the algebras $\mathbb{F}_q[V]$ and $\Gamma(V)$ (though it does not preserve homogeneity) and as operators, we have

$$\mathcal{P} \circ \gamma_q = \gamma_q \circ \mathcal{P} : \Gamma(V) \longrightarrow \Gamma(V)$$
$$\chi(\mathcal{P}) \circ \gamma_q = \gamma_q \circ \chi(\mathcal{P}) : \Gamma(V) \longrightarrow \Gamma(V).$$

PROOF: Introduce the operator

$$\sqrt[q]{} : \mathbb{F}_q[V] \longrightarrow \mathbb{F}_q[V]$$

defined by

$$\sqrt[q]{f} = \begin{cases} h & \text{if } h^q = f \\ 0 & \text{otherwise.} \end{cases}$$

This is well defined because the map $h \longmapsto h^q$ is a linear monomorphism from $\mathbb{F}_q[V]$ to itself. From the definition of the divided power operators γ_k it follows that the two squares

$$\begin{array}{ccc} \Gamma(V) & \xrightarrow{\gamma_q} & \Gamma(V) \\ \mathcal{P}\downarrow & & \downarrow\mathcal{P} \\ \Gamma(V) & \xrightarrow[\gamma_q]{} & \Gamma(V) \end{array} \qquad \begin{array}{ccc} \mathbb{F}[V] & \xleftarrow{\sqrt[q]{\ }} & \mathbb{F}[V] \\ \chi(\mathcal{P})\uparrow & & \uparrow\chi(\mathcal{P}) \\ \mathbb{F}[V] & \xleftarrow[\sqrt[q]{\ }]{} & \mathbb{F}[V] \end{array}$$

are vector space duals to each other. So it suffices to prove the formula

$$\sqrt[q]{} \circ \chi(\mathcal{P}) = \chi(\mathcal{P}) \circ \sqrt[q]{} : \mathbb{F}_q[V] \longrightarrow \mathbb{F}_q[V].$$

Since $\chi(\mathcal{P})$ is an automorphism, $\chi(\mathcal{P})(f)$ is a q-th power if and only if f is a q-th power. If f is a q-th power, say $f = h^q$ then $\chi(\mathcal{P})(f) = \big(\chi(\mathcal{P})(h)\big)^q$ so

$$\sqrt[q]{\chi(\mathcal{P})(f)} = \sqrt[q]{\chi(\mathcal{P})(h^q)} = \sqrt[q]{\big(\chi(\mathcal{P})(h)\big)^q} = \chi(\mathcal{P})(h) = \chi(P)\big(\sqrt[q]{f}\big)$$

as claimed. □

LEMMA III.6.3: *Let V be a finite dimensional vector space over the Galois field $\mathbb{F}_q$ with $q = p^\nu$ elements, p a prime. If $\theta \in \Gamma(V)$ and $f \in \mathbb{F}_q[V]$ then*

$$\Upsilon_q(f \cap \theta) = f^q \cap \Upsilon_q(\theta).$$

PROOF: Let $z_1, \ldots, z_n$ be a basis for V^*. The operator Υ_q is $\mathbb{F}_q$-linear as is the Frobenius operator $f \longmapsto f^q$. It therefore suffices to check this formula for θ a divided power monomial γ_E and f a monomial z^F. If $E - F \notin \mathbb{N}_0^n$ then $qE - qF \notin \mathbb{N}_0^n$ also so

$$\Upsilon_q(z^F \cap \gamma_E) = \Upsilon_q(0) = 0 = z^{qF} \cap \Upsilon_q(\gamma_E).$$

Whereas, if $E - F \in \mathbb{N}_0^n$ we have

$$\Upsilon_q(z^F \cap \gamma_E) = \Upsilon_q(\gamma_{E-F}) = \gamma_{qE-qF} = z^{qF} \cap \gamma_{qF} = z^{qF} \cap \Upsilon_q(\gamma_E)$$

and the result follows. □

THEOREM III.6.4: *Let V be a finite dimensional vector space over the Galois field $\mathbb{F}_q$ with $q = p^\nu$ elements, $p \in \mathbb{N}$ a prime. Suppose that $I \subset \mathbb{F}_q[V]$ is a $\mathcal{P}^*$-invariant $\overline{\mathbb{F}_q[V]}$-primary irreducible ideal. Then*

$$\mathrm{Wu}(I^{[q]}) = \mathrm{Wu}(I)^q \in \mathbb{F}_q[V]/I^{[q]}$$

and

$$\chi\mathrm{Wu}(I^{[q]}) = \chi\mathrm{Wu}(I)^q \in \mathbb{F}_q[V]/I^{[q]}.$$

PROOF: Choose a basis $z_1, \ldots, z_n$ for V^* and let $u_1, \ldots, u_n$ be the dual basis for V. Let $\theta_I \in \Gamma(V)$ generate the dual principal system $I^\perp$ of I. By Theorem II.6.6 the element

$$\theta_{I^{[q]}} = \gamma_{q-1}(u_1) \cdots \gamma_{q-1}(u_n) \Upsilon_q(\theta_I)$$

generates the dual principal system of $I^{[q]}$. Apply $\chi(\mathcal{P})$ to this formula. By Theorem III.3.5, and Lemmas III.6.2 and III.6.3 we get

$$\begin{aligned} \mathrm{Wu}(I^{[q]}) \cap \theta_{I^{[q]}} &= \chi(\mathcal{P})(\theta_{I^{[q]}}) = \chi(\mathcal{P})\big(\gamma_{q-1}(u_1) \cdots \gamma_{q-1}(u_n)\Upsilon_q(\theta_I)\big) \\ &= \gamma_{q-1}(u_1) \cdots \gamma_{q-1}(u_n)\chi(\mathcal{P})(\Upsilon_q(\theta_I)) \end{aligned}$$

(since $\gamma_{q-1}(u_1), \ldots, \gamma_{q-1}(u_n)$ are $\mathscr{P}^*$-invariant)

$$\begin{aligned}&= \gamma_{q-1}(u_1)\cdots\gamma_{q-1}(u_n)\Upsilon_q\big(\chi(\mathcal{P})(\theta_I)\big)\\&= \gamma_{q-1}(u_1)\cdots\gamma_{q-1}(u_n)\Upsilon_q\big(\mathrm{Wu}(I)\cap\theta_I\big)\\&= \gamma_{q-1}(u_1)\cdots\gamma_{q-1}(u_n)\big(\mathrm{Wu}(I)^q\cap\Upsilon_q(\theta_I)\big)\\&= \mathrm{Wu}(I)^q\cap\big(\gamma_{q-1}(u_1)\cdots\gamma_{q-1}(u_n)\Upsilon_q(\theta_I)\big) = \mathrm{Wu}(I)^q\cap\theta_{I^{[q]}}\end{aligned}$$

and the first formula follows. The proof of the second formula is analogous. □

COROLLARY III.6.5: *Let $\mathbb{F}_q$ be the Galois field with $q = p^\nu$ elements, p a prime and V a finite dimensional vector space over $\mathbb{F}_q$. Suppose $I \subset \mathbb{F}_q[V]$ is a $\mathscr{P}^*$-invariant $\overline{\mathbb{F}_q[V]}$-primary irreducible ideal. If the corresponding Poincaré duality quotient algebra has trivial Wu classes, then for any $e \in \mathbb{N}_0$ so does the Poincaré duality quotient algebra $\mathbb{F}_q[V]/I^{[q^e]}$.* □

The *dual* of Corollary III.6.5 can be given a direct computational proof. For the sake of simplicity we confine ourselves to the case $q = 2$.

PROPOSITION III.6.6: *Let $f \in \mathbb{F}_2[z_1, \ldots, z_k]$ be an $\mathscr{A}^*$-indecomposable polynomial. Then $z_1z_2\cdots z_kf^2$ is again indecomposable.*

PROOF: Suppose to the contrary that $z_1z_2\cdots z_kf^2$ is $\mathscr{A}^*$-decomposable. Then it can be written in the form

$$z_1\cdots z_kf^2 = \sum_{i\in I}\mathrm{Sq}^{r_i}\left(z^{M_i}\right)$$

where $z^{M_i} = z_1^{m_{i,1}}\cdots z_k^{m_{i,k}}$ and $r_i > 0$ for $i \in I$. We say that a monomial z^M is of type I if the exponent sequence $M = (m_1, \ldots, m_k)$ consists entirely of odd integers. Otherwise we say it is of type II. Since

$$\mathrm{Sq}^i(f^{2j}) = \big(\mathrm{Sq}^{i/2}(f)\big)^2,$$

where by convention $\mathrm{Sq}^{i/2} = 0$ if i is odd, it follows from the Cartan formula that applying a squaring operation to a monomial of type II gives a sum of monomials of the same type.

On the other hand, if z^M is of type I then $z^M = z_1\cdots z_k(z^N)^2$ where $N = (n_1, \ldots, n_k)$. Again the Cartan formula says,

$$\begin{aligned}\mathrm{Sq}^r(z_1\cdots z_k(z^N)^2) &= z_1\cdots z_k\mathrm{Sq}^r((z^N)^2) + \sum z_1^{1+\varepsilon_1}\cdots z_k^{1+\varepsilon_k}\mathrm{Sq}^s((z^N)^2)\\&= z_1\cdots z_k(\mathrm{Sq}^{r/2}(z^N))^2 + \sum z_1^{1+\varepsilon_1}\cdots z_k^{1+\varepsilon_k}\big(\mathrm{Sq}^{\frac{s}{2}}(z^N)\big)^2,\end{aligned}$$

where the sum runs over all sequences $(\varepsilon_1, \ldots, \varepsilon_k, s)$ with $s \in \mathbb{N}_0$, $\varepsilon_j \in$

$\{0, 1\}$, $r = s + \sum_1^k \varepsilon_i$, and at least one of the ε_j is 1. The terms of this sum are all of type II.

If we write

$$z_1 \cdots z_k f^2 = \sum_{M_i \text{ type I}} \mathrm{Sq}^{r_i}\left(z^{M_i}\right) + \sum_{M_i \text{ type II}} \mathrm{Sq}^{r_i}\left(z^{M_i}\right)$$

then each term of the first sum has the form $z_1 \cdots z_k \left(\mathrm{Sq}^{r_i/2}(z^{N_i})\right)^2$ plus a sum of terms of type II. Hence we may rewrite this equation in the form

$$z_1 \cdots z_k f^2 - \sum_{N_i \text{ type I}} z_1 \cdots z_k \left(\mathrm{Sq}^{r_i/2}(z^{N_i})\right)^2 = \sum_{L_i \text{ type II}} z^{L_i}.$$

Then the left hand side of this equation is a sum of monomials of type I and the right hand side a sum of monomials of type II. Therefore both sides must be zero. This gives us

$$z_1 \cdots z_k f^2 = \sum_{N_i \text{ type I}} z_1 \cdots z_k \left(\mathrm{Sq}^{r_i/2}(z^{N_i})\right)^2 = z_1 \cdots z_k \left(\sum_{M_i \text{ type I}} \mathrm{Sq}^{r_i/2}(z^{N_i}) \right)^2,$$

so dividing by $z_1 \cdots z_k$ and extracting square roots

$$f = \sum_{N_i \text{ type I}} \mathrm{Sq}^{r_i/2}(z^{N_i})$$

contrary to the fact that f is $\mathscr{A}^*$-indecomposable. □

Remark: The converse of Proposition III.6.6 does not hold. As an example of rank 3 consider the monomial $x^5 y^5 z^5 \in \mathbb{F}_2[x, y, z]$. It is indecomposable (modulo hit elements it is equivalent to xy^2z^{12}, which appears in Boardman's list [8] in the last family). But, the monomial $x^2y^2z^2$ obviously is decomposable, and $xyz(x^2y^2z^2)^2 = x^5y^5z^5$.

In their study of the *Hit Problem* for $\mathbb{F}_2[z_1, \ldots, z_n]$ in [16] Section 2 M. C. Crabb and J. R. Hubbuck make a key observation concerning the periodicity operator $\mathbf{P}$. Namely for any integer d it follows from the *Boundedness Conjecture*, which is proved in [12], that the periodicity map

$$\Gamma(u_1, \ldots, u_n)^{\mathscr{A}^*}_{d\cdot 2^t + n(2^t-1)} \xrightarrow{\mathbf{P}} \Gamma(u_1, \ldots, u_n)^{\mathscr{A}^*}_{d\cdot 2^{t+1} + n(2^{t+1}-1)}$$

is in fact a bijection for all sufficiently large $t \in \mathbb{N}$, i.e., there is a $t(n) \in \mathbb{N}$ such that the indicated map is a bijection for all $t \geq t(n)$. In view of Theorems II.6.6 and III.4.1 we therefore obtain the following result.

Theorem III.6.7: *Let $d, n \in \mathbb{N}$. The assignment $I \rightsquigarrow I^{[2]}$ defines a map from the set of $\mathscr{P}^*$-invariant $\mathfrak{m}$-primary irreducible ideals with trivial Wu classes $I \subset \mathbb{F}_2[z_1, \ldots, z_n]$ whose Poincaré duality quotient algebra is of* *formal dimension* $d \cdot 2^t + n(2^t - 1)$ *to the set of* $\mathscr{P}^*$*-invariant* $\mathfrak{m}$*-primary irreducible ideals* $J \subset \mathbb{F}_2[z_1, \ldots, z_n]$ *with trivial Wu classes and with Poincaré duality quotient algebra of formal dimension* $d \cdot 2^{t+1} + n(2^{t+1} - 1)$*. For sufficiently large* t*, depending only on* n*, not on* d*, this map is a bijection.* □

III.7 Examples

In this section we examine in some detail special ideals arising in connection with the *Hit Problem* for $\mathbb{F}_2[x, y, z]$. As usual when working mod 2 we use the notation $\mathscr{A}^*$ for the Steenrod algebra instead of $\mathscr{P}^*$. These examples give ample opportunity to illustrate many of the results developed up to this point in a nontrivial way.

The first example arose by studying the tables and lists in the three variable case [39] in the interpretation of [8]. There are two obvious sources of $\mathscr{A}^*$-indecomposables in $\mathbb{F}_2[x, y, z]$, to wit, *spikes* (see [82]) and $\mathscr{A}^*$-indecomposables from the one and two variable case regarded as being in three variables via tensor products (see Proposition III.4.2). The first degree in which these considerations do not suffice to find a generating set for $\left(\mathbb{F}_2[x, y, z]_{\mathscr{A}^*}\right)_k$ is $k = 7$. In retrospect here is one explanation, from our viewpoint, of what is going on.

EXAMPLE 1: As noted in Example 1 of Section III.1 the element $xyz^5 \in \mathbb{F}_2[x, y, z]$ is $\mathscr{A}^*$-indecomposable. The vector space $\mathbb{F}_2[x, y, z]_7$ has dimension 36, so there are 36 linearly independent monomials. According to [39] the dimension of $\left(\mathbb{F}_2[x, y, z]_{\mathscr{A}^*}\right)_7$ is 10.

Recall that two monomials $z^{E'}, z^{E''} \in \mathbb{F}_2[z_1, \ldots, z_n]_k$ are $\mathscr{A}^*$**-equivalent** if they have the same image in $\mathbb{F}_2[z_1, \ldots, z_n]_{\mathscr{A}^*}$. A bit of computation shows that the $\mathscr{A}^*$-equivalence class of the element xyz^5 in $\mathbb{F}_2[x, y, z]$ is the set of monomials

$$\mathscr{D} = \Big\{xyz^5,\ xy^5z,\ x^5yz,\ xy^2z^4,\ x^2yz^4,\ xy^4z^2,\ x^2y^4z,\ x^4yz^2,\ x^4y^2z,$$
$$x^2y^2z^3,\ x^2y^3z^2,\ x^3y^2z^2\Big\}.$$

This set contains 12 monomials.

The spikes (see [82]) $x^7, y^7, z^7, xy^3z^3, x^3yz^3, x^3y^3z$ give us 6 equivalence classes of $\mathscr{A}^*$-indecomposable monomials in $\mathbb{F}_2[x, y, z]_7$. From direct computation one sees that the 6 monomials

$$xy^6,\ x^2y^5,\ x^3y^4,\ x^4y^3,\ x^5y^2,\ x^6y \in \mathbb{F}_2[x, y]$$

form a single $\mathscr{A}^*$-equivalence class. These elements remain $\mathscr{A}^*$-indecomposable in $\mathbb{F}_2[x, y, z]$ and continue to form a single $\mathscr{A}^*$-equivalence class. By symmetry we deduce that

$$xz^6,\ x^2z^5,\ x^3z^4,\ x^4z^3,\ x^5z^2,\ x^6z$$

and

$$yz^6,\ y^2z^5,\ y^3z^4,\ y^4z^3,\ y^5z^2,\ y^6z$$

each form additional $\mathscr{A}^*$-equivalence classes. These account for 24 of the 36 linearly independent monomials in $\mathbb{F}_2[x, y, z]_7$, and give 9 linearly independent elements in $\left(\mathbb{F}_2[x, y, z]_{\mathscr{A}^*}\right)_7$. The remaining 12 elements of $\mathbb{F}_2[x, y, z]_7$ fall into the single $\mathscr{A}^*$-equivalence class $\mathscr{D}$. This gives 10 $\mathscr{A}^*$-equivalence classes of monomials in $\mathbb{F}_2[x, y, z]_7$ which project under the canonical projection to a basis for $\left(\mathbb{F}_2[x, y, z]_{\mathscr{A}^*}\right)_7$.

We let $\Delta \in \Gamma(u, v, w)_7$ be the element supported on $\mathscr{D}$, i.e., Δ regarded as an $\mathbb{F}_2$-valued linear form on $\mathbb{F}_2[x, y, z]_7$ takes the value 1 on the monomials in $\mathscr{D}$ and the value 0 on all other monomials of degree 7. By Proposition I.3.2 the corresponding big ancestor ideal $\mathfrak{A}(\ker(\Delta)) = I(\Delta) \subset \mathbb{F}_2[x, y, z]$ is an $\overline{\mathbb{F}_2[x, y, z]}$-primary irreducible ideal.

From the preceding discussion it follows that Δ factors through the canonical projection $\mathbb{F}_2[x, y, z]_7 \longrightarrow \left(\mathbb{F}_2[x, y, z]_{\mathscr{A}^*}\right)_7$, viz., there is a map $\bar{\Delta}$ making the following diagram

$$\begin{array}{ccc} \mathbb{F}_q[V]_d & \xrightarrow{\ \Delta\ } & \mathbb{F}_q \\ {\scriptstyle\pi}\big\downarrow & \nearrow{\scriptstyle\bar{\Delta}} & \\ \left(\mathbb{F}_q[V]_{\mathscr{P}^*}\right)_d & & \end{array}$$

commutative. Hence by Proposition III.5.1, the Poincaré duality quotient algebra $\mathbb{F}_2[x, y, z]/I(\Delta)$ has trivial Wu classes.

Using the method of catalecticant matrices described in Section VI.2 we found that generators for $I(\Delta)$ may be taken to be:

$$f_1 = x^3y + x^3z + xz^3 + yz^3$$
$$f_2 = xy^3 + y^3z + xz^3 + yz^3$$
$$f_3 = xy^2z + xyz^2 + xy^3 + x^3y + xz^3 + x^3z$$
$$f_4 = x^2yz + xy^2z + x^3z + xz^3 + y^3z + yz^3$$
$$f_5 = x^4 + y^4 + z^4 + x^2y^2 + x^2z^2 + y^2z^2 + x^2yz + xy^2z + xyz^2 + x^3y + x^3z.$$

Each of these polynomials has degree 4 and the Poincaré series of the corresponding Poincaré duality quotient $H(\Delta) = \mathbb{F}_2[x, y, z]/I(\Delta)$ is

$$1 + 3t + 6t^2 + 10t^3 + 10t^4 + 6t^5 + 3t^6 + t^7.$$

Corollary III.6.5, respectively Proposition III.6.6, implies that the Wu classes of the Poincaré duality quotients

$$\mathbb{F}_2[x, y, z]/I(\Delta)^{[2^{k-1}]} \qquad k \in \mathbb{N}$$

are trivial, respectively, that the monomials

$$(xyz)^{2^k-1}z^{2^{k+1}} \qquad k \in \mathbb{N}$$

are all $\mathscr{A}^*$-indecomposable. Finally, from Proposition III.4.2 we conclude that

$$(z_1 \cdots z_n)^{2^k-1}z_n^{2^{k+1}} \in \mathbb{F}_2[z_1, \ldots, z_n]$$

are $\mathscr{A}^*$-indecomposable for $n \geq 2$ and $k \in \mathbb{N}$.

The Frobenius powers $I(\Delta), I(\Delta)^{[2]}, \ldots, I(\Delta)^{[2^i]}, \ldots$ are all $\mathscr{A}^*$-invariant $\overline{\mathbb{F}_2[x, y, z]}$-primary irreducible ideals. The formal dimension of the Poincaré duality quotient $H(I(\Delta)^{[2^i]})$ is $3(2^i - 1) + 7 \cdot 2^i$. The periodicity operator $\mathbf{P}$ when iterated i times and applied to $xyz^5 = xyz \cdot z^4$ gives as representative for the fundamental class

$$\mathbf{P}(xyz^5) = (xyz)^{2^i-1}(xyz^5)^{2^i}.$$

So the monomials in the sequence

$$(xyz)^{2^i-1}(xyz^5)^{2^i} = (xyz)^{2^{i+1}-1}z^{2^{i+2}} \qquad i \in \mathbb{N}_0$$

are all $\mathscr{A}^*$-indecomposable. Note that $z^2 = \mathrm{Sq}^1(z)$ is $\mathscr{A}^*$-decomposable, but that the elements $\mathbf{P}^i(z^2)$ are all $\mathscr{A}^*$-indecomposable, $i \in \mathbb{N}$. This is another example to show that the converse of Proposition III.6.6 does not hold.

One conclusion which can be drawn from this example is the following: if the $\mathscr{P}^*$-equivalence classes of monomials in $\mathbb{F}_q[V]_k$ project to a basis for $\left(\mathbb{F}_q[V]_{\mathscr{P}^*}\right)_k$ then a linear form $\gamma : \mathbb{F}_q[V]_k \longrightarrow \mathbb{F}_q$ supported on a union of $\mathscr{P}^*$-equivalence classes of the monomials defines a $\mathscr{P}^*$-invariant $\overline{\mathbb{F}_q[V]}$-primary irreducible ideal with trivial Wu classes. However, it need not be the case that the $\mathscr{P}^*$-equivalence classes of monomials in $\mathbb{F}_q[V]_k$ project to a basis for $\left(\mathbb{F}_q[V]_{\mathscr{P}^*}\right)_k$. A simple example where this fails is $\mathbb{F}_2[x, y, z]_4$ since the monomials x^2yz, xy^2z, xyz^2 have distinct images in $\left(\mathbb{F}_2[x, y, z]_{\mathscr{A}^*}\right)_4$, but

$$\mathrm{Sq}^1(xyz) = x^2yz + xy^2z + xyz^2 \in \mathbb{F}_2[x, y, z],$$

so in $\left(\mathbb{F}_2[x, y, z]_{\mathscr{A}^*}\right)_4$ their images are linearly dependent (cf. Example 2 in Section III.5).

The ideal $L^{[2]}$ of Example 1 in Section III.5 has a Poincaré duality quotient with xyz^5 as a representative for the fundamental class also. However, as noted there the corresponding quotient algebra has nontrivial Wu classes. Although the quotient algebras $\mathbb{F}_2[x, y, z]/I(\Delta)$ and $\mathbb{F}_2[x, y, z]/L^{[2]}$ have the same Poincaré series, and a common representative for a fundamental class, they are *not* isomorphic. Aside from the use already noted of the Wu classes to show this, one could proceed as follows. In degree 4 the ideal $L^{[2]}$ contains a two dimensional subspace consisting of 4-th powers of linear forms. In other words, if Q denotes the ideal (x^4, y^4, z^4) in $\mathbb{F}_2[x, y, z]$, then the intersection $L^{[2]} \cap Q$ is nonzero in degree 4. On the other hand, in degree 4 the ideal $I(\Delta)$ contains no 4-th powers at all. Since any automorphism of $\mathbb{F}_2[x, y, z]$ implemented by an element of $GL(3, \mathbb{F}_2)$ must preserve the space of 4-th powers, it follows there is no automorphism mapping $I(\Delta)$ onto $L^{[2]}$. By Theorem II.4.2 the corresponding Poincaré duality quotient algebras are not isomorphic.

In [16] M. C. Crabb and J. R. Hubbuck develop a method to study the *Hit Problem* for $\mathbb{F}_2[V]$ based on what they call the *ring of lines*. This is a subalgebra of $\Gamma(V)^{\mathscr{A}^*}$ which coincides with $\Gamma(V)^{\mathscr{A}^*}$ for $\dim_{\mathbb{F}_2}(V) = 1$ or 2, and in most homogeneous degrees when $\dim_{\mathbb{F}_2}(V)$ is arbitrary. In what follows we study the first case not fitting into this scheme of things: as we will see in Section V.5 this example arises quite naturally from our point of view. Here is a discussion of this example in a raw state based on Proposition III.5.3.

EXAMPLE 2: An element of minimal degree not fitting into the scheme of [16] occurs in degree 8 in the three variable case. The monomial $xy^2z^5 \in \mathbb{F}_2[x, y, z]_8$ is $\mathscr{A}^*$-indecomposable, but does not arise by dualizing from the ring of lines.[7] In our notation and terminology an element of $\Gamma(u, v, w)^{\mathscr{A}^*}$ which accounts for the $\mathscr{P}^*$-indecomposability of the element xy^2z^5 is the element

$$\begin{aligned} \varkappa = {} & \gamma_1(u)\gamma_2(v)\gamma_1(w)\gamma_4(w) + \gamma_1(u)\gamma_4(v)\gamma_1(w)\gamma_2(w) \\ & + \gamma_1(u)\gamma_1(v)\gamma_2(w)\gamma_4(w) + \gamma_2(u)\gamma_1(v)\gamma_2(v)\gamma_1(w)\gamma_2(w). \end{aligned}$$

This element has support $\{xy^2z^5, xy^4z^3, xyz^6, x^2y^3z^3\}$ and is $\mathscr{A}^*$-invariant. By Macaulay's $\mathscr{A}^*$-Double Annihilator Theorem the ideal $I(\varkappa) \subset \mathbb{F}_2[x, y, z]$ corresponding to $\varkappa$ is $\mathscr{A}^*$-invariant and has a Poincaré duality quotient $H(\varkappa)$ with trivial Wu classes. The fundamental class is represented by the monomials in the support of $\varkappa$. Using the method of catalecticant matrices described in Section VI.2 we computed the ideal $I(\varkappa)$ and found

[7] See [16] Proposition 3.7 or see also [8] Section 5 where we consider the case $s = 2$, $t = 1$; $n = 11 \cdot 2^u - 3$ and set $u = 0$ to obtain the same element up to permutation of the variables. This is not the only element of $\Gamma(u, v, w)^{\mathscr{A}^*}$ with xy^2z^5 in its support. See the discussion of Example 1 in Section V.5.

$I(\varkappa) = (f, g, h)$ where

$$\begin{aligned} f &= x^3 \\ g &= y^4 + xy^3 + x^2y^2 \\ h &= z^4 + xyz^2 + xy^2z + x^2yz + x^2y^2 + x^2z^2. \end{aligned}$$

As we will show in Section V.5 the $\mathscr{A}^*$-indecomposability of the element xy^2z^5 has another completely natural explanation from our viewpoint: it represents the fundamental class of the Poincaré duality quotient with trivial Wu classes $\mathbb{F}_2[x, y, z]/(w_4, w_3, w_2^2)$, where $w_4, w_3, w_2 \in \mathbb{F}_2[x, y, z]$ are the Stiefel–Whitney classes (see Section V.5). The element ω in $\Gamma(u, v, w)$ generating the Macaulay dual of the ideal (w_4, w_3, w_2^2) has xy^2z^5 in its support.

The last example amplifies a number of points already touched on as well as leading to an infinite number of further similar examples.

EXAMPLE 3: Consider again the Poincaré duality quotient algebra $H = \mathbb{F}_2[x, y, z]/K$, where $K = (x^4, y^4, z^4) \subset \mathbb{F}_2[x, y, z]$, introduced in Example 1 of Section III.1. As in that example let $f = x^3 + y^3 + z^3 \in \mathbb{F}_2[x, y, z]$. Then, as we have already seen, $\mathbf{Sq}(f) \equiv f \bmod K$, and so $H/\mathrm{Ann}_H(f)$ is an unstable algebra over the Steenrod algebra. We show that $H/\mathrm{Ann}_H(f)$ has trivial Wu classes by applying Theorem III.3.5 in the following way. Set $L = (K : f) \subset \mathbb{F}_2[x, y, z]$ so $\mathrm{Ann}_H(f) \cong L/K$, and

$$L = (K : f) = (xyz, x^3 + y^3 + z^3, x^4, y^4, z^4) \subset \mathbb{F}_2[x, y, z]$$

is an $\mathscr{A}^*$-invariant ideal. The quotient algebra $\mathbb{F}_2[x, y, z]/L$ has Poincaré series

$$P(\mathbb{F}_2[x, y, z], t) = 1 + 3t + 6t^2 + 8t^3 + 6t^4 + 3t^5 + t^6.$$

By Theorem II.5.1 a generator for the Macaulay dual of L is

$$(x^3 + y^3 + z^3) \cap \gamma_3(u)\gamma_3(v)\gamma_3(w) = \gamma_3(v)\gamma_3(w) + \gamma_3(u)\gamma_3(w) + \gamma_3(u)\gamma_3(v)$$

and the monomials that represent a fundamental class are y^3z^3, x^3z^3, and x^3y^3. The ideal $L \subset \mathbb{F}_2[x, y, z]$ is the start of an infinite family of $\mathscr{A}^*$-invariant ideals $L = L_2, L_3, \ldots$ in $\mathbb{F}_2[x, y, z]$ defined by

$$((x^{2^k}, y^{2^k}, z^{2^k}) : x^{2^k-1} + y^{2^k-1} + z^{2^k-1}) = (xyz, x^{2^k-1} + y^{2^k-1} + z^{2^k-1}, x^{2^k}, y^{2^k}, z^{2^k}),$$

for $k = 2, 3, \ldots$. These ideals all have trivial Wu classes and require five generators.

Part IV
Dickson, symmetric, and other coinvariants

THIS part is concerned with applications of the material developed so far to families of ideals originating in invariant theory. If $\rho : G \hookrightarrow \mathrm{GL}(n, \mathbb{F})$ is a representation of a finite group for which the ring of invariants $\mathbb{F}[V]^G$ is a polynomial algebra, then the corresponding Hilbert ideal $\mathfrak{h}(G)$ is generated by a regular sequence in $\mathbb{F}[V]$ of maximal length (this follows e.g., by combining [87] Corollaries 6.7.13 and 6.4.4).[1] It is therefore irreducible and $\overline{\mathbb{F}[V]}$-primary, so $\mathbb{F}[V]_G = \mathbb{F}[V]/\mathfrak{h}(G)$ is a Poincaré duality algebra ([87] Theorem 6.5.1). Using regular ideals that arise in this way allows us to illustrate many of the ideas introduced up to this point as well as point out some special features of Hilbert ideals.

Often the generators of the Hilbert ideal are polynomials of special interest in their own right, such as the elementary symmetric polynomials or the Dickson polynomials. In this Part we also examine the Poincaré duality quotients of $\mathbb{F}_q[V]$ by ideals generated by powers of such polynomials.

IV.1 Dickson coinvariants

Let $\mathbb{F}_q$ be the Galois field with $q = p^\nu$ elements where p is a prime and set $V = \mathbb{F}_q^n$. The ring of invariants $\mathbb{F}_q[V]^{\mathrm{GL}(n,\mathbb{F}_q)}$ is a polynomial algebra, known as the **Dickson algebra** (see e.g. [87] Section 8.1) which we denote by $\mathbf{D}(n)$ if $\mathbb{F}_q$ is clear from the context. Generators for this ring of invariants

[1] Not every Hilbert ideal which is regular arises in this way (see e.g. [92]) and it is an open problem to determine conditions on a representation $\rho : G \hookrightarrow \mathrm{GL}(n, \mathbb{F})$ in the modular case such that the Hilbert ideal is a regular ideal. In the nonmodular case ρ must be a pseudoreflection representation (see e.g. [98], [40], and [47]) and $\mathbb{F}[V]^G$ is a polynomial algebra.

are called **Dickson polynomials**; we denote them by

$$\mathbf{d}_{n,0}, \ldots, \mathbf{d}_{n,n-1} \in \mathbb{F}_q[z_1, \ldots, z_n].$$

We index these polynomials so that $\deg(\mathbf{d}_{n,i}) = q^n - q^i$ and introduce the notation $\mathfrak{d}(k_0, \ldots, k_{n-1})$ for the ideal of $\mathbb{F}_q[z_1, \ldots, z_n]$ generated by the forms $\mathbf{d}_{n,0}^{k_0}, \ldots, \mathbf{d}_{n,n-1}^{k_{n-1}}$. Except for $p = 2$ the Dickson polynomials are uniquely determined only up to a nonzero scalar, but they can be unambiguously defined by the Stong–Tamagawa formulae (see e.g. [87] Theorem 8.1.6), viz.,

$$\mathbf{d}_{n,i} = \sum_{\substack{W^* \leq V^* \\ \dim(W^*)=i}} \prod_{z \notin W^*} z.$$

With these notations we have the following formulae for the action of the Steenrod operations on the Dickson polynomials (see e.g. [66] Appendix A or [68] Corollary 6.8.2):

$$\mathcal{P}^{p^m}(\mathbf{d}_{n,i}) = 0 \text{ for } i = 0, \ldots, n-1 \text{ unless } p^m = q^k \text{ for some } k$$

in which case

$$(\text{✣}) \qquad \mathcal{P}^{q^k}(\mathbf{d}_{n,i}) = \begin{cases} -\mathbf{d}_{n,i-1} & \text{for } k = i-1 \geq 0 \\ -\mathbf{d}_{n,i}\mathbf{d}_{n,n-1} & \text{for } k = n-1 \geq 0 \\ 0 & \text{otherwise} \end{cases}$$

The following property of the Dickson polynomials seems not to have been observed before, but is a routine matter to verify using these formulae.

PROPOSITION IV.1.1: *Let $a_0, \ldots, a_{n-1} \in \mathbb{N}_0$ be a sequence of integers. Then the ideal $\mathfrak{d}(q^{a_0}, \ldots, q^{a_{n-1}})$ in $\mathbb{F}_q[z_1, \ldots, z_n]$ generated by the elements $\mathbf{d}_{n,0}^{q^{a_0}}, \ldots, \mathbf{d}_{n,n-1}^{q^{a_{n-1}}}$ is closed under the action of the Steenrod algebra if and only if $0 \leq a_0 \leq \cdots \leq a_{n-1}$.*

PROOF: The Steenrod operations $\{\mathcal{P}^{p^m} \mid m \in \mathbb{N}_0\}$ generate $\mathscr{P}^*$ so it is enough to show that the ideal $\mathfrak{d}(q^{a_0}, \ldots, q^{a_{n-1}})$ is closed under the action of these operations, and for this it is enough to check that $\mathcal{P}^{p^m}(\mathbf{d}_{n,i}^{q^{a_i}}) \in \mathfrak{d}(q^{a_0}, \ldots, q^{a_{n-1}})$ for $i = 0, \ldots, n-1$ and all $m \in \mathbb{N}_0$. By the preceding formulae the only cases that are nonzero occur for $p^m = q^k$, where we find by the Cartan formula

$$\mathcal{P}^{q^k}(\mathbf{d}_{n,i}^{q^{a_i}}) = (\mathcal{P}^{q^{k-a_i}}(\mathbf{d}_{n,i}))^{q^{a_i}} = \begin{cases} (-\mathbf{d}_{n,i-1})^{q^{a_i}} & \text{for } k - a_i = i-1 \geq 0 \\ (-\mathbf{d}_{n,i}\mathbf{d}_{n,n-1})^{q^{a_i}} & \text{for } k - a_i = n-1 \geq 0 \\ 0 & \text{otherwise.} \end{cases}$$

If $a_i \geq a_{i-1}$ it follows that $(-\mathbf{d}_{n,i-1})^{q^{a_i}}$ is divisible by $(\mathbf{d}_{n,i-1})^{q^{a_{i-1}}}$ and so belongs to the ideal $\mathfrak{d}(q^{a_0}, \ldots, q^{a_{n-1}})$. Likewise $(-\mathbf{d}_{n,i}\mathbf{d}_{n,n-1})^{q^{a_i}}$ is divisible by $\mathbf{d}_{n,i}^{q^{a_i}}$ so is also in $\mathfrak{d}(q^{a_0}, \ldots, q^{a_{n-1}})$.

Conversely, suppose the ideal $\mathfrak{d}(q^{a_0}, \ldots, q^{a_{n-1}})$ is closed under the action of the Steenrod algebra. As a special case (put $k = 0$) of the formulae (✣) we have

$$\mathcal{P}^1(\mathbf{d}_{n,1}) = -\mathbf{d}_{n,0}$$

so that

$$\mathcal{P}^{q^{a_1}}(\mathbf{d}_{n,1}^{q^{a_1}}) = \left(\mathcal{P}^1(\mathbf{d}_{n,1})\right)^{q^{a_1}} = (-1)^{q^{a_1}} \mathbf{d}_{n,0}^{q^{a_1}}.$$

Since $\mathfrak{d}(q^{a_0}, \ldots, q^{a_{n-1}})$ is closed under the action of the Steenrod algebra this implies $\mathbf{d}_{n,0}^{q^{a_1}} \in \mathfrak{d}(q^{a_0}, \ldots, q^{a_{n-1}})$ so $0 \le a_0 \le a_1$.

The proof is completed by induction on n as follows: consider the natural map

$$\pi : \mathbb{F}_q[z_1, \ldots, z_n] \longrightarrow \mathbb{F}_q[z_1, \ldots, z_{n-1}]$$

defined by $\pi(z_i) = z_i$ for $i = 1, \ldots, n-1$ and $\pi(z_n) = 0$. This map commutes with the action of the Steenrod algebra and satisfies ([87] Theorem 8.1.6)

$$\pi(\mathbf{d}_{n,i}) = \begin{cases} \mathbf{d}_{n-1,i-1}^q & \text{if } i = 1, \ldots, n-1 \\ 0 & \text{if } i = 0. \end{cases}$$

Therefore applying the map π to the ideal $\mathfrak{d}(q^{a_0}, \ldots, q^{a_{n-1}})$ we obtain $\pi\left(\mathfrak{d}(q^{a_0}, \ldots, q^{a_{n-1}})\right) = \mathfrak{d}(q^{a_1+1}, \ldots, q^{a_{n-1}+1})$. So in $\mathbb{F}_q[z_1, \ldots, z_{n-1}]$ the ideal $\mathfrak{d}(q^{a_1+1}, \ldots, q^{a_{n-1}+1})$ is closed under the action of the Steenrod algebra. By the inductive assumption we have $a_1 + 1 \le \cdots \le a_{n-1} + 1$ and the result follows. □

If any of the inequalities in the chain $a_0 \le a_1 \le \cdots \le a_{n-1}$ are strict then it may happen that the Wu classes of the corresponding $\mathcal{P}^*$-Poincaré duality quotient algebra are nontrivial. For the basic ideal $\mathfrak{d}(1, \ldots, 1)$ generated by the Dickson polynomials themselves they are trivial by a result of S. A. Mitchell ([60] Appendix B). Since we intend to reprove Mitchell's theorem in the next section using precisely this fact, we include an independent proof.

Denote by $\mathbf{L}_n$ a Dickson–Euler class (see e.g. [94] or [87] Chapter 2 for a discussion of Euler classes), i.e., for each line in $V^* = \mathrm{Span}_{\mathbb{F}_q}(z_1, \ldots, z_n)$ we choose a nonzero linear form in that line and take the product of the resulting set of linear forms. Then $\mathbf{L}_n^{q-1} \mathrel{\overset{\circ}{=}} \mathbf{d}_{n,0}$ loc. cit., and one may make the choices (see e.g. [9] Chapitre V Exercise 6) so that equality holds. The point of departure is the following lemma that allows us to make use of Theorem III.3.5 to compute the Wu classes of the Dickson coinvariants.

LEMMA IV.1.2: *Let $n \in \mathbb{N}$ and consider the ideals $K = (z_1^{q^n}, \ldots, z_n^{q^n}) \subset \mathbb{F}_q[z_1, \ldots, z_n]$ and $L = (\mathbf{d}_{n,0}, \ldots, \mathbf{d}_{n,n-1}) \subset \mathbb{F}_q[z_1, \ldots, z_n]$. Then $K \subset L$ and $(K : L) = (\mathbf{L}_n) + K$.*

PROOF: Let $\mathcal{P}^{\Delta_k} \in \mathcal{P}^*$ be the Milnor primitive element of degree $q^k - 1$ for $k \in \mathbb{N}$ and define $\mathcal{P}^{\Delta_0}$ by $\mathcal{P}^{\Delta_0}(u) = \deg(u) \cdot u$ for any $u \in \mathbb{F}_q[z_1, \ldots, z_n]$. If $z \in \mathbb{F}_q[z_1, \ldots, z_n]$ is a linear form then $\mathcal{P}^{\Delta_k}(z) = z^{q^k}$ (see e.g. [87] Lemma 10.4.2). Introduce the operator

$$\Delta = \mathcal{P}^{\Delta_n} + \mathbf{d}_{n,n-1}\mathcal{P}^{\Delta_{n-1}} + \cdots + \mathbf{d}_{n,0}\mathcal{P}^{\Delta_0}.$$

Recall that Δ is identically zero on $\mathbb{F}_q[z_1, \ldots, z_n]$ (see e.g. [87] Lemma 10.6.2, or [2] where this result appears implicitly in Section 5). Therefore applying the operator Δ to the variables $z_1, \ldots, z_n$ we obtain after rearrangement

$$\begin{aligned} -z_1^{q^n} &= z_1^{q^{n-1}}\mathbf{d}_{n,n-1} + \cdots + z_1\mathbf{d}_{n,0} \\ \vdots\ &\quad \vdots \\ -z_n^{q^n} &= z_n^{q^{n-1}}\mathbf{d}_{n,n-1} + \cdots + z_n\mathbf{d}_{n,0}, \end{aligned}$$

which shows that $K \subset L$. We rewrite the preceding equations as one matrix equation, viz.,

$$(\maltese) \qquad -\begin{bmatrix} z_1^{q^n} \\ \vdots \\ z_n^{q^n} \end{bmatrix} = \begin{bmatrix} z_1^{q^{n-1}} & \cdots & z_1 \\ \vdots & \cdots & \vdots \\ z_n^{q^{n-1}} & \cdots & z_n \end{bmatrix} \cdot \begin{bmatrix} \mathbf{d}_{n,n-1} \\ \vdots \\ \mathbf{d}_{n,0} \end{bmatrix} = \mathbf{A} \begin{bmatrix} \mathbf{d}_{n,n-1} \\ \cdots \\ \mathbf{d}_{n,0} \end{bmatrix}$$

defining the matrix $\mathbf{A}$. Let $\mathbf{A}^{\mathrm{cof}}$ be the matrix of transposed cofactors of $\mathbf{A}$. Then multiplying both sides of equation $(\maltese)$ by $\mathbf{A}^{\mathrm{cof}}$ and using the fact that $\mathbf{A}^{\mathrm{cof}}\mathbf{A} = \det(\mathbf{A}) \cdot \mathbf{I} = \det(\mathbf{A}^{\mathrm{cof}}) \cdot \mathbf{I} = \mathbf{A}\mathbf{A}^{\mathrm{cof}}$, where $\mathbf{I}$ is the $n \times n$ identity matrix, we obtain

$$\mathbf{A}^{\mathrm{cof}} \cdot \begin{bmatrix} -z_1^{q^n} \\ \vdots \\ -z_n^{q^n} \end{bmatrix} = \det(\mathbf{A}) \cdot \begin{bmatrix} \mathbf{d}_{n,n-1} \\ \vdots \\ \mathbf{d}_{n,0} \end{bmatrix}.$$

Since ([9] Chapitre V, Exercise 6)

$$\mathbf{L}_n \mathrel{\hat{=}} \det \begin{bmatrix} z_1^{q^{n-1}} & \cdots & z_1 \\ \vdots & \cdots & \vdots \\ z_n^{q^{n-1}} & \cdots & z_n \end{bmatrix} = \det(\mathbf{A})$$

we conclude from this that $\mathbf{L}_n \cdot L \subseteq K$ so $\mathbf{L}_n \in (K : L)$. By Theorem I.2.1 $(K : L) = (h) + K$ for some $h \in \mathbb{F}_q[V]$, and Corollary I.2.4 tells us that

$$\begin{aligned} \deg(h) &= \text{f-dim}(\mathbb{F}_q[z_1, \ldots, z_n]/K) - \text{f-dim}(\mathbb{F}_q[z_1, \ldots, z_n]/L) \\ &= n(q^n - 1) - \sum_{s=0}^{n-1}(q^n - q^s - 1) \\ &= \frac{q^n - 1}{q - 1} = \deg(\mathbf{L}_n). \end{aligned}$$

Since $\mathbf{L}_n \notin K$ therefore we may choose $h = \mathbf{L}_n$. □

This result is greatly generalized by K. Kuhnigk in [44] where, amongst other things, she computes a Macaulay dual for the ideal generated by the Dickson polynomials, thereby also determining all the monomials which represent a fundamental class of the Dickson coinvariants.

PROPOSITION IV.1.3: *Let $\mathbb{F}_q$ be the Galois field with $q = p^\nu$ elements, p a prime, and $V = \mathbb{F}_q^n$ the standard n-dimensional vector space over $\mathbb{F}_q$. The Wu classes of the algebra of Dickson coinvariants $\mathbb{F}_q[z_1, \ldots, z_n]_{\mathrm{GL}(n, \mathbb{F}_q)}$ are trivial.*

PROOF: We apply Theorem III.3.5 to the $K \subset L$ paradigm with

$$K = (z_1^{q^n}, \ldots, z_n^{q^n}) \subset \mathbb{F}_q[z_1, \ldots, z_n]$$
$$L = (\mathbf{d}_{n,0}, \ldots, \mathbf{d}_{n,n-1}) \subset \mathbb{F}_q[z_1, \ldots, z_n]$$
$$(K : L) = (\mathbf{L}_n) + K, \text{ and } L = (K : \mathbf{L}_n).$$

An elementary computation, or induction and Theorem III.3.5, shows that the Wu classes of $\mathbb{F}_q[z]/(z^{q^s})$ are trivial for any $s \in \mathbb{N}_0$, so taking tensor products one sees that the Wu classes of $\mathbb{F}_q[z_1, \ldots, z_n]/K$ are trivial. The Dickson–Euler class $\mathbf{L}_n$ is a product of linear forms so it is a Thom class, and hence $\mathcal{P}(\mathbf{L}_n) = \mathbf{L}_n \cdot f \in \mathbb{F}_q[z_1, \ldots, z_n]$ for some inhomogeneous polynomial $f = 1 + f_1 + \cdots$, with $f_i \in \mathbb{F}[z_1, \ldots, z_n]_{i(q-1)}$ for $i \in \mathbb{N}$. Since both $\mathbf{L}_n$ and $\mathcal{P}(\mathbf{L}_n)$ are $\mathrm{SL}(n, \mathbb{F}_q)$-invariant, the homogeneous components of f belong to the Dickson algebra $\mathbf{D}(n)$, so the components of positive degree of $\mathcal{P}(\mathbf{L}_n)$ belong to the ideal L. By Theorem III.3.5 we therefore have

$$\chi\mathrm{Wu}(L) \equiv f \cdot \chi\mathrm{Wu}(K) = 1 \bmod L$$

as claimed. □

Denote by $H(q^{a_0}, \ldots, q^{a_{n-1}})$ the Poincaré duality quotient algebra of $\mathbb{F}_q[z_1, \ldots, z_n]$ by the ideal $\mathfrak{d}(q^{a_0}, \ldots, q^{a_{n-1}})$. If $a_0 = a_1 = \cdots = a_{n-1} = a$ then $H(q^a, \ldots, q^a)$ is obtained by applying the Frobenius operator $-^{[q^a]}$ to the Hilbert ideal $\mathfrak{h}(\mathrm{GL}(n, \mathbb{F}_q)) = (\mathbf{d}_{n,0}, \ldots, \mathbf{d}_{n,n-1})$, and passing to the corresponding Poincaré duality quotient algebra. From Proposition IV.1.3 and Corollary III.6.5 it follows that the Wu classes, apart from Wu_0 of course, of the algebra $H(q^a, \ldots, q^a)$ are zero. Further $\mathcal{P}^*$-invariant ideals with trivial Wu classes can be obtained from these by the procedure used in the proof of Propositions V.1.3 and V.1.4. For $q = 2$ we will give in Section V.4 a condition on the exponents $k_0, \ldots, k_{n-1}$ that assures that the ideal $\mathfrak{d}(k_0, \ldots, k_{n-1})$ is $\mathcal{A}^*$-invariant and has trivial Wu classes. For $n = 2$ or 3 we give independent proofs in Sections V.1 and V.3.

IV.2 Wu classes of algebras of coinvariants

Let $\rho : G \hookrightarrow \mathrm{GL}(n, \mathbb{F}_q)$ be a representation of a finite group such that $\mathbb{F}[V]^G$ is a polynomial algebra. In [60] Appendix B S. A. Mitchell showed that in this case the Wu classes of the algebra of coinvariants $\mathbb{F}_q[V]_G$ are trivial. In fact he proved more: if $\mathbb{F}_q[f_1, \ldots, f_n] \subseteq \mathbb{F}_q[z_1, \ldots, z_n]$ is a $\mathscr{P}^*$-invariant subalgebra and $f_1, \ldots, f_n \in \mathbb{F}_q[z_1, \ldots, z_n]$ is a regular sequence, then the quotient algebra $\mathbb{F}_q[z_1, \ldots, z_n]/(f_1, \ldots, f_n)$ has trivial Wu classes. We are indebted to R. E. Stong for clarifying a number of points in Mitchell's argument which allowed us to construct a variant of the proof that fits nicely with the development at this point. Here is how it goes.

NOTATION: *If $A \longrightarrow B$ is a map of commutative graded connected algebras over the field $\mathbb{F}$ then we denote $\mathbb{F} \otimes_A B$ by $B//A$.*

LEMMA IV.2.1: *Let $n \in \mathbb{N}$ and suppose given a chain of algebras*

$$\mathbb{F}[h_1, \ldots, h_n] \subseteq \mathbb{F}[f_1, \ldots, f_n] \subseteq \mathbb{F}[z_1, \ldots, z_n]$$

where $h_1, \ldots, h_n \in \mathbb{F}[f_1, \ldots, f_n]$ and $f_1, \ldots, f_n \in \mathbb{F}[z_1, \ldots, z_n]$ are regular sequences.[2] Introduce $H = \mathbb{F}[z_1, \ldots, z_n]/(h_1, \ldots, h_n)$ and $H' = \mathbb{F}[f_1, \ldots, f_n]/(h_1, \ldots, h_n)$. Then H is a free H'-module.

PROOF: Let $H'' = \mathbb{F}[z_1, \ldots, z_n]/(f_1, \ldots, f_n)$. Then

$$\mathbb{F} \longrightarrow H' \longrightarrow H \longrightarrow H'' \longrightarrow \mathbb{F}$$

is a coexact sequence, and for the Poincaré series of these algebras we have

$$P(H, t) = P(H', t) \cdot P(H'', t).$$

Since $H'' \cong \mathbb{F} \otimes_{H'} H$ is the $\mathbb{F}$-vector space of H'-indecomposable elements of H we can find an epimorphism of H'-modules $\varphi : H' \otimes H'' \longrightarrow H$ by lifting a vector space basis for H'' to H. Since $H' \otimes H''$ and H have the same Poincaré series φ must also be a monomorphism, and hence is an isomorphism. □

PROPOSITION IV.2.2: *Let $q = p^\nu$ where $p \in \mathbb{N}$ is a prime and $\nu \in \mathbb{N}$. Suppose*

$$\mathbb{F}_q \longrightarrow H' \longrightarrow H \longrightarrow H'' \longrightarrow \mathbb{F}_q$$

is a coexact sequence of $\mathscr{P}^*$-Poincaré duality algebras over $\mathbb{F}_q$ and H is a free H'-module. If $\mathrm{Wu}_k(H) = 0$ for $1 \leq k \leq \left[\frac{\text{f-dim}(H'')}{q-1}\right]$, then $\mathrm{Wu}(H'') = 1$.

---

2 It does not matter if we assume $h_1, \ldots, h_n$ is a regular sequence in $\mathbb{F}[f_1, \ldots, f_n]$ or $\mathbb{F}[z_1, \ldots, z_n]$: the conditions are equivalent.

PROOF: Let $[H] \in H$ be a fundamental class. Choose a lift $h \in H$ of a fundamental class $[H''] \in H''_{\text{f-dim}(H'')}$. It follows from Lemma IV.2.1 that $[H] = [H'] \cdot h$ for a suitable choice of fundamental class $[H']$ of H'. Let $h'' \in H''$ and suppose that $\deg(h'') + k(q-1) = \text{f-dim}(H'')$. Choose a lift $\widetilde{h}'' \in H$ of h''. Then for $k > 0$ the Cartan formula gives

$$<\mathcal{P}^k(h'') \mid [H'']> = <[H'] \cdot \mathcal{P}^k(\widetilde{h}'') \mid [H]> = <\mathcal{P}^k([H'] \cdot \widetilde{h}'') \mid [H]>$$
$$= <\mathrm{Wu}_k(H) \cdot [H'] \cdot \widetilde{h}'' \mid [H]> = 0$$

since $\mathcal{P}^i([H']) = 0 \in H'$ for $i > 0$ and $\mathrm{Wu}_k(H) = 0$. Therefore $\mathrm{Wu}_k(H'') = 0$. Since this holds for any $1 \le k \le \left[\frac{\text{f-dim}(H'')}{q-1}\right]$ the result follows. □

COROLLARY IV.2.3 (S. A. Mitchell): *Let $q = p^\nu$ where $p \in \mathbb{N}$ is a prime and $\nu \in \mathbb{N}$. Suppose $\mathbb{F}_q[f_1, \ldots, f_n] \subseteq \mathbb{F}_q[z_1, \ldots, z_n]$ is a $\mathcal{P}^*$-invariant subalgebra and $f_1, \ldots, f_n \in \mathbb{F}_q[z_1, \ldots, z_n]$ is a regular sequence. Then*

$$\mathrm{Wu}(\mathbb{F}_q[z_1, \ldots, z_n]/(f_1, \ldots, f_n)) = 1.$$

PROOF: The algebra $\mathbb{F}_q[f_1, \ldots, f_n]$ contains a fractal of the Dickson algebra, say $\mathbb{F}_q[\mathbf{d}^{q^s}_{n,0}, \ldots, \mathbf{d}^{q^s}_{n,n-1}] \subseteq \mathbb{F}_q[f_1, \ldots, f_n]$ (see e.g. [53] Appendix B or [66] Theorem 7.4.4). By Lemma IV.2.1 the sequence

$$\mathbb{F}_q \longrightarrow \frac{\mathbb{F}_q[f_1, \ldots, f_n]}{(\mathbf{d}^{q^s}_{n,0}, \ldots, \mathbf{d}^{q^s}_{n,n-1})} \longrightarrow \frac{\mathbb{F}_q[z_1, \ldots, z_n]}{(\mathbf{d}^{q^s}_{n,0}, \ldots, \mathbf{d}^{q^s}_{n,n-1})} \longrightarrow \frac{\mathbb{F}_q[z_1, \ldots, z_n]}{(f_1, \ldots, f_n)} \longrightarrow \mathbb{F}_q$$

is coexact and $\mathbb{F}_q[z_1, \ldots, z_n]/(\mathbf{d}^{q^s}_{n,0}, \ldots, \mathbf{d}^{q^s}_{n,n-1})$ is free as a module over $\mathbb{F}_q[f_1, \ldots, f_n]/(\mathbf{d}^{q^s}_{n,0}, \ldots, \mathbf{d}^{q^s}_{n,n-1})$. By Proposition IV.1.3 the Wu classes of the Dickson coinvariants $\mathbb{F}_q[z_1, \ldots, z_n]/(\mathbf{d}_{n,0}, \ldots, \mathbf{d}_{n,n-1})$ are trivial, so by Theorem III.6.4 $\mathbb{F}_q[z_1, \ldots, z_n]/(\mathbf{d}^{q^s}_{n,0}, \ldots, \mathbf{d}^{q^s}_{n,n-1})$ has trivial Wu classes also, and the result follows from Lemma IV.2.1 and Proposition IV.2.2. □

REMARK: If $f_1, \ldots, f_n \in \mathbb{F}_q[z_1, \ldots, z_n]$ it is *not* enough to assume that the *ideal* $(f_1, \ldots, f_n) \subset \mathbb{F}_q[z_1, \ldots, z_n]$ is $\mathcal{P}^*$-invariant and regular to conclude that the quotient has trivial Wu classes. A simple counterexample is provided by $\mathbb{F}_q[z]/(z^k)$ where k is not a power of q. Nor does the result generalize to unstable algebras other than $\mathbb{F}_q[z_1, \ldots, z_n]$. For example it can fail for quotients of the Dickson algebra, see e.g. Example 1 in Section VI.6.

COROLLARY IV.2.4 (S. A. Mitchell): *Let $\rho : G \hookrightarrow \mathrm{GL}(n, \mathbb{F}_q)$ be a representation of a finite group over the Galois field $\mathbb{F}_q$ with $q = p^\nu$ elements, $p \in \mathbb{N}$ a prime. If $\mathbb{F}_q[V]^G$ is a polynomial algebra, then* $\mathrm{Wu}(\mathbb{F}_q[V]_G) = 1$.
□

IV.3 The Macaulay dual of Dickson coinvariants mod 2

Since the Dickson coinvariants have trivial Wu classes, any monomial that represents a fundamental class is $\mathscr{P}^*$-indecomposable. In this section we determine for $q = 2$ all such monomials, i.e., which monomials z^D in $\mathbb{F}_2[z_1, \ldots, z_n]$ represent the fundamental class of the coinvariant algebra $\mathbb{F}_2[z_1, \ldots, z_n]_{\mathrm{GL}(n, \mathbb{F}_2)}$. It turns out that this set of monomials is the Σ_n-orbit of the single monomial $z_1^{2^n-2^0-1} z_2^{2^n-2^1-1} \cdots z_n^{2^n-2^{n-1}-1}$. To prove this we first compute a generator for the dual principal system $(\mathbf{d}_{n,0}, \ldots, \mathbf{d}_{n,n-1})^\perp$ using the $K \subset L$ paradigm. This arrangement of the material is due to K. Kuhnigk and replaces our original proofs which were considerably longer, more cumbersome, and less intuitive. We are indebted to her for permission to use it here. In her doctoral dissertation [44] she also determines a generator of the Macaulay dual of $(\mathbf{d}_{n,0}, \ldots, \mathbf{d}_{n,n-1})$ for arbitrary q, as well as certain ideals of the form $(\mathbf{d}_{n,0}^{a_0}, \ldots, \mathbf{d}_{n,n-1}^{a_{n-1}})$, where $a_0, \ldots, a_{n-1} \in \mathbb{N}$.

NOTATION: *The monomial* $z_1^{2^n-2^0-1} z_2^{2^n-2^1-1} \cdots z_n^{2^n-2^{n-1}-1}$ *plays an important role in this section and we reserve the notation* z^{D_n} *for it. The entries in the sequence* D_n *are* $(\deg(\mathbf{d}_{n,0}) - 1, \ldots, \deg(\mathbf{d}_{n,n-1}) - 1)$. *The orbit of* $D_n \in \mathbb{N}_0^n$ *under the natural* Σ_n*-action will be denoted by* $\mathfrak{D}_n$.

THEOREM IV.3.1: *The generator of the dual principal system to the ideal* $(\mathbf{d}_{n,0}, \ldots, \mathbf{d}_{n,n-1}) \subset \mathbb{F}_2[z_1, \ldots, z_n]$ *is*

$$\sum_{D \in \mathfrak{D}_n} \gamma_D \in \Gamma(u_1, \ldots, u_n)$$

where $\mathfrak{D}_n \subset \mathbb{N}_0^n$ *is the* Σ_n*-orbit of the index sequence* D_n.

PROOF: Recall the identity (see e.g. [87] Section 8.1)

$$\Phi(X) = \prod_{0 \neq z \in V^*} (X + z) = \sum_{i=0}^{n} \mathbf{d}_{n,i} X^{2^i - 1}$$

that is often used to define the Dickson polynomials. Note that $\Phi(z_i) = 0$ for $i = 1, \ldots, n$. Writing these n equations as a single matrix equation gives after some rearrangement

$$\begin{bmatrix} z_1^{2^n-1} \\ \vdots \\ z_n^{2^n-1} \end{bmatrix} = \begin{bmatrix} 1 & z_1^{2-1} & \cdots & z_1^{2^{n-1}-1} \\ \vdots & \vdots & \vdots \\ 1 & z_n^{2-1} & \cdots & z_n^{2^{n-1}-1} \end{bmatrix} \begin{bmatrix} \mathbf{d}_{n,0} \\ \cdots \\ \mathbf{d}_{n,n-1} \end{bmatrix}.$$

From this we see that $(z_1^{2^n-1}, \ldots, z_n^{2^n-1}) \subset (\mathbf{d}_{n,0}, \ldots, \mathbf{d}_{n,n-1})$. If we apply Cramer's rule to this matrix equation we find as in the proof of Lemma

IV.1.2 that

$$
\begin{aligned}
\big((z_1^{2^n-1}, \ldots, z_n^{2^n-1}) : (\mathbf{d}_{n,0}, \ldots, \mathbf{d}_{n,n-1})\big)
&= \det \begin{bmatrix} 1 & z_1^{2-1} & \ldots & z_1^{2^{n-1}-1} \\ \vdots & \vdots & & \vdots \\ 1 & z_n^{2-1} & \ldots & z_n^{2^{n-1}-1} \end{bmatrix} + (z_1^{2^n-1}, \ldots, z_n^{2^n-1}) \\
&= \Big(\sum_{E \in \mathfrak{E}_n} z^E\Big) + (z_1^{2^n-1}, \ldots, z_n^{2^n-1}),
\end{aligned}
$$

where $\mathfrak{E}_n$ is the Σ_n-orbit of $(2^0-1, 2^1-1, \ldots, 2^{n-1}-1) \in \mathbb{N}_0^n$. The generator of the dual principal system $(z_1^{2^n-1}, \ldots, z_n^{2^n-1})^\perp$ is the divided power monomial $\gamma_{2^n-2}(u_1)\cdots\gamma_{2^n-2}(u_n) \in \Gamma(u_1, \ldots, u_n)$. By Theorem II.5.1 it follows that

$$
\Big(\sum_{E \in \mathfrak{E}_n} z^E\Big) \cap \gamma_{2^n-2}(u_1)\cdots\gamma_{2^n-2}(u_n) = \sum_{D \in \mathfrak{D}_n} \gamma_D \in \Gamma(u_1, \ldots, u_n)
$$

generates the dual principal system $(\mathbf{d}_{n,0}, \ldots, \mathbf{d}_{n,n-1})^\perp$ as claimed. □

COROLLARY IV.3.2: *If a monomial $z^D \in \mathbb{F}_2[z_1, \ldots, z_n]$ represents a fundamental class of $\mathbb{F}_2[z_1, \ldots, z_n]_{\mathrm{GL}(n, \mathbb{F}_2)}$ then D is a permutation of D_n.*

PROOF: The support of the generator

$$
\sum_{D \in \mathfrak{D}_n} \gamma_D \in \Gamma(u_1, \ldots, u_n)
$$

of the dual principal system $(\mathbf{d}_{n,0}, \ldots, \mathbf{d}_{n,n-1})^\perp$ consists precisely of the monomials z^D with $D \in \mathfrak{D}_n$. □

IV.4 Symmetric coinvariants

Denote by $\tau : \Sigma_{n+1} \hookrightarrow \mathrm{GL}(n+1, \mathbb{F})$ the tautological representation of the symmetric group Σ_{n+1} over the field $\mathbb{F}$. The action of Σ_{n+1} on $\mathbb{F}^{n+1}$ stabilizes the subspace V of vectors with coordinate sum 0, so by restriction induces a faithful representation $\tilde{\tau} : \Sigma_{n+1} \hookrightarrow \mathrm{GL}(n, \mathbb{F})$. The ring of invariants of $\tilde{\tau}$ is a polynomial algebra. To wit, if $z_0, z_1, \ldots, z_n$ is the canonical basis for the linear forms on $\mathbb{F}^{n+1}$, then the inclusion $\iota : V \hookrightarrow \mathbb{F}^{n+1}$ induces a map between the rings of invariants $\iota^* : \mathbb{F}[z_0, z_1, \ldots, z_n]^{\Sigma_{n+1}} \longrightarrow \mathbb{F}[V]^{\Sigma_{n+1}}$. One knows by the fundamental theorem on symmetric polynomials, [87] Theorem 1.1.1, that $\mathbb{F}[z_0, z_1, \ldots, z_n]^{\Sigma_{n+1}} \cong \mathbb{F}[e_1, \ldots, e_{n+1}]$, where the generators, although not unique, may be chosen to be the elementary symmetric polynomials $e_1, e_2, \ldots, e_{n+1} \in \mathbb{F}[z_0, z_1, \ldots, z_{n+1}]$. The kernel of ι^* is the principal ideal generated by e_1. If we set $w_i = \iota^*(e_i)$ for $i = 2, \ldots, n+1$

then it is easy to see that $\mathbb{F}[V]^{\Sigma_{n+1}} \cong \mathbb{F}[w_2, \ldots, w_{n+1}]$ using [87] Theorem 5.5.5. Note that $\mathbb{F}[z_0, z_1, \ldots, z_n]_{\Sigma_{n+1}} \xrightarrow[\cong]{\iota^*} \mathbb{F}[V]/(w_2, \ldots, w_{n+1})$. If $\mathbb{F}$ is the Galois field $\mathbb{F}_q$ this is an isomorphism of $\mathscr{P}^*$-algebras. From topological considerations it is clear that the Wu classes of $\mathbb{F}_q[V]/(w_2, \ldots, w_{n+1})$ must be trivial if the ground field is finite: this also follows from S. A. Mitchell's theorem, Corollary IV.2.3. For the sake of illustration we include a simple algebraic proof that makes use of the $K \subset L$ paradigm.

Lemma IV.4.1 (R. E. Stong): *Let $k \in \mathbb{N}$, G be a finite group, and $\rho : G \hookrightarrow \mathrm{GL}(k, \mathbb{F})$ a representation. If $I \subseteq \mathbb{F}[V]$ is stable under the action of G then the action of G on $\mathbb{F}[V]$ passes down to the quotient algebra $\mathbb{F}[V]/I$. If $\mathbb{F} = \mathbb{F}_q$ is a Galois field and I is a $\mathscr{P}^*$-invariant irreducible $\overline{\mathbb{F}_q[V]}$-primary ideal then $\chi\mathrm{Wu}(I) \in \left(\mathbb{F}_q[V]/I\right)^G$.*

Proof: Choose a fundamental class $[H]$ for the Poincaré duality quotient algebra $H = \mathbb{F}_q[V]/I$. Let $u \in \mathbb{F}_q[V]$ and $g \in G$ be arbitrary. Then one has

$$\begin{aligned} &<(g \cdot \chi\mathrm{Wu}(I)) \cdot u \mid [H]> = <g \cdot (\chi\mathrm{Wu}(I) \cdot (g^{-1} \cdot u)) \mid [H]> \\ &= <\chi\mathrm{Wu}(I) \cdot (g^{-1} \cdot u) \mid g^{-1} \cdot [H]> = <\chi(\mathcal{P})(g^{-1} \cdot u) \mid g^{-1} \cdot [H]> \\ &= <g^{-1} \cdot \chi(\mathcal{P})(u) \mid g^{-1} \cdot [H]> = <\chi(\mathcal{P})(u)) \mid gg^{-1} \cdot [H]> \\ &= <\chi(\mathcal{P})(u) \mid [H]> = <\chi\mathrm{Wu}(I) \cdot u \mid [H]>. \end{aligned}$$

Since $u \in \mathbb{F}_q[V]$ was arbitrary this means $g \cdot \chi\mathrm{Wu}(I) - \chi\mathrm{Wu}(I) = 0$ and since $g \in G$ was also arbitrary the result follows. □

Proposition IV.4.2: *Let $\mathbb{F}_q$ be the Galois field with q elements, where $q = p^\nu$ and $p \in \mathbb{N}$ is a prime. Then, with the preceding notations, the Wu classes of the Poincaré duality quotient $\mathbb{F}_q[V]/(w_2, \ldots, w_{n+1})$ are trivial.*

Proof: The map $\iota^* : \mathbb{F}_q[z_0, z_1, \ldots, z_n]_{\Sigma_{n+1}} \longrightarrow \mathbb{F}_q[V]/(w_2, \ldots, w_{n+1})$ is an isomorphism, so it is equivalent to show that $\mathbb{F}_q[z_0, z_1, \ldots, z_n]_{\Sigma_{n+1}}$ has trivial Wu classes. To this end recall the identity

$$\xi(t) = t^{n+1} + e_1 t^n + \cdots + e_n t + e_{n+1} = \prod_{i=0}^{n} (t + z_i),$$

which is often used to define the elementary symmetric polynomials. If we evaluate $\xi(t)$ at $-z_i$ for $i = 0, 1, \ldots, n$ we obtain zero. Hence rearranging gives us the equations

$$-(-z_i)^{n+1} = e_1(-z_i)^n + \cdots + e_n(-z_i) + e_{n+1} \quad i = 0, 1, \ldots, n.$$

These in turn can be regarded as a system of linear equations, viz.,

$$-\begin{bmatrix}(-z_0)^{n+1}\\ \vdots \\ (-z_n)^{n+1}\end{bmatrix} = \begin{bmatrix}(-z_0)^n & \cdots & -z_0 & 1\\ \vdots & \vdots & \vdots & \vdots \\ (-z_n)^n & \cdots & -z_n & 1\end{bmatrix} \cdot \begin{bmatrix} e_1 \\ \vdots \\ e_{n+1}\end{bmatrix}.$$

This implies that $(z_0^{n+1}, z_1^{n+1}, \ldots, z_n^{n+1}) \subseteq (e_1, \ldots, e_{n+1})$. If we multiply both sides of this equation by the matrix of transposed cofactors we obtain $(\Delta_{n+1}e_1, \ldots, \Delta_{n+1}e_{n+1}) \subset (z_0^{n+1}, z_1^{n+1}, \ldots, z_n^{n+1})$, where

$$\Delta_{n+1} = \det \begin{bmatrix} z_0^n & \cdots & z_0 & 1 \\ \vdots & \vdots & \vdots & \vdots \\ z_n^n & \cdots & z_n & 1\end{bmatrix} = \prod_{i<j}(z_i - z_j),$$

by the rule for expanding Vandermonde determinants. So, if we set $K = (z_0^{n+1}, z_1^{n+1}, \ldots, z_n^{n+1})$ and $L = (e_1, \ldots, e_{n+1})$ we have a $K \subset L$ paradigm. Reasoning as in Lemma IV.1.2 (see also Proposition VI.3.1) $(K : L) = (\Delta_{n+1}) + K$. Since

$$\begin{aligned}\mathcal{P}(\Delta_{n+1}) &= \mathcal{P}\Big(\prod_{i<j}(z_i - z_j)\Big) = \prod_{i<j}\mathcal{P}(z_i - z_j) = \prod_{i<j}\Big((z_i - z_j) + (z_i - z_j)^q\Big)\\ &= \prod_{i<j}(z_i - z_j)\cdot \prod_{i<j}\Big(1 + (z_i - z_j)^{q-1}\Big)\end{aligned}$$

we find that

$$\frac{\mathcal{P}(\Delta_{n+1})}{\Delta_{n+1}} = \prod_{i<j}\Big(1 + (z_i - z_j)^{q-1}\Big).$$

Note that in the fraction on the left hand side of this formula both the numerator and the denominator are Σ_{n+1} det-relative invariants, so the fraction is a Σ_{n+1} invariant. Therefore this form is congruent to 1 modulo the Hilbert ideal $\mathfrak{h}(\Sigma_{n+1})$ and by Theorem III.3.5 we obtain

$$\chi\mathrm{Wu}(L) = \prod_{i<j}\Big(1 + (z_i - z_j)^{q-1}\Big) \cdot \chi\mathrm{Wu}(K) \equiv \chi\mathrm{Wu}(K) \bmod L.$$

By Lemma IV.4.1 $\chi\mathrm{Wu}(K) \equiv 1 \bmod L$ so we conclude $\chi\mathrm{Wu}(L) \equiv 1 \bmod L$ and the result is established. □

Part V
The Hit Problem mod 2

THEOREM III.4.1 presented a very unexpected connection between what started out as a study of Poincaré duality quotients of $\mathbb{F}[V]$ for the sake of invariant theory and the *Hit Problem*, a problem involving the action of the Steenrod algebra on $\mathbb{F}_q[V]$. This Part is devoted to the exploitation of that result: we apply what has been developed in Parts I–IV to the problem of computing the $\mathscr{A}^*$-indecomposable elements of $\mathbb{F}_2[z_1, \ldots, z_n]$. Again, more than one surprise occurs.

The objects of study in this Part are the $\mathscr{A}^*$-invariant Poincaré duality quotients of $\mathbb{F}_2[z_1, \ldots, z_n]$ with trivial Wu classes. We examine the cases $n = 2$ and $n = 3$ in some detail. For $n = 2$ we account for the known results on the *Hit Problem* in terms of spikes[1] in the sense of [82] and the fundamental classes of ideals generated by powers of Dickson polynomials. In the case $n = 3$ we need to bring in the ideals generated by powers of Stiefel–Whitney classes. These account for some of the $\mathscr{A}^*$-indecomposable monomials that up to this juncture had somewhat mysterious computational origins. We include in this Part a complete list of the ideals $(\mathbf{d}_{n,0}^{a_0}, \ldots, \mathbf{d}_{n,n-1}^{a_{n-1}})$ which are $\mathscr{A}^*$-invariant, as well as a list of the quotients $\mathbb{F}_2[z_1, \ldots, z_n]/(\mathbf{d}_{n,0}^{a_0}, \ldots, \mathbf{d}_{n,n-1}^{a_{n-1}})$ with trivial Wu classes. There are also partial results for the ideals generated by powers of Stiefel–Whitney classes.

In this Part we work almost exclusively over the Galois field $\mathbb{F}_2$. We employ the classical notation $\mathscr{A}^*$ for the mod 2 Steenrod algebra, reserving $\mathscr{P}^*$ for results that are valid for any prime power. We write Sq^k for the squaring operations in $\mathscr{A}^*$, and $\mathbf{Sq} = 1 + \mathrm{Sq}^1 + \cdots$ for the total Steenrod squaring

[1] These are in essence the elements that come from the one variable case. They are the products $z_1^{q^{a_1}-1} \cdots z_n^{q^{a_n}-1} \in \mathbb{F}_q[z_1, \ldots, z_n]$ where $a_1, \ldots, a_n \in \mathbb{N}_0^n$.

operation in the mod 2 case. Many of the results in this Part could be extended to any Galois field at the expense of a more complex notation and argumentation. We refrain from doing so as it appears to us that no new ideas occur if 2 is replaced by q.

Since there is a very large literature on *Hit Problems* we suggest the reader consult the bibliographies of [108] and [109].

V.1 Powers of Dickson polynomials in 2 variables

Let $\mathbf{d}_{2,0} = x^2y + xy^2$, $\mathbf{d}_{2,1} = x^2 + xy + y^2 \in \mathbb{F}_2[x, y]$ be the two Dickson polynomials. They form a regular sequence, as do $\mathbf{d}^a_{2,0}, \mathbf{d}^b_{2,1}$ for any $a, b \in \mathbb{N}$. Write $\mathfrak{d}(a, b) = (\mathbf{d}^a_{2,0}, \mathbf{d}^b_{2,1})$ for the ideal in $\mathbb{F}_2[x, y]$ generated by $\mathbf{d}^a_{2,0}, \mathbf{d}^b_{2,1}$. Then the quotient algebra $H(a, b) = \mathbb{F}_2[x, y]/\mathfrak{d}(a, b)$ satisfies Poincaré duality. In this section we will determine which of the ideals $\mathfrak{d}(a, b)$ are $\mathscr{A}^*$-invariant, and which of the quotient algebras $H(a, b)$ have trivial Wu classes. Here is the final result.

THEOREM V.1.1: *An ideal $\mathfrak{d}(a, b) = (\mathbf{d}^a_{2,0}, \mathbf{d}^b_{2,1})$ is invariant under the action of the mod 2 Steenrod algebra $\mathscr{A}^*$ if and only if $b = 2^\ell \cdot c$ with c odd and $a \le 2^\ell$. The quotient algebra $H(a, b) = \mathbb{F}_2[x, y]/\mathfrak{d}(a, b)$ has trivial Wu classes if and only if $a = 2^t$ and $b = 2^s - 2^t$ with $s > t$.*

The proof of this theorem proceeds in stages. To begin with, the ideal $\mathfrak{d}(1, 1)$ has trivial Wu classes: this is an easy computation using the formulae (see also Proposition IV.1.3)

$$\mathbf{Sq}(\mathbf{d}_{2,1}) = \mathbf{d}_{2,1} + \mathbf{d}_{2,0} + \mathbf{d}^2_{2,1}$$
$$\mathbf{Sq}(\mathbf{d}_{2,0}) = \mathbf{d}_{2,0}(1 + \mathbf{d}_{2,1} + \mathbf{d}_{2,0})$$

and the structure of the algebra $H(1, 1)$. One way to visualize this algebra is with the aid of Figure V.1.1. The nodes • indicate basis vectors for the homogeneous component whose degree is equal to the number of rows in the diagram above the node labeled 1, which is assigned the degree zero. The monomials representing a fundamental class, viz., x^2y and xy^2 are an $\mathscr{A}^*$-equivalence class and $\mathscr{A}^*$-indecomposable, so there is no possible squaring operation into the top nonzero degree. By applying Frobenius powers to $\mathfrak{d}(1, 1)$ we conclude the following result from Corollary III.6.5.

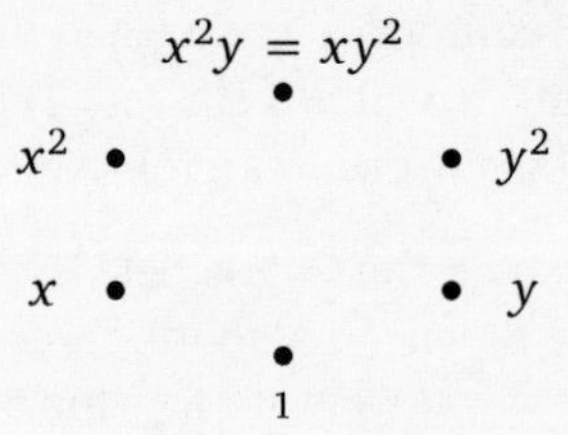

FIGURE V.1.1: $H(1, 1) = \mathbb{F}_2[x, y]/\mathfrak{d}(1, 1)$

PROPOSITION V.1.2: *Let $t \in \mathbb{N}_0$. Then the ideals $\mathfrak{d}(2^t, 2^t) = \mathfrak{d}(1, 1)^{[2^t]}$ are $\mathscr{A}^*$-invariant and have trivial Wu classes.* □

Note that the ideals $\mathfrak{d}(2^t, 2^t)$ have $\mathbf{d}_{2,0}^{2^t}$ as the generator of larger degree. To gain an idea of what the Poincaré duality quotient algebra $H(2^t, 2^t)$ looks like, we computed its Poincaré polynomial $P(H(2^t, 2^t), u)$ and found the following formula:

$$\begin{aligned} P(H(2^t, 2^t), u) &= (1 + u + u^2 + \cdots + u^{2^{t+1}+2^t-1})(1 + u + u^2 + \cdots + u^{2^{t+1}-1}) \\ &= 1 + 2u + 3u^2 + \cdots + 2^{t+1}u^{2^{t+1}-1} + 2^{t+1}u^{2^{t+1}} + \cdots + 2^{t+1}u^{2^{t+1}+2^t-1} \\ &\quad +(2^{t+1} - 1)u^{2^{t+1}+2^t} + \cdots + 3u^{2^{t+2}+2^t-4} + 2u^{2^{t+2}+2^t-3} + u^{2^{t+2}+2^t-2}. \end{aligned}$$

When pictured graphically $H(2^t, 2^t)$ looks a bit like a lighthouse standing with its base completely covering a small island as shown in the accompanying graphic. In the graphic the degree of a homogeneous component of $H(2^t, 2^t)$ is the height above the apex at the bottom labeled 0, with the width of the graphic representing the dimension of that homogeneous component. Notice the maximum width, by which we mean the maximum dimension of a homogeneous component of $H(2^t, 2^t)$, is equal to 2^{t+1}. In [56] we discuss which of these algebras can occur as the mod 2-cohomology of a topological space.

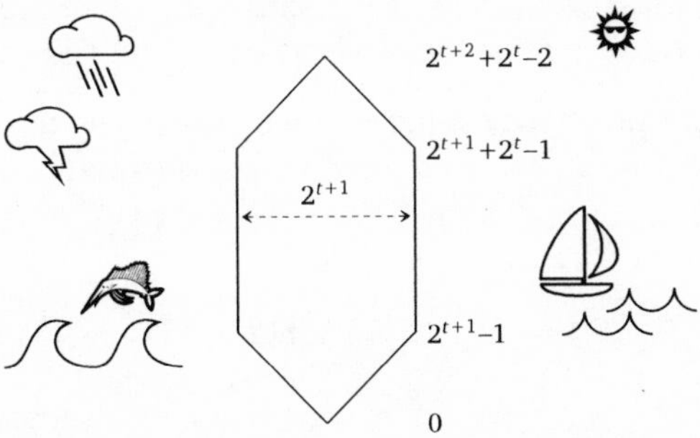

The next step in the proof of Theorem V.1.1 is to determine which of the ideals $\mathfrak{d}(a, b)$ is $\mathscr{A}^*$-invariant. This we do by an interpolation process. The main tools involved are Lemma III.1.3 and Theorem III.3.5.

PROPOSITION V.1.3: *Let $a, b \in \mathbb{N}$ and write $b = 2^\ell \cdot c$ where c is odd. Suppose that the ideal $\mathfrak{d}(a, b) \subset \mathbb{F}_2[x, y]$ is $\mathscr{A}^*$-invariant. Then $a \leq 2^\ell$. In particular, if b is odd then $a = 1$. Conversely, if $a \leq 2^\ell$ then the ideal $\mathfrak{d}(a, b)$ is $\mathscr{A}^*$-invariant.*

PROOF: Note that $\mathbf{Sq}(\mathbf{d}_{2,1}) = \mathbf{d}_{2,1} + \mathbf{d}_{2,0} + \mathbf{d}_{2,1}^2$ so

$$\mathrm{Sq}^{2^\ell}(\mathbf{d}_{2,1}^b) = (\mathrm{Sq}^1(\mathbf{d}_{2,1}^c))^{2^\ell} = (c \cdot \mathbf{d}_{2,1}^{c-1} \cdot \mathbf{d}_{2,0})^{2^\ell} = (\mathbf{d}_{2,1}^{c-1} \cdot \mathbf{d}_{2,0})^{2^\ell}$$

since c is odd. If $\mathfrak{d}(a, b) = (\mathbf{d}_{2,0}^a, \mathbf{d}_{2,1}^b)$ is $\mathscr{A}^*$-invariant then $(\mathbf{d}_{2,1}^{c-1} \cdot \mathbf{d}_{2,0})^{2^\ell} \in (\mathbf{d}_{2,0}^a, \mathbf{d}_{2,1}^b)$ so $2^\ell \geq a$ as claimed.

Conversely, suppose $a \leq 2^\ell$. Note that the ideal $\mathfrak{d}(1, c) = (\mathbf{d}_{2,0}, \mathbf{d}_{2,1}^c)$ is $\mathscr{A}^*$-invariant because $\mathbf{d}_{2,0}$ is a Thom class in $\mathbb{F}_2[x, y]$ and

$$\begin{aligned}\mathbf{Sq}(\mathbf{d}_{2,1}^c) &= (\mathbf{d}_{2,1} + \mathbf{d}_{2,0} + \mathbf{d}_{2,1}^2)^c \\ &\equiv (\mathbf{d}_{2,1} + \mathbf{d}_{2,1}^2)^c = \mathbf{d}_{2,1}^c(1 + \mathbf{d}_{2,1})^c = 0 \bmod (\mathbf{d}_{2,0}, \mathbf{d}_{2,0}^c).\end{aligned}$$

Therefore by Proposition III.6.1 the ideal $\mathfrak{d}(2^\ell, 2^\ell \cdot c) = (\mathbf{d}_{2,0}^{2^\ell}, \mathbf{d}_{2,1}^b)$ is also $\mathscr{A}^*$-invariant. Set $K = (\mathbf{d}_{2,0}^{2^\ell}, \mathbf{d}_{2,1}^b)$ and $L = (\mathbf{d}_{2,0}^a, \mathbf{d}_{2,1}^b)$ so $(K : L) = (\mathbf{d}_{2,0}^{2^\ell - a}) + K$ and $L = (K : \mathbf{d}_{2,0}^{2^\ell - a})$. Since $\mathbf{d}_{2,0}^{2^\ell - a} \in \mathbb{F}_2[x, y]$ is a Thom class Lemma III.1.3 implies that L is $\mathscr{A}^*$-invariant as claimed. □

The final stage in the proof of Theorem V.1.1 is to determine which of the $\mathscr{A}^*$-invariant ideals $\mathfrak{d}(a, b)$ have trivial Wu classes. We do this in two steps: first we show that the ideals $\mathfrak{d}(2^t, 2^s - 2^t)$ for $s > t \in \mathbb{N}_0$ all have trivial Wu classes, and then that the other $\mathscr{A}^*$-invariant ideals $\mathfrak{d}(a, b)$ have nontrivial Wu classes.

PROPOSITION V.1.4: *Suppose that $s > t \in \mathbb{N}_0$. Then $\mathfrak{d}(2^t, 2^s - 2^t)$ is an $\mathscr{A}^*$-invariant $\mathfrak{m}$-primary irreducible ideal and has trivial Wu classes.*

PROOF: By Frobenius periodicity, Corollary III.6.5, it suffices to consider the case $t = 0$. The ideal $\mathfrak{d}(2^s, 2^s)$ is $\mathscr{A}^*$-invariant and has trivial Wu classes by Proposition V.1.2. By Proposition V.1.3 the ideal $\mathfrak{d}(1, 2^s - 1)$ is also $\mathscr{A}^*$-invariant. Note that

$$(\mathfrak{d}(2^s, 2^s) : \mathfrak{d}(1, 2^s - 1)) = (\mathbf{d}_{2,0}^{2^s-1}\mathbf{d}_{2,1}) + \mathfrak{d}(2^s, 2^s).$$

By Proposition II.6.7 the ideal $(\mathbf{d}_{2,0}^{2^s-1}\mathbf{d}_{2,1}) + \mathfrak{d}(2^s, 2^s)$ is $\mathscr{A}^*$-invariant, and moreover we have by Theorem I.2.1

$$(\mathfrak{d}(2^s, 2^s) : \mathbf{d}_{2,0}^{2^s-1}\mathbf{d}_{2,1}) = \mathfrak{d}(1, 2^s - 1))$$

so if we set

$$\begin{aligned}K &= \mathfrak{d}(2^s, 2^s) \\ L &= \mathfrak{d}(1, 2^s - 1)\end{aligned}$$

and

$$h = \mathbf{d}_{2,0}^{2^s-1}\mathbf{d}_{2,1}$$

the hypotheses of Theorem III.3.5 are fulfilled. Hence we may compute $\mathrm{Wu}(\mathfrak{d}(1, 2^s - 1))$ and $\chi\mathrm{Wu}(\mathfrak{d}(1, 2^s - 1))$ by means of that theorem. Our strategy is to show $\chi\mathrm{Wu}(\mathfrak{d}(1, 2^s - 1)$ is trivial.

To this end we note that

$$\begin{aligned}\mathbf{Sq}(\mathbf{d}_{2,0}^{2^s-1}) &= (\mathbf{d}_{2,0} + \mathbf{d}_{2,0}\mathbf{d}_{2,1} + \mathbf{d}_{2,0}^2)^{2^s-1} \\ \mathbf{Sq}(\mathbf{d}_{2,1}) &= (\mathbf{d}_{2,1} + \mathbf{d}_{2,0} + \mathbf{d}_{2,1}^2)\end{aligned}$$

and hence if we rewrite $\mathbf{d}_{2,1} + \mathbf{d}_{2,0} + \mathbf{d}_{2,1}^2$ as a sum of two terms, viz., $(\mathbf{d}_{2,1}(1 + \mathbf{d}_{2,1})) + \mathbf{d}_{2,0}$, and multiply out we obtain

$$\begin{aligned}\mathbf{Sq}(\mathbf{d}_{2,0}^{2^s-1}\mathbf{d}_{2,1}) &= (\mathbf{d}_{2,0} + \mathbf{d}_{2,0}\mathbf{d}_{2,1} + \mathbf{d}_{2,0}^2)^{2^s-1} \cdot (\mathbf{d}_{2,1} + \mathbf{d}_{2,0} + \mathbf{d}_{2,1}^2)\\ &= \mathbf{d}_{2,0}^{2^s-1}\mathbf{d}_{2,1}(1 + \mathbf{d}_{2,1} + \mathbf{d}_{2,0})^{2^s-1}(1 + \mathbf{d}_{2,1}) + \mathbf{d}_{2,0}^{2^s}(1 + \mathbf{d}_{2,1} + \mathbf{d}_{2,0})^{2^s-1}\\ &\equiv \mathbf{d}_{2,0}^{2^s-1}\mathbf{d}_{2,1}(1 + \mathbf{d}_{2,1} + \mathbf{d}_{2,0})^{2^s-1}(1 + \mathbf{d}_{2,1}) \bmod K.\end{aligned}$$

Therefore dividing by $\mathbf{d}_{2,0}^{2^s-1}\mathbf{d}_{2,1}$ and reducing modulo L we find

$$\begin{aligned}\chi\mathrm{Wu}(\mathfrak{d}(1, 2^s - 1)) &\equiv (1 + \mathbf{d}_{2,1} + \mathbf{d}_{2,0})^{2^s-1}(1 + \mathbf{d}_{2,1})\\ &\equiv (1 + \mathbf{d}_{2,1})^{2^s-1}(1 + \mathbf{d}_{2,1}) \equiv (1 + \mathbf{d}_{2,1})^{2^s} \equiv 1 + \mathbf{d}_{2,1}^{2^s}\\ &\equiv 1 \bmod \mathfrak{d}(1, 2^s - 1).\end{aligned}$$

Hence the Wu classes of $\mathfrak{d}(1, 2^s - 1)$ are trivial as claimed. □

PROPOSITION V.1.5: *Let $k \in \mathbb{N}$ and choose $m \in \mathbb{N}_0$ such that $2^m \le k \le 2^{m+1} - 1$. Then*

$$\begin{aligned}\mathbf{Sq}(\mathrm{Wu}(\mathfrak{d}(1, k))) &= (1 + \mathbf{d}_{2,1})^{k+1}\\ \chi\mathrm{Wu}(\mathfrak{d}(1, k)) &= (1 + \mathbf{d}_{2,1})^{2^{m+1}-(k+1)}.\end{aligned}$$

PROOF: Consider the decreasing chain of ideals

$$\mathfrak{d}(1, k) \supsetneqq \mathfrak{d}(1, k + 1) \supsetneqq \cdots \supsetneqq \mathfrak{d}(1, 2^{m+1} - 1).$$

Each ideal in this chain is $\mathscr{A}^*$-invariant and we may therefore apply Theorem III.3.5 to adjacent pairs of ideals. To this end note that

$$\mathfrak{d}(1, j) = (\mathfrak{d}(1, j + 1) : \mathbf{d}_{2,1})$$

and recall

$$\mathbf{Sq}(\mathbf{d}_{2,1}) = \mathbf{d}_{2,0} + \mathbf{d}_{2,1} + \mathbf{d}_{2,1}^2 \equiv \mathbf{d}_{2,1}(1 + \mathbf{d}_{2,1}) \bmod \mathfrak{d}(1, j).$$

The recipe of Theorem III.3.5 therefore yields

$$\chi\mathrm{Wu}(\mathfrak{d}(1, j)) = (1 + \mathbf{d}_{2,1}) \cdot \chi\mathrm{Wu}(\mathfrak{d}(1, j + 1)). \tag{$\because$}$$

By Proposition V.1.4 the ideal $\mathfrak{d}(1, 2^{m+1} - 1)$ has trivial Wu classes. Inducting down from this case using the formula (∵) we arrive at the formula

$$\chi\mathrm{Wu}(\mathfrak{d}(1, k)) = (1 + \mathbf{d}_{2,1})^{2^{m+1}-(k+1)}.$$

In the quotient algebra $\mathbb{F}_2[x, y]/\mathfrak{d}(1, k)$ we have $\mathbf{d}_{2,1}^{2^{m+1}} = 0$. Therefore

$$(1 + \mathbf{d}_{2,1})^{2^{m+1}-(k+1)} \cdot (1 + \mathbf{d}_{2,1})^{(k+1)} = (1 + \mathbf{d}_{2,1})^{2^{m+1}} = (1 + \mathbf{d}_{2,1}^{2^{m+1}}) = 1$$

and the formula for $\mathbf{Sq}(\mathrm{Wu}(\mathfrak{d}(1, k)))$ follows from Lemma III.3.3. □

PROPOSITION V.1.6: *Let $a, b \in \mathbb{N}$ and write $b = 2^\ell \cdot c$ with c odd. Suppose that the ideal $\mathfrak{d}(a, b)$ is $\mathscr{A}^*$-invariant with trivial Wu classes. Then $a = 2^\ell$ and $c + 1$ is a power of 2.*

PROOF: Since $\mathfrak{d}(a, b)$ is $\mathscr{A}^*$-invariant $a \leq 2^\ell$ by Proposition V.1.3. The ideal $\mathfrak{d}(2^\ell, 2^\ell \cdot c)$ is the ℓ-th Frobenius power of the ideal $\mathfrak{d}(1, c)$. Choose $m \in \mathbb{N}_0$ so that $2^m \leq c \leq 2^{m+1} - 1$. (N.b. Since c is odd, $2^m < c$ unless $c = 1$, in which case $m = 0$ and the entire inequality collapses to an equality.) Then Proposition V.1.5 gives

$$\chi\mathrm{Wu}(\mathfrak{d}(1, c) = (1 + \mathbf{d}_{2,1})^{2^{m+1}-(c+1)}$$

so by Theorem III.6.4

$$\chi\mathrm{Wu}(\mathfrak{d}(2^\ell, 2^\ell \cdot c)) = (1 + \mathbf{d}_{2,1}^{2^\ell})^{2^{m+1}-(c+1)}.$$

If we set $K = \mathfrak{d}(2^\ell, 2^\ell \cdot c)$ and $L = \mathfrak{d}(a, b)$ then $(K : L) = (\mathbf{d}_{2,0}^{2^\ell - a}) + K$ and $L = (K : \mathbf{d}_{2,0}^{2^\ell - a})$. Both K and L are $\mathscr{A}^*$-invariant so again Theorem III.3.5 provides a recipe to compute $\chi\mathrm{Wu}(L)$ from $\chi\mathrm{Wu}(K)$. To apply this we note

$$\mathrm{Sq}(\mathbf{d}_{2,0}^{2^\ell - a}) = \mathbf{d}_{2,0}^{2^\ell - a} \cdot (1 + \mathbf{d}_{2,1} + \mathbf{d}_{2,0})^{2^\ell - a},$$

so the recipe gives

$$(\maltese) \qquad \chi\mathrm{Wu}(L) = (1 + \mathbf{d}_{2,1} + \mathbf{d}_{2,0})^{2^\ell - a} \cdot (1 + \mathbf{d}_{2,1}^{2^\ell})^{2^{m+1}-(c+1)}.$$

We have

$$(1 + \mathbf{d}_{2,1} + \mathbf{d}_{2,0})^{2^\ell - a} = 1 + \cdots + \mathbf{d}_{2,1}^{2^\ell - a} + \cdots$$

and, since $2^{m+1} - (c + 1)$ is even, $\mathbf{d}_{2,1}^{2^{\ell+1}}$ is the least positive power of $\mathbf{d}_{2,1}$ occurring in the expansion of $(1 + \mathbf{d}_{2,1}^{2^\ell})^{2^{m+1}-(c+1)}$, so we also have

$$\chi\mathrm{Wu}(L) = 1 + \cdots + \mathbf{d}_{2,1}^{2^\ell - a} + \cdots.$$

Since $b = 2^\ell \cdot c$ and $a \leq 2^\ell$ it follows that $b > 2^\ell - a$, and therefore if $a \neq 2^\ell$ one has $\chi\mathrm{Wu}_{2(2^\ell - a)}(L) = \mathbf{d}_{2,1}^{2^\ell - a} + \cdots \not\equiv 0 \bmod L$ contrary to assumption. So $a = 2^\ell$ and $\chi\mathrm{Wu}(L) = (1 + \mathbf{d}_{2,1}^{2^\ell})^{2^{m+1}-(c+1)}$.

Next we show that $c + 1 = 2^{m+1}$. Note that $2^{\ell+m+1} > 2^\ell \cdot c$ from the way we chose m, so in $\mathbb{F}_2[x, y]/L$ we have

$$(1 + \mathbf{d}_{2,1}^{2^\ell})^{2^{m+1}-(c+1)} \cdot (1 + \mathbf{d}_{2,1}^{2^\ell})^{c+1} = (1 + \mathbf{d}_{2,1}^{2^\ell})^{2^{m+1}} = 1 + \mathbf{d}_{2,1}^{2^{\ell+m+1}} = 1$$

and therefore

$$\mathbf{Sq}(\mathrm{Wu}(L)) = (1 + \mathbf{d}_{2,1}^{2^\ell})^{c+1}.$$

If $c + 1$ is not a power of 2 then the dyadic expansion of $c + 1$ is of the form

$$c + 1 = c_{m+1}2^{m+1} + \cdots c_1 2 + c_0$$

where $c_j \neq 0$ for some $j < m + 1$. Therefore the binomial coefficient $\binom{c+1}{2^j}$ is odd, so the homogeneous component of $\mathbf{Sq}(\mathrm{Wu}(L))$ of degree $2^{\ell+j+1}$ is $\mathbf{d}_{2,1}^{2^\ell \cdot 2^j} = \mathbf{d}_{2,1}^{2^{\ell+j}}$ which is nonzero. This contradicts the assumption that $L = \mathfrak{d}(a, b)$ has trivial Wu classes, so $c + 1 = 2^{m+1}$ completing the proof. □

PROOF OF THEOREM V.1.1: Combine Propositions V.1.3, V.1.4, and V.1.6. □

In the proof of Proposition V.1.6 we developed a formula, namely (✠), for the conjugate Wu classes of any $\mathscr{A}^*$-invariant ideal $\mathfrak{d}(a, b)$ in $\mathbb{F}_2[x, y]$ that will be of use in the sequel. We record it next.

COROLLARY V.1.7: *Let $a, b \in \mathbb{N}$ and write $b = 2^\ell \cdot c$ with c odd. Assume that $a \le 2^\ell$ so that the ideal $\mathfrak{d}(a, b)$ is $\mathscr{A}^*$-invariant. Choose $m \in \mathbb{N}_0$ so that $2^m \le c \le 2^{m+1} - 1$. Then*

$$\chi\mathrm{Wu}(\mathfrak{d}(a, b)) = (1 + \mathbf{d}_{2,1} + \mathbf{d}_{2,0})^{2^\ell - a} \cdot (1 + \mathbf{d}_{2,1}^{2^\ell})^{2^{m+1}-(c+1)}. \quad \square$$

The proof of Theorem V.1.1 involves only two basic constructions and the initial $\mathscr{A}^*$-invariant ideal with trivial Wu classes $\mathfrak{d}(1, 1)$. These suffice to yield all of the ideals generated by powers of the two Dickson polynomials $\mathbf{d}_{2,0}$ and $\mathbf{d}_{2,1}$ which are $\mathscr{A}^*$-invariant and have trivial Wu classes. To wit, the ideals $\mathfrak{d}(2^t, 2^s - 2^t)$ arise as follows:

(1) first Frobenius periodicity is used to obtain the ideals $\mathfrak{d}(2^r, 2^r) = \mathfrak{d}(1, 1)^{[2^r]}$ which have trivial Wu classes,
(2) then the operator $(- : \mathbf{d}_{2,0}^{2^r-1}\mathbf{d}_{2,1})$ is applied to the ideal $\mathfrak{d}(2^r, 2^r)$ (and Theorem III.6.4 is used) to obtain the examples of ideals $\mathfrak{d}(1, 2^r - 1)$ with trivial Wu classes, and finally
(3) Frobenius periodicity is applied to $\mathfrak{d}(1, 2^r - 1)$ to yield all the remaining examples.

The portion of the proof that shows that the other $\mathscr{A}^*$-invariant ideals of the form $\mathfrak{d}(a, b)$ have nontrivial Wu classes involves the same two basic constructions.

The ideals $\mathfrak{d}(2^t, 2^s - 2^t)$ with $2^s - 2^t = 2^t$ were discussed following Proposition V.1.2: they are just the ideals $\mathfrak{d}(2^t, 2^t) = \mathfrak{d}(1, 1)^{[2^t]}$ obtained as Frobenius powers of the basic ideal $\mathfrak{d}(1, 1)$. The ideals $\mathfrak{d}(2^t, 2^s - 2^t)$ for which $2^s - 2^t > 2^t$, i.e., for which $s > t + 1$, have $\mathbf{d}_{2,0}^{2^t}$ as the generator of lowest degree (in contrast to the ideals $\mathfrak{d}(2^t, 2^t)$ which have $\mathbf{d}_{2,0}^{2^t}$ as the generator of largest degree). The corresponding Poincaré duality algebra has Poincaré series

$$\begin{aligned} P(H(2^t, 2^s - 2^t), u) &= (1 + u + u^2 + \cdots + u^{2^{t+1}+2^t-1})(1 + u + u^2 + \cdots + u^{2^{s+1}-2^{t+1}-1}) \\ &= 1 + 2u + 3u^2 + \cdots + (2^{t+1} + 2^t)u^{2^{t+1}+2^t-1} + (2^{t+1} + 2^t)u^{2^{t+1}+2^t} + \cdots + \\ &(2^{t+1} + 2^t)u^{2^{s+1}-2^{t+1}-1} + (2^{t+1} + 2^t - 1)u^{2^{s+1}-2^{t+1}} + \cdots + 2u^{2^{s+1}-2^t-3} + u^{2^{s+1}-2^t-2}. \end{aligned}$$

Again, if pictured graphically this looks a bit like a lighthouse. This

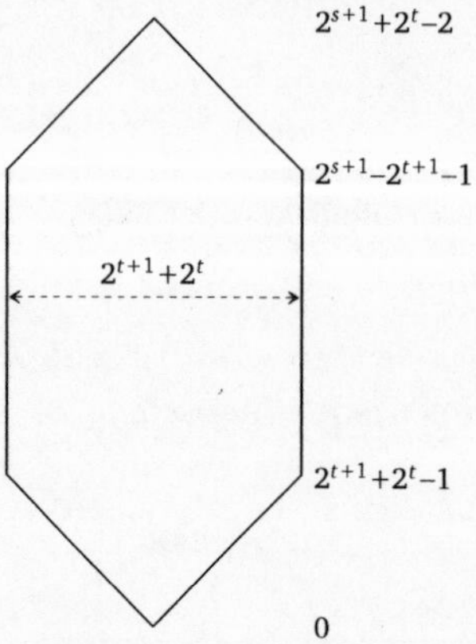

time the maximum width is $2^{t+1}+2^t$, which is independent of $s \in \mathbb{N}$. The appearance of this graphic is a special feature of the case of two variables: for more than a small number of variables it can happen that the coefficients of the Poincaré polynomial of a Poincaré duality quotient of $\mathbb{F}[z_1, \ldots, z_n]$ do not form a nondecreasing sequence up to the middle dimension (see [95] Theorem 4.2 and the example that follows it).

Setting $m = s - t$ we see that there are two periodicity operators acting on the family of ideals

$$\mathfrak{d}(2^t, 2^t(2^m-1)) \qquad t \in \mathbb{N}_0,\ m \in \mathbb{N};$$

the one increases t by 1 and the other m by 1. Increasing t corresponds to raising both Dickson polynomials to the next highest power of 2, so is related in an obvious way to the operator Φ prevalent in the literature on the Steenrod algebra (see for example [76]) and which also occurred in Proposition II.6.7. The operator corresponding to increasing m is more mysterious, and depends on a special property of Dickson polynomials.

V.2 $\mathscr{A}^*$-Indecomposable elements in $\mathbb{F}_2[x, y]$

Our purpose in this section is to make more explicit the connection discussed in Section III.4 between $\mathscr{P}^*$-invariant Poincaré duality quotients of $\mathbb{F}_q[z_1, \ldots, z_n]$ and a solution to the *Hit Problem* for $\mathbb{F}_q[z_1, \ldots, z_n]$. For $q = 2$ and $n = 2$ this problem was solved by F. P. Peterson in [74] who gave a set of monomials that generate $\mathbb{F}_2[x, y]$ as a module over the Steenrod algebra $\mathscr{A}^*$. We intend to show that the fundamental classes of the naturally occurring families of Poincaré duality quotients of $\mathbb{F}_2[x, y]$ with an unstable $\mathscr{A}^*$-algebra structure studied in Section V.1 can be used to provide a basis for $\mathbb{F}_2[x, y]_{\mathscr{A}^*}$.

It is actually much more convenient to describe a vector space basis for $\Gamma(u, v)^{\mathscr{A}^*}$ using our work on $\mathscr{A}^*$-invariant Poincaré duality quotients of $\mathbb{F}_2[x, y]$. We then establish a precise relationship between our basis and those already in the literature, e.g., [75], [82], [3], and [8]. The first step is to find explicit formulae for the[2] Macaulay duals of the $\mathscr{A}^*$-invariant

---

[2] We are working over $\mathbb{F}_2$ where a Macaulay dual is actually unique.

ideals $(\mathbf{d}_{2,0}^{a}, \mathbf{d}_{2,1}^{b})$ with trivial Wu classes (see also [44]).

NOTATION: *Denote by $\Delta_{s,t}$ the Macaulay dual of the ideal $(\mathbf{d}_{2,0}^{2^t}, \mathbf{d}_{2,1}^{2^s-2^t})$.*

We begin by computing the special cases $\Delta_{r+1,r}$, $r \in \mathbb{N}_0$, which we then use to determine the Macaulay duals of the remaining cases.

LEMMA V.2.1: *Let $r \in \mathbb{N}_0$. The Macaulay dual of the ideal $(\mathbf{d}_{2,0}^{2^r}\mathbf{d}_{2,1}^{2^r}) \subset \mathbb{F}_2[x, y]$ is*

$$\Delta_{r+1,r} = \gamma_{2^{r+1}-1}(u)\gamma_{2^{r+1}+2^r-1}(v) + \gamma_{2^{r+1}+2^r-1}(u)\gamma_{2^{r+1}-1}(v),$$

where $u, v \in \Gamma(u, v)$ are dual to $x, y \in \mathbb{F}_2[x, y]$.

PROOF: Recall that $(x^3, y^3) \subset (\mathbf{d}_{2,0}, \mathbf{d}_{2,1})$ and

$$\big((x^3, y^3) : (\mathbf{d}_{2,0}, \mathbf{d}_{2,1})\big) = (x + y) + (x^3, y^3)$$

(see e.g. Section II.5 Example 1). Taking Frobenius powers and using Proposition II.6.3 we find

$$\big((x^{2^{r+1}+2^r}, y^{2^{r+1}+2^r}) : (\mathbf{d}_{2,0}^{2^r}, \mathbf{d}_{2,1}^{2^r})\big) = (x^{2^r} + y^{2^r}) + (x^{2^{r+1}+2^r}, y^{2^{r+1}+2^r})$$

and therefore by Theorem II.6.6 the Macaulay dual of $(\mathbf{d}_{2,0}^{2^r}, \mathbf{d}_{2,1}^{2^r})$ is

$$\begin{aligned}\Delta_{r+1,r} &= (x^{2^r} + y^{2^r}) \cap \gamma_{2^{r+1}+2^r-1}(u)\gamma_{2^{r+1}+2^r-1}(v)\\ &= \gamma_{2^{r+1}-1}(u)\gamma_{2^{r+1}+2^r-1}(v) + \gamma_{2^{r+1}+2^r-1}(u)\gamma_{2^{r+1}-1}(v)\end{aligned}$$

as claimed. □

LEMMA V.2.2: *Let $s > t \in \mathbb{N}_0$ and set $r = s - t$. The Macaulay dual of the ideal $(\mathbf{d}_{2,0}^{2^t}, \mathbf{d}_{2,1}^{2^s-2^t})$ is*

$$\Delta_{s,t} = \sum_{k=1}^{2^{r+1}-2} \gamma_{2^{s+1}-1-k2^t}(u)\gamma_{2^t-1+k2^t}(v),$$

where $u, v \in \Gamma(u, v)$ are dual to $x, y \in \mathbb{F}_2[x, y]$.

PROOF: We have $(\mathbf{d}_{2,0}^{2^s}, \mathbf{d}_{2,1}^{2^s}) \subset (\mathbf{d}_{2,0}^{2^t}, \mathbf{d}_{2,1}^{2^s-2^t})$ and

$$\big((\mathbf{d}_{2,0}^{2^s}, \mathbf{d}_{2,1}^{2^s}) : (\mathbf{d}_{2,0}^{2^t}, \mathbf{d}_{2,1}^{2^s-2^t})\big) = (\mathbf{d}_{2,0}^{2^s-2^t}\mathbf{d}_{2,1}^{2^t}) + (\mathbf{d}_{2,0}^{2^s}, \mathbf{d}_{2,1}^{2^s}).$$

We next develop a formula for the transition element $(\mathbf{d}_{2,0}^{2^r-1}\mathbf{d}_{2,1})^{2^t}$. We have

$$\begin{aligned}\mathbf{d}_{2,0}^{2^r-1} &= \big(xy(x+y)\big)^{2^r-1} = x^{2^r-1}y^{2^r-1}(x+y)^{2^r-1}\\ &= x^{2^r-1}y^{2^r-1}\sum_{i+j=2^r-1} x^i y^j = \sum_{i+j=2^r-1} x^{2^r+i-1}y^{2^r+j-1}\end{aligned}$$

and

$$\mathbf{d}_{2,0}^{2^r-1}\mathbf{d}_{2,1} = \Big(\sum_{i+j=2^r-1} x^{2^r+i-1}y^{2^r+j-1}\Big)(x^2 + xy + y^2)$$

$$= \sum_{i+j=2^r-1} x^{2^r+i+1}y^{2^r+j-1} + \sum_{i+j=2^r-1} x^{2^r+i}y^{2^r+j} + \sum_{i+j=2^r-1} x^{2^r+i-1}y^{2^r+j+1}.$$

With the exception of two terms each, the terms of the first and third sums cancel each other, and we are left with the following formula for $\mathbf{d}_{2,0}^{2^r-1}\mathbf{d}_{2,1}$:

$$\sum_{i+j=2^r-1} x^{2^r+i}y^{2^r+j} + x^{2^r-1}y^{2^{r+1}} + x^{2^r}y^{2^{r+1}-1} + x^{2^{r+1}-1}y^{2^r} + x^{2^{r+1}}y^{2^r-1}.$$

Hence $\mathbf{d}_{2,0}^{2^s-2^t}\mathbf{d}_{2,1}^{2^t}$ is equal to

$$\sum_{i+j=2^r-1} x^{2^s+i2^t}y^{2^s+j2^t} + x^{2^s-2^t}y^{2^{s+1}} + x^{2^s}y^{2^{s+1}-2^t} + x^{2^{s+1}-2^t}y^{2^s} + x^{2^{s+1}}y^{2^s-2^t}.$$

From Lemma V.2.1 we have that

$$\Delta_{s+1,s} = \gamma_{2^{s+1}-1}(u)\gamma_{2^{s+1}+2^s-1}(v) + \gamma_{2^{s+1}+2^s-1}(u)\gamma_{2^{s+1}-1}(v)$$

is a Macaulay dual for the ideal $(\mathbf{d}_{2,0}^{2^s}, \mathbf{d}_{2,1}^{2^s})$. Hence by Theorem II.5.1 the Macaulay dual of the ideal $(\mathbf{d}_{2,0}^{2^t}, \mathbf{d}_{2,1}^{2^s-2^t})$ is

$$\Delta_{s,t} = \mathbf{d}_{2,0}^{2^s-2^t}\mathbf{d}_{2,1}^{2^t} \cap \left(\gamma_{2^{s+1}-1}(u)\gamma_{2^{s+1}+2^s-1}(v) + \gamma_{2^{s+1}+2^s-1}(u)\gamma_{2^{s+1}-1}(v)\right).$$

Substituting $\mathbf{d}_{2,0}^{2^s-2^t}\mathbf{d}_{2,1}^{2^t}$ with the preceding formula we find after some calculation

$$\Delta_{s,t} = \sum_{i+j=2^r-1} \gamma_{2^s-i2^t-1}(u)\gamma_{2^{s+1}-j2^t-1}(v) + \sum_{i+j=2^r-1} \gamma_{2^{s+1}-i2^t-1}(u)\gamma_{2^s-j2^t-1}(v)$$
$$+ \gamma_{2^{s+1}-1}(u)\gamma_{2^t-1}(v) + \gamma_{2^t-1}(u)\gamma_{2^{s+1}-1}(v).$$

In the first sum of the previous formula the indices $2^s - i2^t - 1$ of the divided powers of u satisfy

$$2^t - 1 = 2^s - (2^r - 1)2^t - 1 \le 2^s - i2^t - 1 \le 2^s - 1$$

while in the second sum the indices $2^{s+1} - i2^t - 1$ satisfy

$$2^s + 2^t - 1 = 2^{s+1} - (2^r - 1)2^t - 1 \le 2^{s+1} - i2^t - 1 \le 2^{s+1} - 1.$$

Hence we can combine the two sums and the remaining terms into a single sum, viz.,

$$\Delta_{s,t} = \sum_{k=1}^{2^{r+1}-2} \gamma_{2^{s+1}-k2^t-1}(u)\gamma_{2^t+k2^t-1}(v)$$

as claimed. □

NOTATION: *For $k \in \mathbb{N}_0$ write $k = \sum_{i=0}^{\infty} \alpha_i(k)2^i$ for the 2-adic expansion of k. So $\alpha_i(k)$ is the i-th binary digit of k.*

LEMMA V.2.3: *Let $d = 2^{s+1} + 2^t - 2$ with $s + 1 \geq t \geq 0$. Assume that we have two nonnegative integers a and b with $a + b = d$. Suppose there exists n with $0 \leq n < t$ such that*

$$\alpha_i(a) = \begin{cases} 1 & 0 \leq i < n \\ 0 & i = n. \end{cases}$$

Then

$$\alpha_i(b) = \begin{cases} 1 & 0 \leq i < n \\ 0 & i = n \end{cases}$$

as well. If $a > 2^n$ or $b > 2^n$ it follows that the element $x^a y^b \in \mathbb{F}_2[x, y]$ is $\mathscr{A}^*$-decomposable.

PROOF: Denote the difference $a - (2^n - 1)$ by $\overline{a}$. Then $\alpha_i(\overline{a}) = 0$ for all $0 \leq i \leq n$. We have

$$\begin{aligned} b &= d - a = d - (a - (2^n - 1)) + (2^n - 1) = d - \overline{a} - (2^n - 1) \\ &= (2^{s+1} + 2^t - 2^n - 1) - \overline{a} \end{aligned}$$

and since by assumption $n < t \leq s + 1$ the claim about $\alpha_i(b)$ follows easily from this.

To prove that $x^a y^b$ is $\mathscr{A}^*$-decomposable if $a > 2^n$ or $b > 2^n$ we procede by induction on n. Without loss of generality we assume $a > 2^n$. If $n = 0$ then $\alpha_0(a) = \alpha_0(b) = 0$, so $\mathrm{Sq}^1(x^{a-1}y^b) = x^a y^b$.

Next assume that $n > 0$. We have

$$\begin{aligned} \mathrm{Sq}^{2^n}(x^{a-2^n} y^b) &= x^a y^b + \sum_{j=0}^{2^n-1} \mathrm{Sq}^j(x^{a-2^n}) \mathrm{Sq}^{2^n-j}(y^b) \\ &= x^a y^b + \sum_{j=0}^{2^n-1} x^{a-2^n+j} y^{b+2^n-j} \end{aligned}$$

and for each $0 \leq j \leq 2^n - 1$ there exists $0 \leq h(j) < n$ such that

$$\alpha_i(x^{a-2^n+j}) = \begin{cases} 1 & 0 \leq i < h(j) \\ 0 & i = h(j). \end{cases}$$

Since $a > 2^n$ and $j \geq 2^{h(j)}$, we have that $a - 2^n + j > 2^{h(j)}$. Therefore we can use induction to conclude that all the terms in the sum $\sum_{j=0}^{2^n-1} x^{a-2^n+j} y^{b+2^n-j}$ are $\mathscr{A}^*$-decomposable. Hence the same holds for $x^a y^b$. □

PROPOSITION V.2.4: *Let $x^a y^b$, $a, b \in \mathbb{N}_0$ be a monomial of degree d in $\mathbb{F}_2[x, y]$.*

(i) If $d = 2^{t+1} - 2$, $t \geq 0$, then $x^a y^b$ is $\mathscr{A}^*$-indecomposable if and only if $a = 2^t - 1 = b$.

(ii) *If $d = 2^{s+1} + 2^t - 2$, $s \geq t \geq 0$, then $x^a y^b$ is $\mathscr{A}^*$-indecomposable if and only if $a = 2^{s+1} - 1 - k2^t$ and $b = 2^t - 1 + k2^t$ with $0 \leq k \leq 2^{s-t+1} - 1$.*

There are no $\mathscr{A}^*$-indecomposable monomials in any other degrees.

Proof: It is known that the only degrees in which $\mathscr{A}^*$-indecomposable monomials can be found are those of the form $d = (2^\alpha - 1) + (2^\beta - 1)$, $\alpha, \beta \in \mathbb{N}_0$ (see e.g. [74] or [82]). So the two cases for d listed above cover all possible degrees that need to be considered. We therefore find the following.

(i) In the first case, the given monomial $x^{2^t-1}y^{2^t-1}$ is a spike and thus indecomposable. All other monomials $x^a y^b$ fulfill the hypothesis of Lemma V.2.3 and are thus decomposable.

(ii) In the second case, the monomials corresponding to $k = 0$ and $k = 2^{s-t+1} - 1$ are the two spikes $x^{2^{s+1}-1}y^{2^t-1}$ and $x^{2^t-1}y^{2^{s+1}-1}$. If $s > t$ then the other monomials given form the support of $\Delta_{s,t}$. All other monomials in this degree satisfy the hypothesis of Lemma V.2.3 and are therefore decomposable.

This yields the desired conclusion. □

Proposition V.2.5: *$\Gamma(u, v)_d^{\mathscr{A}^*} = 0$ unless d has one of the following three forms, in which case the indicated divided power forms provide an $\mathbb{F}_2$-basis for $\Gamma(u, v)_d^{\mathscr{A}^*}$.*

degree	$\mathbb{F}_2$-basis for $\Gamma(u, v)_d^{\mathscr{A}^*}$
$d = 2^{t+1} - 2,\ t \geq 0$	$\gamma_{2^t-1}(u)\gamma_{2^t-1}(v)$
$d = 2^{t+1} + 2^t - 2,\ t \geq 0$	$\gamma_{2^{t+1}-1}(u)\gamma_{2^t-1}(v),\ \gamma_{2^t-1}(u)\gamma_{2^{t+1}-1}(v)$
$d = 2^{s+1} + 2^t - 2,\ s > t \geq 0$	$\gamma_{2^{s+1}-1}(u)\gamma_{2^t-1}(v),\ \gamma_{2^t-1}(u)\gamma_{2^{s+1}-1}(v),\ \Delta_{s,t}$

In particular it follows that the dimension of $\Gamma(u, v)_d^{\mathscr{A}^*}$ is at most 3 as has already been shown in [12].

Proof: It is clear from Proposition V.2.4 that $\Gamma(u, v)_d^{\mathscr{A}^*} = 0$ in degrees other than those that are listed and that the given divided power forms span $\Gamma(u, v)_d^{\mathscr{A}^*}$ in the degrees where it is not trivial. Since their supports are disjoint they are also linearly independent and so provide a basis. □

To establish a concordance between Proposition V.2.5 and the results of [3] and [16] we observe that in our notation, the basis for $\Gamma(u, v)_{2^{s+1}-1+2^t-1}^{\mathscr{A}^*}$ that is chosen there consists of the two spikes

$$\gamma_{2^{s+1}-1}(u)\gamma_{2^t-1}(v),\ \gamma_{2^t-1}(u)\gamma_{2^{s+1}-1}(v)$$

and the form

$$\gamma_{2^{s+1}-1}(u+v)\gamma_{2^t-1}(v).$$

Although the last form seems to be asymmetric in u and v it is not: to see this expand $\gamma_{2^{s+1}-1}(u+v)$ using the *beginner's binomial theorem* to obtain

$$\gamma_{2^{s+1}-1}(u+v) = \sum_{i+j=2^{s+1}-1} \gamma_i(u)\gamma_j(v).$$

Then, note that

$$\gamma_{2^t-1}(v) = \gamma_1(v)\gamma_2(v)\cdots\gamma_{2^{t-1}}(v)$$

annihilates $\gamma_j(v)$ for j not of the form $k \cdot 2^t$ and

$$\gamma_{2^t-1}(v)\gamma_{k\cdot 2^t}(v) = \gamma_{2^t-1+k\cdot 2^t}(v).$$

So we find that

$$\begin{aligned}\gamma_{2^{s+1}-1}(u+v)\gamma_{2^t-1}(v) &= \sum_{k=0}^{2^{s+1-t}-2^t} \gamma_{2^{s+1}-1-k2^t}(u)\gamma_{2^t-1+k2^t}(v)\\ &= \gamma_{2^{s+1}-1}(u)\gamma_{2^t-1}(v) + \Delta_{s,t} + \gamma_{2^t-1}(u)\gamma_{2^{s+1}-1}(v).\end{aligned}$$

From this concordance with [3] the reader can establish similar concordances with the bases for $\left(\mathbb{F}_2[x, y]_{\mathscr{A}^*}\right)_k$ of [74], [8], and [39].

V.3 Powers of Dickson polynomials in 3 variables

In this section we consider the quotients of $\mathbb{F}_2[x, y, z]$ by the ideals generated by powers of the Dickson polynomials $\mathfrak{d}(a_0, a_1, a_2) = (\mathbf{d}_{3,0}^{a_0}, \mathbf{d}_{3,1}^{a_1}, \mathbf{d}_{3,2}^{a_2})$. We determine which of these ideals are $\mathscr{A}^*$-invariant as well as which of the corresponding Poincaré duality quotient algebras $\mathbb{F}_2[x, y, z]/\mathfrak{d}(a_0, a_1, a_2)$ have trivial Wu classes. The methods employed are the same as in Section V.1 so we will be a little less complete in giving the details of computations than we gave in that section.

The action of the total Steenrod operation **Sq** on the Dickson polynomials in three variables over $\mathbb{F}_2$ is summarized in the following formulae (see e.g. [66] Section A.2):

$$(\divideontimes)\qquad \begin{aligned}\mathbf{Sq}(\mathbf{d}_{3,2}) &= (\mathbf{d}_{3,0} + \mathbf{d}_{3,1}) + \mathbf{d}_{3,2}(\mathbf{d}_{3,2} + 1)\\ \mathbf{Sq}(\mathbf{d}_{3,1}) &= \mathbf{d}_{3,0}(1 + \mathbf{d}_{3,2}) + \mathbf{d}_{3,1}(\mathbf{d}_{3,1} + \mathbf{d}_{3,2} + 1)\\ \mathbf{Sq}(\mathbf{d}_{3,0}) &= \mathbf{d}_{3,0}(\mathbf{d}_{3,0} + \mathbf{d}_{3,1} + \mathbf{d}_{3,2} + 1).\end{aligned}$$

We have chosen this way of writing things to emphasize the deviation of $\mathbf{d}_{3,2}$ and $\mathbf{d}_{3,1}$ from being Thom classes.

LEMMA V.3.1: *For positive integers a_0, a_1, a_2 write a_i for $i = 0, 1$, and 2 in the form $a_i = 2^{s_i} \cdot b_i$ where b_i is odd. Then the ideal $\mathfrak{d}(a_0, a_1, a_2) = (\mathbf{d}_{3,0}^{a_0}, \mathbf{d}_{3,1}^{a_1}, \mathbf{d}_{3,2}^{a_2}) \subset \mathbb{F}_2[x, y, z]$ is $\mathscr{A}^*$-invariant if and only if $2^{s_1} \geq a_0$ and $2^{s_2} \geq a_1$.*

PROOF: If $2^{s_1} < a_0$ then

$$\mathrm{Sq}^{2^{s_1}}(\mathbf{d}_{3,1}^{a_1}) = \mathrm{Sq}^{2^{s_1}}(\mathbf{d}_{3,1}^{b_1 \cdot 2^{s_1}}) = \left(\mathrm{Sq}^1(\mathbf{d}_{3,1}^{b_1})\right)^{2^{s_1}} = \left(\mathbf{d}_{3,1}^{b_1-1} \cdot \mathbf{d}_{3,0}\right)^{2^{s_1}}$$

since b_1 is odd. But this is not in $\mathfrak{d}(a_0, a_1, a_2) = (\mathbf{d}_{3,0}^{a_0}, \mathbf{d}_{3,1}^{a_1}, \mathbf{d}_{3,2}^{a_2})$ since $2^{s_1} < a_0$ and $2^{s_1}(b_1 - 1) < a_1$. Likewise if $2^{s_2} < a_1$ then

$$\mathrm{Sq}^{2^{s_2+1}}(\mathbf{d}_{3,2}^{a_2}) = \mathrm{Sq}^{2 \cdot 2^{s_2}}(\mathbf{d}_{3,2}^{s_2 \cdot b_2}) = \left(\mathrm{Sq}^2(\mathbf{d}_{3,2}^{b_2})\right)^{2^{s_2}} = \left(\mathbf{d}_{3,2}^{b_2-1}\mathbf{d}_{3,1}\right)^{2^{s_2}}$$

since b_2 is odd and $\mathrm{Sq}^1(\mathbf{d}_{3,2}) = 0$. This is not in the ideal $\mathfrak{d}(a_0, a_1, a_2) = (\mathbf{d}_{3,0}^{a_0}, \mathbf{d}_{3,1}^{a_1}, \mathbf{d}_{3,2}^{a_2})$ since $2^{s_2} < a_1$ and $2^{s_2}(b_2 - 1) < a_2$. So the condition for $\mathscr{A}^*$-invariance of the ideal $\mathfrak{d}(a_0, a_1, a_2) = (\mathbf{d}_{3,0}^{a_0}, \mathbf{d}_{3,1}^{a_1}, \mathbf{d}_{3,2}^{a_2}) \subset \mathbb{F}_2[x, y, z]$ is necessary.

Conversely, if $2^{s_1} \geq a_0$ and $2^{s_2} \geq a_1$, then

$$\begin{aligned}
\mathbf{Sq}(\mathbf{d}_{3,2}^{a_2}) &= \left(\mathbf{d}_{3,0} + \mathbf{d}_{3,1} + \mathbf{d}_{3,2}(\mathbf{d}_{3,2} + 1)\right)^{a_2} \\
&= \left((\mathbf{d}_{3,0} + \mathbf{d}_{3,1})^{2^{s_2}} + \mathbf{d}_{3,2}^{2^{s_2}}(\mathbf{d}_{3,2} + 1)^{2^{s_2}}\right)^{b_2} \\
&\equiv \mathbf{d}_{3,2}^{a_2}(\mathbf{d}_{3,2} + 1)^{a_2} \equiv 0 \bmod \mathfrak{d}(a_0, a_1, a_2)
\end{aligned}$$

since $2^{s_2} \geq a_1 \geq a_0$. Likewise

$$\begin{aligned}
\mathbf{Sq}(\mathbf{d}_{3,1}^{a_1}) &= \left(\mathbf{d}_{3,0} + \mathbf{d}_{3,0}\mathbf{d}_{3,2} + \mathbf{d}_{3,1}(\mathbf{d}_{3,1} + \mathbf{d}_{3,2} + 1)\right)^{a_1} \\
&= \left(\mathbf{d}_{3,0}^{2^{s_1}} + \mathbf{d}_{3,0}^{2^{s_1}}\mathbf{d}_{3,2}^{2^{s_1}} + \mathbf{d}_{3,1}^{2^{s_1}}(\mathbf{d}_{3,1} + \mathbf{d}_{3,2} + 1)^{2^{s_1}}\right)^{b_1} \\
&\equiv \mathbf{d}_{3,1}^{a_1}(\mathbf{d}_{3,1} + \mathbf{d}_{3,2} + 1)^{a_1} \equiv 0 \bmod \mathfrak{d}(a_0, a_1, a_2)
\end{aligned}$$

as $2^{s_1} \geq a_0$, whence the ideal $\mathfrak{d}(a_0, a_1, a_2) = (\mathbf{d}_{3,0}^{a_0}, \mathbf{d}_{3,1}^{a_1}, \mathbf{d}_{3,2}^{a_2}) \subset \mathbb{F}_2[x, y, z]$ is $\mathscr{A}^*$-invariant. □

Our next task is to determine which of the $\mathscr{A}^*$-invariant ideals of the form $\mathfrak{d}(a_0, a_1, a_2) = (\mathbf{d}_{3,0}^{a_0}, \mathbf{d}_{3,1}^{a_1}, \mathbf{d}_{3,2}^{a_2}) \subset \mathbb{F}_2[x, y, z]$ have trivial Wu classes.

PROPOSITION V.3.2: *Let $a_0, a_1, a_2 \in \mathbb{N}$ and suppose that the ideal $\mathfrak{d}(a_0, a_1, a_2) = (\mathbf{d}_{3,0}^{a_0}, \mathbf{d}_{3,1}^{a_1}, \mathbf{d}_{3,2}^{a_2}) \subset \mathbb{F}_2[x, y, z]$ is $\mathscr{A}^*$-invariant. Write $a_i = 2^{s_i} \cdot b_i$ with b_i odd, $i = 0, 1, 2$. If for some $s_3 \in \mathbb{N}$ with $s_3 > s_2$ one has $b_2 = 2^{s_3 - s_2} - 1$, $b_1 = 2^{s_2 - s_1} - 1$, $s_0 = s_1$, and $b_0 = 1$, then the Wu classes of the corresponding quotient algebra $\mathbb{F}_2[x, y, z]/\mathfrak{d}(a_0, a_1, a_2)$ are trivial.*

PROOF: Note that $s_2 \geq s_1$ and, since $b_0 = 1$ and $s_0 = s_1$,

$$\begin{aligned}
(\mathbf{d}_{3,0}^{a_0}, \mathbf{d}_{3,1}^{a_1}, \mathbf{d}_{3,2}^{a_2}) &= (\mathbf{d}_{3,0}^{2^{s_0}}, \mathbf{d}_{3,1}^{2^{s_0}(2^{s_2-s_0}-1)}, \mathbf{d}_{3,2}^{2^{s_2}(2^{s_3-s_2}-1)}) \\
&= \left(\mathbf{d}_{3,0}, \mathbf{d}_{3,1}^{2^{s_2-s_0}-1}, \mathbf{d}_{3,2}^{2^{s_2-s_0}(2^{s_3-s_2}-1)}\right)^{[2^{s_0}]}.
\end{aligned}$$

Hence by Theorem III.6.4 it is enough to consider the ideals of the form $(\mathbf{d}_{3,0}, \mathbf{d}_{3,1}^{2^s-1}, \mathbf{d}_{3,2}^{2^s(2^t-1)})$ (one sets $s = s_2 - s_0$ and $t = s_3 - s_2$). To show these have trivial Wu classes we employ Corollary III.6.5. We set

$$K = (\mathbf{d}_{3,0}^{2^{s+t}}, \mathbf{d}_{3,1}^{2^{s+t}}, \mathbf{d}_{3,2}^{2^{s+t}})$$
$$L = (\mathbf{d}_{3,0}, \mathbf{d}_{3,1}^{2^s-1}, \mathbf{d}_{3,2}^{2^s(2^t-1)})$$
$$h = \mathbf{d}_{3,0}^{2^{s+t}-1}\mathbf{d}_{3,1}^{2^s(2^t-1)+1}\mathbf{d}_{3,2}^{2^s}.$$

Here is a short resumé of the relevant portion of the computation. It makes use of the formulae (⁘) , as well as the fact that $\mathbf{d}_{3,0}$ and $\mathbf{d}_{3,1}^{2^s}$ belong to the ideal $\mathfrak{d}(1, 2^s-1, 2^s(2^t-1))$. We use the notation $\equiv$ to denote equality in the quotient algebra $\mathbb{F}_2[x, y, z]/\mathfrak{d}(1, 2^s-1, 2^s(2^t-1))$:

$$\begin{aligned}
&\chi\mathrm{Wu}(\mathfrak{d}(1, 2^s-1, 2^s(2^t-1))\\
&\quad\equiv (\mathbf{d}_{3,0} + \mathbf{d}_{3,1} + \mathbf{d}_{3,2} + 1)^{2^{s+t}-1}(\mathbf{d}_{3,1} + \mathbf{d}_{3,2} + 1)^{2^s(2^t-1)+1}(\mathbf{d}_{3,2} + 1)^{2^s}\\
&\quad\equiv (\mathbf{d}_{3,1} + \mathbf{d}_{3,2} + 1)^{2^{s+t}-1}(\mathbf{d}_{3,1}^{2^s} + \mathbf{d}_{3,2}^{2^s} + 1)^{2^t-1}(\mathbf{d}_{3,1} + \mathbf{d}_{3,2} + 1)(\mathbf{d}_{3,2} + 1)^{2^s}\\
&\quad\equiv (\mathbf{d}_{3,1} + \mathbf{d}_{3,2} + 1)^{2^{s+t}}(\mathbf{d}_{3,2} + 1)^{2^{s+t}-2^s}(\mathbf{d}_{3,2} + 1)^{2^s}\\
&\quad\equiv (\mathbf{d}_{3,1}^{2^{s+t}} + \mathbf{d}_{3,2}^{2^{s+t}} + 1)(\mathbf{d}_{3,2}^{2^{s+t}} + 1) \equiv 1. \quad \square
\end{aligned}$$

The final task in this section is to show that there are no other examples of quotients $\mathbb{F}_2[x, y, z]/\mathfrak{d}(a_0, a_1, a_2)$ with trivial Wu classes apart from those given by Proposition V.3.2 and their Frobenius powers. We will use the notations introduced in Lemma V.3.1 and write $a_i = 2^{s_i} \cdot b_i$ where b_i is odd for $i = 0, 1, 2$. Let us consider an ideal $(\mathbf{d}_{3,0}^{a_0}, \mathbf{d}_{3,1}^{a_1}, \mathbf{d}_{3,2}^{a_2}) \subset \mathbb{F}_2[x, y, z]$ that is $\mathscr{A}^*$-invariant. By Lemma V.3.1 we may write

$$(\mathbf{d}_{3,0}^{a_0}, \mathbf{d}_{3,1}^{a_1}, \mathbf{d}_{3,2}^{a_2}) = (\mathbf{d}_{3,0}^{b_0}, \mathbf{d}_{3,1}^{a_1/2^{s_0}}, \mathbf{d}_{3,2}^{a_2/2^{s_0}})^{[2^{s_0}]}.$$

So to compute the Wu classes of $\mathscr{A}^*$-invariant ideals in $\mathbb{F}_2[x, y, z]$ generated by powers of Dickson polynomials we may restrict our attention to those where $a_0 = b_0$ is odd, i.e., $s_0 = 0$. If $a_0 = b_0 \neq 1$ then $\mathscr{A}^*$-invariance implies that $s_2 \geq s_1 \geq 2$. Choose a large power of 2, say 2^r, with $2^r \geq \max\{a_0, a_1, a_2\}$ and apply Theorem III.6.4 to the situation

$$K = (\mathbf{d}_{3,0}^{2^r}, \mathbf{d}_{3,1}^{2^r}, \mathbf{d}_{3,2}^{2^r})$$
$$L = (\mathbf{d}_{3,0}^{a_0}, \mathbf{d}_{3,1}^{a_1}, \mathbf{d}_{3,2}^{a_2})$$
$$h = \mathbf{d}_{3,0}^{2^r-a_0} \cdot \mathbf{d}_{3,1}^{2^r-a_1} \cdot \mathbf{d}_{3,2}^{2^r-a_2}.$$

Note that $2^r - a_0$ is odd, and $2^r - a_1$ and $2^r - a_2$ are even. We find

$$\begin{aligned}
&\chi\mathrm{Wu}(\mathbf{d}_{3,0}^{a_0}, \mathbf{d}_{3,1}^{a_1}, \mathbf{d}_{3,2}^{a_2})\\
(✠)\quad &\equiv (1 + \mathbf{d}_{3,2} + \mathbf{d}_{3,1} + \mathbf{d}_{3,0})^{2^r-a_0} \cdot (1 + \mathbf{d}_{3,2} + \mathbf{d}_{3,1})^{2^r-a_1} \cdot (1 + \mathbf{d}_{3,2})^{2^r-a_2}\\
&\equiv (1 + \mathbf{d}_{3,2} + \mathbf{d}_{3,1} + \mathbf{d}_{3,0} + \cdots) \cdot (1 + \mathbf{d}_{3,2}^2 + \mathbf{d}_{3,1}^2)^{\frac{2^r-a_1}{2}} \cdot (1 + \mathbf{d}_{3,2}^2)^{\frac{2^r-a_2}{2}},
\end{aligned}$$

where $\equiv$ means equality in $\mathbb{F}_2[x, y, z]/(\mathbf{d}_{3,0}^{a_0}, \mathbf{d}_{3,1}^{a_1}, \mathbf{d}_{3,2}^{a_2})$. Hence we have shown the following.

LEMMA V.3.3: *Let $a_0, a_1, a_2 \in \mathbb{N}$. If $a_0 = b_0 \cdot 2^{s_0}$ with b_0 odd and not equal to 1 and the ideal $(\mathbf{d}_{3,0}^{a_0}, \mathbf{d}_{3,1}^{a_1}, \mathbf{d}_{3,2}^{a_2}) \subset \mathbb{F}_2[x, y, z]$ is $\mathscr{A}^*$-invariant, then*

$$\begin{aligned}\chi\mathrm{Wu}(\mathbf{d}_{3,0}^{a_0}, \mathbf{d}_{3,1}^{a_1}, \mathbf{d}_{3,2}^{a_2}) &= 1 + \mathbf{d}_{3,2}^{2^{s_0}} + \mathbf{d}_{3,1}^{2^{s_0}} + \mathbf{d}_{3,2}^{2^{s_0}} + \cdots \\ &\neq 1 \in \mathbb{F}_2[x, y, z]/(\mathbf{d}_{3,0}^{a_0}, \mathbf{d}_{3,1}^{a_1}, \mathbf{d}_{3,2}^{a_2}).\end{aligned}$$

In particular if $(\mathbf{d}_{3,0}^{a_0}, \mathbf{d}_{3,1}^{a_1}, \mathbf{d}_{3,2}^{a_2})$ has trivial Wu classes then $b_0 = 1$. □

So applying Theorem III.6.4 allows us to reduce consideration to ideals of the form $(\mathbf{d}_{3,0}^{a_0}, \mathbf{d}_{3,1}^{a_1}, \mathbf{d}_{3,2}^{a_2})$ which are $\mathscr{A}^*$-invariant where $b_0 = 1$ and $s_0 = 0$, i.e., $a_0 = 1$.

LEMMA V.3.4: *Let $a_1, a_2 \in \mathbb{N}$. If the ideal $(\mathbf{d}_{3,0}, \mathbf{d}_{3,1}^{a_1}, \mathbf{d}_{3,2}^{a_2})$ is $\mathscr{A}^*$-invariant and has trivial Wu classes then $a_1 = 2^s - 1$ and $a_2 = 2^s(2^t - 1)$ for some $s, t \in \mathbb{N}$.*

PROOF: Choose r so that $2^r > \max\{a_1, a_2\}$ and apply Theorem III.6.4 to the situation

$$\begin{aligned}K &= (\mathbf{d}_{3,0}^{2^r}, \mathbf{d}_{3,1}^{2^r}, \mathbf{d}_{3,2}^{2^r}) \\ L &= (\mathbf{d}_{3,0}, \mathbf{d}_{3,1}^{a_1}, \mathbf{d}_{3,2}^{a_2}) \\ h &= \mathbf{d}_{3,0}^{2^r-1} \cdot \mathbf{d}_{3,1}^{2^r-a_1} \cdot \mathbf{d}_{3,2}^{2^r-a_2}.\end{aligned}$$

Then, since $\mathbf{d}_{3,0} \in L$, a short computation gives

$$\begin{aligned}&\chi\mathrm{Wu}(\mathbf{d}_{3,0}, \mathbf{d}_{3,1}^{a_1}, \mathbf{d}_{3,2}^{a_2}) \\ &\underset{\mathrm{mod} L}{\equiv} (1 + \mathbf{d}_{3,2} + \mathbf{d}_{3,1})^{2^r-1} \cdot (1 + \mathbf{d}_{3,2} + \mathbf{d}_{3,1})^{2^r-a_1} \cdot (1 + \mathbf{d}_{3,2})^{2^r-a_2} \\ &\underset{\mathrm{mod} L}{\equiv} (1 + \mathbf{d}_{3,2} + \mathbf{d}_{3,1})^{2^{r+1}-(a_1+1)}(1 + \mathbf{d}_{3,2})^{2^r-a_2}.\end{aligned}$$

Terms involving $\mathbf{d}_{3,1}$ in the expansion of $(1 + \mathbf{d}_{3,2} + \mathbf{d}_{3,1})^{2^{r+1}-(a_1+1)}$ are only trivial modulo L if $a_1 + 1 = 2^s$ for some $s \in \mathbb{N}$. In this case we find that

$$\begin{aligned}\chi\mathrm{Wu}(\mathbf{d}_{3,0}, \mathbf{d}_{3,1}^{a_1}, \mathbf{d}_{3,2}^{a_2}) &\underset{\mathrm{mod} L}{\equiv} (1 + \mathbf{d}_{3,2} + \mathbf{d}_{3,1})^{2^{r+1}-2^s} \cdot (1 + \mathbf{d}_{3,2})^{2^r-a_2} \\ &\underset{\mathrm{mod} L}{\equiv} (1 + \mathbf{d}_{3,2})^{2^{r+1}-2^s}(1 + \mathbf{d}_{3,2})^{2^r-a_2} \\ &\underset{\mathrm{mod} L}{\equiv} (1 + \mathbf{d}_{3,2})^{2^{r+1}+2^r-(2^s+a_2)}.\end{aligned}$$

Again, the terms involving $\mathbf{d}_{3,2}$ in the expansion of $(1 + \mathbf{d}_{3,2})^{2^{r+1}+2^r-(2^s+a_2)}$ only vanish modulo L if $2^s + a_2 = 2^{t+s}$ for some $t \in \mathbb{N}$, so that $a_2 = 2^s(2^t - 1)$ as desired. □

Combining Lemmas V.3.3 and V.3.4 with Proposition V.3.2 we obtain the following theorem.

THEOREM V.3.5: *Let $a_0, a_1, a_2 \in \mathbb{N}$ and write $a_i = 2^{s_i} \cdot b_i$ with b_i odd, $i = 0, 1, 2$. The ideal $\mathfrak{d}(a_0, a_1, a_2) = (\mathbf{d}_{3,0}^{a_0}, \mathbf{d}_{3,1}^{a_1}, \mathbf{d}_{3,2}^{a_2}) \subset \mathbb{F}_2[x, y, z]$ is $\mathscr{A}^*$-invariant if and only if $2^{s_1} \geq a_0$ and $2^{s_2} \geq a_1$. It has trivial Wu classes if and only if in addition for some $s_3 \in \mathbb{N}$ with $s_3 > s_2 > s_1$ one has $a_0 = 2^{s_1}$, $a_1 = 2^{s_1}(2^{s_2-s_1} - 1)$,* and $a_2 = 2^{s_2}(2^{s_3-s_2} - 1)$. □

Prior to Lemma V.3.3 we developed a formula (✠) for the conjugate Wu classes of an $\mathscr{A}^*$-invariant ideal $\mathfrak{d}(a_0, a_1, a_2)$ in $\mathbb{F}_2[x, y, z]$. It comes in handy in the sequel, so we record it here.

COROLLARY V.3.6: *Let $a_0, a_1, a_2 \in \mathbb{N}$ and write a_i for $i = 0, 1,$ and 2 in the form $a_i = 2^{s_i} \cdot b_i$ where b_i is odd. Assume $2^{s_1} \geq a_0$ and $2^{s_2} \geq a_1$ so the ideal $\mathfrak{d}(a_0, a_1, a_2) = (\mathbf{d}_{3,0}^{a_0}, \mathbf{d}_{3,1}^{a_1}, \mathbf{d}_{3,2}^{a_2}) \subset \mathbb{F}_2[x, y, z]$ is $\mathscr{A}^*$-invariant. Then the conjugate Wu classes of $\mathfrak{d}(a_0, a_1, a_2)$ are given by*

$$\begin{aligned}&\chi\mathrm{Wu}(\mathbf{d}_{3,0}^{a_0}, \mathbf{d}_{3,1}^{a_1}, \mathbf{d}_{3,2}^{a_2})\\ &= (1 + \mathbf{d}_{3,2} + \mathbf{d}_{3,1} + \mathbf{d}_{3,0})^{2^r - a_0} \cdot (1 + \mathbf{d}_{3,2} + \mathbf{d}_{3,1})^{2^r - a_1} \cdot (1 + \mathbf{d}_{3,2})^{2^r - a_2}.\end{aligned}$$ □

V.4 Powers of Dickson polynomials in many variables

In this section we study the $\mathfrak{m}$-primary irreducible ideals in $\mathbb{F}_2[z_1, \ldots, z_n]$ generated by powers of Dickson polynomials. Write $\mathfrak{d}(a_0, \ldots, a_{n-1})$ for the ideal in the algebra $\mathbb{F}_2[z_1, \ldots, z_n]$ generated by $\mathbf{d}_{n,0}^{a_0}, \ldots, \mathbf{d}_{n,n-1}^{a_{n-1}}$, where $a_0, \ldots, a_{n-1} \in \mathbb{N}$. We have three goals in this section. First, to provide a necessary and sufficient condition for $\mathfrak{d}(a_0, \ldots, a_{n-1})$ to be $\mathscr{A}^*$-invariant. Then, to develop formulae for the conjugate Wu classes of an $\mathscr{A}^*$-invariant ideal $\mathfrak{d}(a_0, \ldots, a_{n-1})$, and finally to find the conditions under which they vanish.

We begin by determining which n-tuples $a_0, \ldots, a_{n-1} \in \mathbb{N}^n$ index an $\mathscr{A}^*$-invariant ideal $\mathfrak{d}(a_0, \ldots, a_{n-1})$. To do so we will make use of the following formula for the action of the total squaring operation on the Dickson polynomials: Namely, for $k = 0, \ldots, n-1$ one has

$$(\ast)\qquad \begin{aligned}\mathbf{Sq}(\mathbf{d}_{n,k}) &= (1 + \mathbf{d}_{n,n-1} + \cdots + \mathbf{d}_{n,k+1}) \cdot (\mathbf{d}_{n,k-1} + \cdots + \mathbf{d}_{n,0})\\ &\quad + \mathbf{d}_{n,k} \cdot (1 + \mathbf{d}_{n,n-1} + \cdots + \mathbf{d}_{n,k}).\end{aligned}$$

Note that this formula expresses $\mathbf{Sq}(\mathbf{d}_{n,k})$ as a sum of two parts: a Thom class portion

$$\mathbf{d}_{n,k} \cdot (1 + \mathbf{d}_{n,n-1} + \cdots + \mathbf{d}_{n,k})$$

and an obstruction to being a Thom class

$$(1 + \mathbf{d}_{n,n-1} + \cdots + \mathbf{d}_{n,k+1}) \cdot (\mathbf{d}_{n,k-1} + \cdots + \mathbf{d}_{n,0}).$$

The formula (⁂) follows for example from the formulae lists in [66] Section A.2 and some gymnastics with the Adem–Wu relations. The following proposition answers the question: which of the ideals $\mathfrak{d}(a_0, \dots, a_{n-1})$ are $\mathscr{A}^*$-invariant?

PROPOSITION V.4.1: *Let $n, a_0, \dots, a_{n-1} \in \mathbb{N}$ and for $k = 0, \dots, n-1$ write $a_k = b_k \cdot 2^{s_k}$ with b_k odd. Then the ideal*

$$\mathfrak{d}(a_0, \dots, a_{n-1}) = \left(\mathbf{d}_{n,0}^{a_0}, \mathbf{d}_{n,1}^{a_1}, \dots, \mathbf{d}_{n,n-1}^{a_{n-1}}\right) \subset \mathbb{F}_2[z_1, \dots, z_n]$$

is $\mathscr{A}^*$-invariant if and only if $2^{s_k} \geq a_{k-1}$ for $k = 1, \dots, n-1$.

PROOF: First of all we note that formula (⁂) gives that $\mathrm{Sq}^i(\mathbf{d}_{n,k}) = 0$ for $0 < i < 2^{k-1}$. Therefore if we apply the Cartan formula to

$$\mathrm{Sq}^{2^{k-1}}(\mathbf{d}_{n,k}^b) = \mathrm{Sq}^{2^{k-1}}(\underbrace{\mathbf{d}_{n,k} \cdots \mathbf{d}_{n,k}}_{\longleftarrow\, b \,\longrightarrow})$$

we find that the only nonzero terms in the resulting sum are of the form

$$\mathbf{d}_{n,k} \cdots \mathbf{d}_{n,k} \cdot \mathrm{Sq}^{2^{k-1}}(\mathbf{d}_{n,k}) \cdot \mathbf{d}_{n,k} \cdots \mathbf{d}_{n,k} = \mathbf{d}_{n,k} \cdots \mathbf{d}_{n,k} \cdot \mathbf{d}_{n,k-1} \cdot \mathbf{d}_{n,k} \cdots \mathbf{d}_{n,k}.$$

Since there are exactly b of these terms we conclude that for b odd

$$\mathrm{Sq}^{2^{k-1}}(\mathbf{d}_{n,k}^b) = \mathbf{d}_{n,k}^{b-1}\mathbf{d}_{n,k-1}.$$

Suppose that for some k, $1 \leq k \leq n-1$, we have $2^{s_k} < a_{k-1}$. Then

$$\begin{aligned} \mathrm{Sq}^{2^{k-1+s_k}}(\mathbf{d}_{n,k}^{a_k}) &= \mathrm{Sq}^{2^{k-1}\cdot 2^{s_k}}(\mathbf{d}_{n,k}^{2^{s_k}\cdot b_k}) = \left(\mathrm{Sq}^{2^{k-1}}(\mathbf{d}_{n,k}^{b_k})\right)^{2^{s_k}} \\ &= \left(\mathbf{d}_{n,k-1} \cdot \mathbf{d}_{n,k}^{b_k-1}\right)^{2^{s_k}} = \mathbf{d}_{n,k-1}^{2^{s_k}} \cdot \mathbf{d}_{n,k}^{2^{s_k}(b_k-1)}. \end{aligned}$$

Since $2^{s_k} < a_{k-1}$ and $2^{s_k}(b_k - 1) < a_k$ this term is not in $\mathfrak{d}(a_0, \dots, a_{n-1})$, so the ideal $\mathfrak{d}(a_0, \dots, a_{n-1})$ is not $\mathscr{A}^*$-invariant.

On the other hand, suppose $2^{s_k} \geq a_{k-1}$ for $k = 1, \dots, n-1$. To show that $\mathfrak{d}(a_0, \dots, a_{n-1})$ is $\mathscr{A}^*$-invariant it is enough to verify that $\mathbf{Sq}(\mathbf{d}_{n,k}^{a_k})$ belongs[3] to $\mathfrak{d}(a_0, \dots, a_{n-1})$ for $k = 0, \dots, n-1$. From formula (⁂) we obtain for $\mathbf{Sq}(\mathbf{d}_{n,k}^{a_k})$

$$\begin{aligned} &\left((1 + \mathbf{d}_{n,n-1} + \cdots + \mathbf{d}_{n,k+1}) \cdot (\mathbf{d}_{n,k-1} + \cdots + \mathbf{d}_{n,0}) + \mathbf{d}_{n,k} \cdot (1 + \mathbf{d}_{n,n-1} + \cdots + \mathbf{d}_{n,k})\right)^{b_k \cdot 2^{s_k}} \\ &= \left((1 + \mathbf{d}_{n,n-1}^{2^{s_k}} + \cdots + \mathbf{d}_{n,k+1}^{2^{s_k}}) \cdot (\mathbf{d}_{n,k-1}^{2^{s_k}} + \cdots + \mathbf{d}_{n,0}^{2^{s_k}}) + \mathbf{d}_{n,k}^{2^{s_k}} \cdot (1 + \mathbf{d}_{n,n-1}^{2^{s_k}} + \cdots + \mathbf{d}_{n,k}^{2^{s_k}})\right)^{b_k}. \end{aligned}$$

Since $2^{s_k} \geq a_{k-1}$ we have $2^{s_k} \geq a_i$ for $i = 0, \dots, k-1$. Hence the elements $\mathbf{d}_{n,0}^{2^{s_k}}, \dots, \mathbf{d}_{n,k-1}^{2^{s_k}}$ belong to $\mathfrak{d}(a_0, \dots, a_{n-1})$, and therefore

$$(1 + \mathbf{d}_{n,n-1}^{2^{s_k}} + \cdots + \mathbf{d}_{n,k+1}^{2^{s_k}}) \cdot (\mathbf{d}_{n,k-1}^{2^{s_k}} + \cdots + \mathbf{d}_{n,0}^{2^{s_k}}) \in \mathfrak{d}(a_0, \dots, a_{n-1}).$$

---

3 What is meant is that the homogeneous components of $\mathbf{Sq}(\mathbf{d}_{n,k}^{a_k})$ belong to $\mathfrak{d}(a_0, \dots, a_{n-1})$. This imprecision of phrase recurs several times in the text.

Therefore modulo the ideal $\mathfrak{d}(a_0, \ldots, a_{n-1})$, we see that $\mathbf{Sq}(\mathbf{d}_{n,k}^{a_k})$ is equivalent to

$$\mathbf{d}_{n,k}^{2^{s_k}\cdot b_k}(1 + \mathbf{d}_{n,n-1} + \cdots + \mathbf{d}_{n,k})^{2^{s_k}\cdot b_k} = \mathbf{d}_{n,k}^{a_k}(1 + \mathbf{d}_{n,n-1} + \cdots + \mathbf{d}_{n,k})^{a_k},$$

so belongs to $\mathfrak{d}(a_0, \ldots, a_{n-1})$ since $\mathbf{d}_{n,k}^{a_k}$ does. □

Our next goal is to determine a formula for the conjugate Wu classes of the $\mathscr{A}^*$-invariant ideals given by Proposition V.4.1. To facilitate this we introduce two number theoretic functions, $\nu(-)$ and $\omega(-)$, by the requirement that $\omega(-)$ takes only odd integer values and $n = 2^{\nu(n)} \cdot \omega(n)$, $\forall\, n \in \mathbb{N}$. Thus $\nu(n)$ is the 2-adic valuation[4] of n and $\omega(n)$ is the largest odd divisor of n.

If $\mathfrak{d}(a_0, \ldots, a_{n-1})$ is $\mathscr{A}^*$-invariant then Proposition V.4.1 implies

$$a_{i-1} \le 2^{\nu(a_i)}, \quad \text{for } i = 1, \ldots, n-1,$$

and therefore

$$2^{\nu(a_0)} \le 2^{\nu(a_1)} \le \cdots \le 2^{\nu(a_{n-1})}.$$

Thus $\nu(a_0) \le \nu(a_1) \le \cdots \le \nu(a_{n-1})$. There may of course be repetitions amongst $\nu(a_0) \le \cdots \le \nu(a_{n-1})$. For example Proposition V.4.1 implies that the ideals $\mathfrak{d}(2^t, \ldots, 2^t, 2^t \cdot u)$, where $t, u \in \mathbb{N}_0$ and u is odd are all $\mathscr{A}^*$-invariant. In this case $a_0, \ldots, a_{n-1}$ all have the same 2-adic valuation.

Let $t_0 < \ldots < t_\ell$ be the distinct elements amongst the exponent sequence $\nu(a_0) \le \cdots \le \nu(a_{n-1})$ and $\mu_0, \ldots, \mu_\ell$ their multiplicities, so $\mu_0, \ldots, \mu_\ell$ is an ordered partition of n and

$$(\nu(a_0), \ldots, \nu(a_{n-1})) = (\underbrace{t_0, \ldots, t_0}_{\longleftarrow\ \mu_0\ \longrightarrow}, \underbrace{t_1, \ldots, t_1}_{\longleftarrow\ \mu_1\ \longrightarrow}, \ldots, \underbrace{t_\ell, \ldots, t_\ell}_{\longleftarrow\ \mu_\ell\ \longrightarrow}).$$

It is also convenient to introduce the notation $\lambda_i = \mu_i - 1$ for $i = 0, \ldots, \ell$.

If $c_1, \ldots, c_\mu$ is a string of natural numbers satisfying

$$\nu(c_1) = \cdots = \nu(c_\mu) = t,$$

so $c_1, \ldots, c_\mu$ all have the same 2-adic valuation t, and

$$c_{i-1} \le 2^{\nu(c_i)} = 2^t \text{ for } i = 2, \ldots, \mu,$$

then from the inequalities

$$\omega(c_{i-1})2^t = c_{i-1} \le 2^t \text{ for } i = 2, \ldots, \mu$$

we deduce that

$$\omega(c_1) = \cdots = \omega(c_{\mu-1}) = 1.$$

[4] A more precise notation would be $\nu_2(n)$, but 2 is the only prime that occurs in this section so we use the simpler notation to avoid multiple subscripts.

In other words

$$(c_1, \ldots, c_\mu) = (2^t, \ldots, 2^t, c_\mu).$$

Note carefully that this does *not* exclude the possibility that $c_\mu = 2^t$ also.

By Proposition V.4.1 this means that a sequence of integers $(a_0, \ldots, a_{n-1})$ indexing an $\mathscr{A}^*$-invariant ideal $\mathfrak{d}(a_0, \ldots, a_{n-1})$ falls into blocks corresponding with the distinct 2-adic valuations $t_0 < \cdots < t_\ell$, viz.,

$$(\underbrace{2^{t_0}, \ldots, 2^{t_0}}_{\lambda_0}, a_{\sigma_0 - 1}, \underbrace{2^{t_1}, \ldots, 2^{t_1}}_{\lambda_1}, a_{\sigma_1-1}, \ldots, \underbrace{2^{t_\ell}, \ldots, 2^{t_\ell}}_{\lambda_\ell}, a_{\sigma_\ell - 1})$$

where $\sigma_i = \mu_0 + \cdots + \mu_i$ for $i = 0, \ldots, \ell$. N.b., $\sigma_\ell - 1 = n - 1$.

We fix this notation[5] for the rest of this section. Specifically, we define the following.

NOTATION: *Let $(a_0, \ldots, a_{n-1})$ be an index sequence for an $\mathscr{A}^*$-invariant ideal $\mathfrak{d}(a_0, \ldots, a_{n-1})$. Then we define the following quantities:*

$\ell + 1$	*is the number of distinct 2-adic valuations in the chain $\nu(a_0) \leq \cdots \leq \nu(a_{n-1})$,*
$t_0 < \cdots < t_\ell$	*their distinct values,*
$\mu_0, \ldots, \mu_\ell$	*their multiplicities,*
$\lambda_i = \mu_i - 1$	*for $i = 0, \ldots, \ell$,*
$\sigma_i = \mu_0 + \cdots + \mu_i$	*for $i = 1, \ldots, \ell$, and,*
$\vartheta_i = \vartheta(a_{\sigma_i - 1})$	*for $i = 0, \ldots, \ell$.*

So the last entry of the i-th block is $2^{t_i} \cdot \vartheta_i = a_{\sigma_i - 1}$. Note in particular that $\sigma_\ell = n$ and the $\mathscr{A}^*$-invariance conditions of Proposition V.4.1 *imply not only that $t_{i+1} > t_i$ but in addition that $2^{t_{i+1}} > a_{\sigma_i - 1} = 2^{t_i} \cdot \vartheta_i$ for $i = 0, \ldots, \ell - 1$.*

Using this notation and the preceding discussion we can strengthen Proposition V.4.1 to include more complete information about the index sequences $(a_0, \ldots, a_{n-1})$ of the $\mathscr{A}^*$-invariant ideals $\mathfrak{d}(a_0, \ldots, a_{n-1})$ in the algebra $\mathbb{F}_2[z_1, \ldots, z_n]$.

PROPOSITION V.4.2: *Let $n, a_0, \ldots, a_{n-1} \in \mathbb{N}$. Then the ideal*

$$\mathfrak{d}(a_0, \ldots, a_{n-1}) = (\mathbf{d}_{n,0}^{a_0}, \mathbf{d}_{n,1}^{a_1}, \ldots, \mathbf{d}_{n,n-1}^{a_{n-1}}) \subset \mathbb{F}_2[z_1, \ldots, z_n]$$

is $\mathscr{A}^*$-invariant if and only if $2^{\nu(a_k)} \geq a_{k-1}$ for $k = 1, \ldots, n-1$. Therefore, if $(a_0, \ldots, a_{n-1})$ indexes an $\mathscr{A}^*$-invariant ideal in $\mathbb{F}_2[z_1, \ldots, z_n]$ then $\nu(a_0) \leq \cdots \leq \nu(a_{n-1})$. If $t_0 < \cdots < t_\ell$ are the distinct values among these 2-adic valuations then $(a_0, \ldots, a_{n-1})$ falls into $\ell + 1$ blocks of consecutive

[5] It would be more accurate to indicate the dependence of these quantities on the sequence $(a_0, \ldots, a_{n-1})$, but as we will only discuss one sequence at a time we have chosen not to encumber the notation with further indices.

entries. The j-th such block has the form

$$\underbrace{2^{t_j}, \ldots, 2^{t_j}, \wp_j \cdot 2^{t_j}}_{\mu_j}$$

where $\wp_j$ *is an odd natural number, and* $2^{t_{i+1}} > a_{\sigma_i - 1} = 2^{t_i} \cdot \wp_i$ *for* $i = 0, \ldots, \ell - 1$. □

Fix a sequence $(a_0, \ldots, a_{n-1}) \in \mathbb{N}^n$ indexing an $\mathscr{A}^*$-invariant ideal. We intend to apply the $K \subset L$ paradigm and Theorem III.3.5 to compute the conjugate Wu classes of $\mathfrak{d}(a_0, \ldots, a_{n-1})$. To do so we choose an integer $t_{\ell+1} \in \mathbb{N}$ that satisfies $2^{t_{\ell+1}} \geq a_{n-1}$. Then the ideal $\mathfrak{d}(2^{t_{\ell+1}}, \ldots, 2^{t_{\ell+1}})$ is $\mathscr{A}^*$-invariant, has trivial Wu classess, and is contained in $\mathfrak{d}(a_0, \ldots, a_{n-1})$. If $\wp(a_{n-1}) = 1$, i.e., the entry a_{n-1} is a power of 2, then we may choose $t_{\ell+1} = t_\ell$. A transition element for the ideal $\mathfrak{d}(a_0, \ldots, a_{n-1})$ regarded as an over ideal of $\mathfrak{d}(2^{t_{\ell+1}}, \ldots, 2^{t_{\ell+1}})$ is the product

$$\begin{aligned} &\mathbf{d}_0^{2^{t_{\ell+1}}-2^{t_0}} \cdots \mathbf{d}_{\sigma_0-2}^{2^{t_{\ell+1}}-2^{t_0}} \cdot \mathbf{d}_{\sigma_0-1}^{2^{t_{\ell+1}}-\wp_0 2^{t_0}} \\ &\quad \vdots \\ &\quad \cdot \mathbf{d}_{\sigma_i-1}^{2^{t_{\ell+1}}-2^{t_i}} \cdots \mathbf{d}_{\sigma_i-2}^{2^{t_{\ell+1}}-2^{t_i}} \cdot \mathbf{d}_{\sigma_i-1}^{2^{t_{\ell+1}}-\wp_i 2^{t_i}} \\ &\qquad \vdots \\ &\qquad \cdot \mathbf{d}_{\sigma_\ell-1}^{2^{t_{\ell+1}}-2^{t_\ell}} \cdots \mathbf{d}_{\sigma_\ell-2}^{2^{t_{\ell+1}}-2^{t_\ell}} \cdot \mathbf{d}_{\sigma_\ell-1}^{2^{t_{\ell+1}}-\wp_\ell 2^{t_\ell}}. \end{aligned}$$

We need to compute the value of **Sq** on this product and rewrite the result in the quotient algebra $\mathbf{D}(n)/(\mathbf{d}_{n,0}^{2^{t_{\ell+1}}}, \ldots, \mathbf{d}_{n,n-1}^{2^{t_{\ell+1}}})$ in the form appearing in Theorem III.3.5 to compute[6] the Wu classes. Since **Sq** is multiplicative its value on this product is a similar product where $\mathbf{d}_{n,i}$ is replaced with $\mathrm{Sq}(\mathbf{d}_{n,i})$ using the formulae (⁘). The strategy is to work block by block through the product and within each block to reduce modulo the ideal $(\mathbf{d}_{n,0}^{2^{t_{\ell+1}}}, \ldots, \mathbf{d}_{n,n-1}^{2^{t_{\ell+1}}})$ after each multiplication.

To reduce the number of indices[7] we write $\mathbf{d}_{n,k}$ in the shorter form $\mathbf{d}_k$ (as n remains fixed in the following discussion) and introduce the polynomials

$$\mathbf{Q}_i = \left[(1 + \mathbf{d}_{n-1} + \cdots + \mathbf{d}_{i+1})(\mathbf{d}_{i-1} + \cdots + \mathbf{d}_0) + \mathbf{d}_i(1 + \mathbf{d}_{n-1} + \cdots + \mathbf{d}_i)\right]$$

as a short-hand for $\mathbf{Sq}(\mathbf{d}_i)$ for $i = 0, \ldots, n-1$ (cf. formula (⁘)). It is also convenient in order to avoid extra discussions of extreme cases to agree that $\mathbf{d}_n = 1$ and $\mathbf{d}_{-1} = 0$.

---

[6] The entire computation that follows takes place in quotients of the Dickson algebra and *not* of $\mathbb{F}_2[z_1, \ldots, z_n]$ as one might expect. See the remark following Theorem V.4.3 for an explanation of why this works.

[7] As well as a startling number of `Overfull \hboxes` produced by TEX.

With these notations, the total Steenrod operation applied to the chosen transition element for the ideal $\mathfrak{d}(a_0, \ldots, a_{n-1})$ over $\mathfrak{d}(2^{t_{\ell+1}}, \ldots, 2^{t_{\ell+1}})$ is the product

$$(\maltese)\quad \begin{aligned} &\mathbf{Q}_0^{2^{t_{\ell+1}}-2^{t_0}} \cdots \mathbf{Q}_{\sigma_0-2}^{2^{t_{\ell+1}}-2^{t_0}} \cdot \mathbf{Q}_{\sigma_0-1}^{2^{t_{\ell+1}}-\omega_0 2^{t_0}} \\ &\quad \vdots \\ &\quad \cdot \mathbf{Q}_{\sigma_{i-1}}^{2^{t_{\ell+1}}-2^{t_i}} \cdots \mathbf{Q}_{\sigma_i-2}^{2^{t_{\ell+1}}-2^{t_i}} \cdot \mathbf{Q}_{\sigma_i-1}^{2^{t_{\ell+1}}-\omega_i 2^{t_i}} \\ &\qquad \vdots \\ &\qquad \cdot \mathbf{Q}_{\sigma_{\ell-1}}^{2^{t_{\ell+1}}-2^{t_\ell}} \cdots \mathbf{Q}_{\sigma_\ell-2}^{2^{t_{\ell+1}}-2^{t_\ell}} \cdot \mathbf{Q}_{\sigma_\ell-1}^{2^{t_{\ell+1}}-\omega_\ell 2^{t_\ell}}. \end{aligned}$$

We start the computation with the first element of the first block, viz.,

$$\begin{aligned} \mathbf{Q}_0^{2^{t_{\ell+1}}-2^{t_0}} &= \left[\mathbf{d}_0(1+\mathbf{d}_{n-1}+\cdots+\mathbf{d}_0)\right]^{2^{t_{\ell+1}}-2^{t_0}} \\ &= \mathbf{d}_0^{2^{t_{\ell+1}}-2^{t_0}}\left[1+\mathbf{d}_{n-1}^{2^{t_0}}+\cdots+\mathbf{d}_0^{2^{t_0}}\right]^{2^{s_0}-1}, \end{aligned}$$

where $s_0 = t_{\ell+1} - t_0$. Expand $\left[1+\mathbf{d}_{n-1}^{2^{t_0}}+\cdots+\mathbf{d}_0^{2^{t_0}}\right]^{2^{s_0}-1}$ as a polynomial in $\mathbf{d}_0^{2^{t_0}}$ whose coefficients are polynomials in $\mathbf{d}_{n-1}^{2^{t_0}}, \ldots, \mathbf{d}_1^{2^{t_0}}$. Note that

$$\mathbf{d}_0^{2^{t_{\ell+1}}-2^{t_0}} \cdot \mathbf{d}_0^{j\cdot 2^{t_0}} \in (\mathbf{d}_0^{2^{t_{\ell+1}}}, \ldots, \mathbf{d}_{n-1}^{2^{t_{\ell+1}}}) \quad \text{for } j > 0,$$

so only the coefficient of $\mathbf{d}_0^{0\cdot 2^{t_0}}$ contributes to the product modulo the ideal $(\mathbf{d}_0^{2^{t_{\ell+1}}}, \ldots, \mathbf{d}_{n-1}^{2^{t_{\ell+1}}})$, and,

$$\mathbf{Q}_0^{2^{t_{\ell+1}}-2^{t_0}} \equiv \mathbf{d}_0^{2^{t_{\ell+1}}-2^{t_0}}\left(1+\mathbf{d}_{n-1}^{2^{t_0}}+\cdots+\mathbf{d}_1^{2^{t_0}}\right)^{2^{s_0}-1}$$

modulo $(\mathbf{d}_0^{2^{t_{\ell+1}}}, \ldots, \mathbf{d}_{n-1}^{2^{t_{\ell+1}}})$.

Next we examine $\mathbf{Q}_1^{2^{t_{\ell+1}}-2^{t_0}}$ and find

$$\begin{aligned} \mathbf{Q}_1^{2^{t_{\ell+1}}-2^{t_0}} &= \left[(1+\mathbf{d}_{n-1}+\cdots+\mathbf{d}_2)\mathbf{d}_0+\mathbf{d}_1(1+\mathbf{d}_{n-1}+\cdots+\mathbf{d}_1)\right]^{2^{t_{\ell+1}}-2^{t_0}} \\ &= \left[(1+\mathbf{d}_{n-1}^{2^{t_0}}+\cdots+\mathbf{d}_2^{2^{t_0}})\mathbf{d}_0^{2^{t_0}}+\mathbf{d}_1^{2^{t_0}}(1+\mathbf{d}_{n-1}^{2^{t_0}}+\cdots+\mathbf{d}_1^{2^{t_0}})\right]^{2^{s_0}-1}. \end{aligned}$$

The terms of the expansion are of the form

$$\left[(1+\mathbf{d}_{n-1}^{2^{t_0}}+\cdots+\mathbf{d}_2^{2^{t_0}})\mathbf{d}_0^{2^{t_0}}\right]^{\alpha} \cdot \left[\mathbf{d}_1^{2^{t_0}}(1+\mathbf{d}_{n-1}^{2^{t_0}}+\cdots+\mathbf{d}_1^{2^{t_0}})\right]^{\beta}$$

where $\alpha, \beta \in \mathbb{N}_0$ and $\alpha+\beta = 2^{s_0}-1$. If $\alpha > 0$ then $\mathbf{d}_0^{2^{t_{\ell+1}}-2^{t_0}} \cdot \mathbf{d}_0^{\alpha\cdot 2^{t_0}}$ is in $(\mathbf{d}_0^{2^{t_{\ell+1}}}, \ldots, \mathbf{d}_{n-1}^{2^{t_{\ell+1}}})$ so, modulo this ideal we have

$$\begin{aligned} &\mathbf{Q}_0^{2^{t_{\ell+1}}-2^{t_0}} \mathbf{Q}_1^{2^{t_{\ell+1}}-2^{t_0}} \\ &\quad \equiv \mathbf{d}_0^{2^{t_{\ell+1}}-2^{t_0}}\left(1+\mathbf{d}_{n-1}^{2^{t_0}}+\cdots+\mathbf{d}_1^{2^{t_0}}\right)^{2^{s_0}-1} \cdot \left[\mathbf{d}_1^{2^{t_0}}\left(1+\mathbf{d}_{n-1}^{2^{t_0}}+\cdots+\mathbf{d}_1^{2^{t_0}}\right)\right]^{2^{s_0}-1} \\ &\quad \equiv \mathbf{d}_0^{2^{t_{\ell+1}}-2^{t_0}}\mathbf{d}_1^{2^{t_{\ell+1}}-2^{t_0}}\left(1+\mathbf{d}_{n-1}^{2^{t_0}}+\cdots+\mathbf{d}_1^{2^{t_0}}\right)^{2(2^{s_0}-1)}. \end{aligned}$$

If we expand $\left(1+\mathbf{d}_{n-1}^{2^{t_0}}+\cdots+\mathbf{d}_1^{2^{t_0}}\right)^{2(2^{s_0}-1)}$ as a polynomial in $\mathbf{d}_1^{2^{t_0}}$ with coefficients that are polynomials in $\mathbf{d}_{n-1}^{2^{t_0}}, \ldots, \mathbf{d}_2^{2^{t_0}}$, then since the product

$\mathbf{d}_1^{2^{t_{\ell+1}}-2^{t_0}} \cdot \mathbf{d}_1^{j \cdot 2^{t_0}}$ belongs to the ideal $(\mathbf{d}_0^{2^{t_{\ell+1}}}, \ldots, \mathbf{d}_{n-1}^{2^{t_{\ell+1}}})$ for $j > 0$ only the coefficient of $\mathbf{d}_1^{0 \cdot 2^{t_0}}$ makes a nonzero contribution to the product modulo this ideal. So we have

$$\underset{\sim}{\mathbf{Q}}_0^{2^{t_{\ell+1}}-2^{t_0}} \underset{\sim}{\mathbf{Q}}_1^{2^{t_{\ell+1}}-2^{t_0}} \equiv \mathbf{d}_0^{2^{t_{\ell+1}}-2^{t_0}} \mathbf{d}_1^{2^{t_{\ell+1}}-2^{t_0}} \left(1 + \mathbf{d}_{n-1}^{2^{t_0}} + \cdots + \mathbf{d}_2^{2^{t_0}}\right)^{2(2^{s_0}-1)}$$

modulo $(\mathbf{d}_0^{2^{t_{\ell+1}}}, \ldots, \mathbf{d}_{n-1}^{2^{t_{\ell+1}}})$. If we continue in this way we find that the product of all the terms in the first block except for the last term yields

$$\begin{aligned} &\underset{\sim}{\mathbf{Q}}_0^{2^{t_{\ell+1}}-2^{t_0}} \cdots \underset{\sim}{\mathbf{Q}}_{\sigma_0-2}^{2^{t_{\ell+1}}-2^{t_0}} \\ &\quad \equiv \mathbf{d}_0^{2^{t_{\ell+1}}-2^{t_0}} \cdots \mathbf{d}_{\sigma_0-2}^{2^{t_{\ell+1}}-2^{t_0}} \cdot \left(1 + \mathbf{d}_{n-1}^{2^{t_0}} + \cdots + \mathbf{d}_{\sigma_0-1}^{2^{t_0}}\right)^{(\mu_0-1)(2^{s_0}-1)}. \end{aligned}$$

To complete the analysis of the first block (i.e., the product of the terms corresponding to the 2-adic valuation t_0 in the product (⊞)) note

$$\begin{aligned} \underset{\sim}{\mathbf{Q}}_{\sigma_0-1}^{2^{t_{\ell+1}}-\wp_0 2^{t_0}} &= \Big[(1 + \mathbf{d}_{n-1} + \cdots + \mathbf{d}_{\sigma_0})(\mathbf{d}_{\sigma_0-2} + \cdots \mathbf{d}_0) \\ &\qquad + \mathbf{d}_{\sigma_0-1} \cdot (1 + \mathbf{d}_{n-1} + \cdots + \mathbf{d}_{\sigma_0-1})\Big]^{2^{t_{\ell+1}}-\wp_0 2^{t_0}} \end{aligned}$$

$$= \Big[(1 + \mathbf{d}_{n-1}^{2^{t_0}} + \cdots + \mathbf{d}_{\sigma_0}^{2^{t_0}})(\mathbf{d}_{\sigma_0-2}^{2^{t_0}} + \cdots \mathbf{d}_0^{2^{t_0}}) + \mathbf{d}_{\sigma_0-1}^{2^{t_0}} \cdot (1 + \mathbf{d}_{n-1}^{2^{t_0}} + \cdots + \mathbf{d}_{\sigma_0-1}^{2^{t_0}})\Big]^{2^{s_0}-\wp_0}.$$

The terms of this expansion have the form

$$\Big[(1 + \mathbf{d}_{n-1}^{2^{t_0}} + \cdots + \mathbf{d}_{\sigma_0}^{2^{t_0}})(\mathbf{d}_{\sigma_0-2}^{2^{t_0}} + \cdots \mathbf{d}_0^{2^{t_0}})\Big]^{\alpha} \cdot \Big[\mathbf{d}_{\sigma_0-1}^{2^{t_0}} \cdot (1 + \mathbf{d}_{n-1}^{2^{t_0}} + \cdots + \mathbf{d}_{\sigma_0-1}^{2^{t_0}})\Big]^{\beta}$$

where $\alpha, \beta \in \mathbb{N}_0$ and $\alpha + \beta = 2^{s_0} - \wp_0$. For $\alpha > 0$ every monomial in the expansion of

$$\Big[(1 + \mathbf{d}_{n-1}^{2^{t_0}} + \cdots + \mathbf{d}_{\sigma_0}^{2^{t_0}})(\mathbf{d}_{\sigma_0-2}^{2^{t_0}} + \cdots \mathbf{d}_0^{2^{t_0}})\Big]^{\alpha}$$

is divisible by $\mathbf{d}_j^{2^{t_0}}$ for some $0 \leq j \leq \sigma_0 - 2$. The product of this term with the rest of the first block is divisible by $\mathbf{d}_j^{2^{t_{\ell+1}}-2^{t_0}} \cdot \mathbf{d}_j^{2^{t_0}}$ which belongs to the ideal $(\mathbf{d}_0^{2^{t_{\ell+1}}}, \ldots, \mathbf{d}_{n-1}^{2^{t_{\ell+1}}})$. Therefore none of these terms contributes to the product of the elements of the first block modulo this ideal. Hence for the product of all the elements of the first block we find

$$\begin{aligned} &\underset{\sim}{\mathbf{Q}}_0^{2^{t_{\ell+1}}-2^{t_0}} \cdots \underset{\sim}{\mathbf{Q}}_{\sigma_0-2}^{2^{t_{\ell+1}}-2^{t_0}} \cdot \underset{\sim}{\mathbf{Q}}_{\sigma_0-1}^{2^{t_{\ell+1}}-\wp_0 2^{t_0}} \equiv \mathbf{d}_0^{2^{t_{\ell+1}}-2^{t_0}} \cdots \mathbf{d}_{\sigma_0-2}^{2^{t_{\ell+1}}-2^{t_0}} \cdot \\ &\left(1 + \mathbf{d}_{n-1}^{2^{t_0}} + \cdots + \mathbf{d}_{\sigma_0-1}^{2^{t_0}}\right)^{(\mu_0-1)(2^{s_0}-1)} \cdot \mathbf{d}_{\sigma_0-1}^{2^{t_{\ell+1}}-\wp_0 2^{t_0}} \cdot \left(1 + \mathbf{d}_{n-1}^{2^{t_0}} + \cdots + \mathbf{d}_{\sigma_0-1}^{2^{t_0}}\right)^{2^{s_0}-\wp_0} \\ &\quad \equiv \mathbf{d}_0^{2^{t_{\ell+1}}-2^{t_0}} \cdots \mathbf{d}_{\sigma_0-2}^{2^{t_{\ell+1}}-2^{t_0}} \cdot \mathbf{d}_{\sigma_0-1}^{2^{t_{\ell+1}}-\wp_0 2^{t_0}} \cdot \left(1 + \mathbf{d}_{n-1}^{2^{t_0}} + \cdots + \mathbf{d}_{\sigma_0-1}^{2^{t_0}}\right)^{\mu_0 \cdot 2^{s_0}-(\lambda_0+\wp_0)} \end{aligned}$$

in $\mathbf{D}(n)/(\mathbf{d}_0^{2^{t_{\ell+1}}}, \ldots, \mathbf{d}_{n-1}^{2^{t_{\ell+1}}})$.

We next examine the product of $\underset{\sim}{\mathbf{Q}}_{\sigma_0}^{2^{t_{\ell+1}}-2^{t_1}}$ (the first element associated with the second block) with $\underset{\sim}{\mathbf{Q}}_0^{2^{t_{\ell+1}}-2^{t_0}} \cdots \underset{\sim}{\mathbf{Q}}_{\sigma_0-2}^{2^{t_{\ell+1}}-2^{t_0}} \cdot \underset{\sim}{\mathbf{Q}}_{\sigma_0-1}^{2^{t_{\ell+1}}-\wp_0 2^{t_0}}$ (the product of all the elements associated with the first block) which we just computed. We have

$$\begin{aligned} \underset{\sim}{\mathbf{Q}}_{\sigma_0}^{2^{t_{\ell+1}}-2^{t_1}} &= \Big[(1 + \mathbf{d}_{n-1} + \cdots + \mathbf{d}_{\sigma_0+1}) \cdot (\mathbf{d}_{\sigma_0-1} + \cdots + \mathbf{d}_0) \\ &\qquad + \mathbf{d}_{\sigma_0}(1 + \mathbf{d}_{n-1} + \cdots + \mathbf{d}_{\sigma_0})\Big]^{2^{t_{\ell+1}}-2^{t_1}} \end{aligned}$$

which expands to a sum of terms of the form

$$\left[(1+\mathbf{d}_{n-1}^{2^{t_1}}+\cdots+\mathbf{d}_{\sigma_0+1}^{2^{t_1}})\cdot(\mathbf{d}_{\sigma_0-1}^{2^{t_1}}+\cdots+\mathbf{d}_0^{2^{t_1}})\right]^{\alpha}\cdot\left[\mathbf{d}_{\sigma_0}^{2^{t_1}}(1+\mathbf{d}_{n-1}^{2^{t_1}}+\cdots+\mathbf{d}_{\sigma_0}^{2^{t_1}})\right]^{\beta}$$

where $\alpha,\ \beta\in\mathbb{N}_0$, $\alpha+\beta=2^{s_1}-1$, and $s_1=t_{\ell+1}-t_1$. Note that by Proposition V.4.2 we have $2^{t_1}>a_{\sigma_0-1}=\omega_0 2^{t_0}\geq 2^{t_0}$ since we have assumed that the ideal $\mathfrak{d}(a_0,\ldots,a_{n-1})$ is $\mathscr{A}^*$-invariant. If $\alpha>0$ then each term of the expansion of

$$\left[(1+\mathbf{d}_{n-1}^{2^{t_1}}+\cdots+\mathbf{d}_{\sigma_0+1}^{2^{t_1}})\cdot(\mathbf{d}_{\sigma_0-1}^{2^{t_1}}+\cdots+\mathbf{d}_0^{2^{t_1}})\right]^{\alpha}$$

is divisible by some $\mathbf{d}_j^{2^{t_1}}$ for $0\leq j\leq\sigma_0-1$. For such j the products $\mathbf{d}_j^{2^{t_{\ell+1}}-2^{t_0}}\mathbf{d}_j^{2^{t_1}}$ and $\mathbf{d}_j^{2^{t_{\ell+1}}-\omega_0 2^{t_0}}\mathbf{d}_j^{2^{t_1}}$ belong to the ideal $(\mathbf{d}_0^{2^{t_{\ell+1}}},\ldots,\mathbf{d}_{n-1}^{2^{t_{\ell+1}}})$. It follows that none of these terms makes a nonzero contribution to the product $\mathbf{Q}_0^{2^{t_{\ell+1}}-2^{t_0}}\cdots\mathbf{Q}_{\sigma_0-2}^{2^{t_{\ell+1}}-2^{t_0}}\cdot\mathbf{Q}_{\sigma_0-1}^{2^{t_{\ell+1}}-\omega_0 2^{t_0}}\cdot\mathbf{Q}_{\sigma_0}^{2^{t_{\ell+1}}-2^{t_1}}$ modulo $(\mathbf{d}_0^{2^{t_{\ell+1}}},\ldots,\mathbf{d}_{n-1}^{2^{t_{\ell+1}}})$. Hence modulo this ideal we have

$$\begin{aligned}&\mathbf{Q}_0^{2^{t_{\ell+1}}-2^{t_0}}\cdots\mathbf{Q}_{\sigma_0-2}^{2^{t_{\ell+1}}-2^{t_0}}\cdot\mathbf{Q}_{\sigma_0-1}^{2^{t_{\ell+1}}-\omega_0 2^{t_0}}\cdot\mathbf{Q}_{\sigma_0}^{2^{t_{\ell+1}}-2^{t_1}}\\ &\quad\equiv\mathbf{d}_0^{2^{t_{\ell+1}}-2^{t_0}}\cdots\mathbf{d}_{\sigma_0-2}^{2^{t_{\ell+1}}-2^{t_0}}\cdot\mathbf{d}_{\sigma_0-1}^{2^{t_{\ell+1}}-\omega_0 2^{t_0}}\\ &\quad\cdot(1+\mathbf{d}_{n-1}^{2^{t_0}}+\cdots+\mathbf{d}_{\sigma_0-1}^{2^{t_0}})^{\mu_0 2^{s_0}-(\lambda_0+\omega_0)}\cdot\mathbf{d}_{\sigma_0}^{2^{t_{\ell+1}}-2^{t_1}}(1+\mathbf{d}_{n-1}^{2^{t_1}}+\cdots+\mathbf{d}_{\sigma_0}^{2^{t_1}})^{2^{s_1}-1}.\end{aligned}$$

Expanding $(1+\mathbf{d}_{n-1}^{2^{t_1}}+\cdots+\mathbf{d}_{\sigma_0}^{2^{t_1}})^{2^{s_1}-1}$ in powers of $\mathbf{d}_{\sigma_0}^{2^{t_1}}$ we see that since $\mathbf{d}_{\sigma_0}^{2^{t_{\ell+1}}-2^{t_1}}\mathbf{d}_{\sigma_0}^{j2^{t_1}}$ belongs to $(\mathbf{d}_0^{2^{t_{\ell+1}}},\ldots,\mathbf{d}_{n-1}^{2^{t_{\ell+1}}})$ for $j>0$ only the term with $j=0$ in the expansion is relevant for our computation. Hence

$$\begin{aligned}&\mathbf{Q}_0^{2^{t_{\ell+1}}-2^{t_0}}\cdots\mathbf{Q}_{\sigma_0-2}^{2^{t_{\ell+1}}-2^{t_0}}\cdot\mathbf{Q}_{\sigma_0-1}^{2^{t_{\ell+1}}-\omega_0 2^{t_0}}\cdot\mathbf{Q}_{\sigma_0}^{2^{t_{\ell+1}}-2^{t_1}}\\ &\quad\equiv\mathbf{d}_0^{2^{t_{\ell+1}}-2^{t_0}}\cdots\mathbf{d}_{\sigma_0-2}^{2^{t_{\ell+1}}-2^{t_0}}\cdot\mathbf{d}_{\sigma_0-1}^{2^{t_{\ell+1}}-\omega_0 2^{t_0}}\cdot\mathbf{d}_{\sigma_0}^{2^{t_{\ell+1}}-2^{t_1}}\\ &\quad\cdot(1+\mathbf{d}_{n-1}^{2^{t_0}}+\cdots+\mathbf{d}_{\sigma_0-1}^{2^{t_0}})^{\mu_0\cdot 2^{s_0}-(\lambda_0+\omega_0)}\\ &\quad\cdot(1+\mathbf{d}_{n-1}^{2^{t_1}}+\cdots+\mathbf{d}_{\sigma_0+1}^{2^{t_1}})^{2^{s_1}-1}\end{aligned}$$

modulo $(\mathbf{d}_0^{2^{t_{\ell+1}}},\ldots,\mathbf{d}_{n-1}^{2^{t_{\ell+1}}})$. Continuing in this way we find for the product of the elements associated to the first two blocks that

$$\begin{aligned}&\mathbf{Q}_0^{2^{t_{\ell+1}}-2^{t_0}}\cdots\mathbf{Q}_{\sigma_0-2}^{2^{t_{\ell+1}}-2^{t_0}}\cdot\mathbf{Q}_{\sigma_0-1}^{2^{t_{\ell+1}}-\omega_0 2^{t_0}}\cdot\mathbf{Q}_{\sigma_0}^{2^{t_{\ell+1}}-2^{t_1}}\cdots\mathbf{Q}_{\sigma_1-2}^{2^{t_{\ell+1}}-2^{t_1}}\cdot\mathbf{Q}_{\sigma_1-1}^{2^{t_{\ell+1}}-\omega_1 2^{t_1}}\\ &\equiv\mathbf{d}_0^{2^{t_{\ell+1}}-2^{t_0}}\cdots\mathbf{d}_{\sigma_0-2}^{2^{t_{\ell+1}}-2^{t_0}}\cdot\mathbf{d}_{\sigma_0-1}^{2^{t_{\ell+1}}-\omega_0 2^{t_0}}\cdot\mathbf{d}_{\sigma_0}^{2^{t_{\ell+1}}-2^{t_1}}\cdots\mathbf{d}_{\sigma_1-2}^{2^{t_{\ell+1}}-2^{t_1}}\cdot\mathbf{d}_{\sigma_1-1}^{2^{t_{\ell+1}}-\omega_1 2^{t_1}}\\ &\quad\cdot(1+\mathbf{d}_{n-1}^{2^{t_0}}+\cdots+\mathbf{d}_{\sigma_0-1}^{2^{t_0}})^{\mu_0\cdot 2^{s_0}-(\lambda_0+\omega_0)}\\ &\quad\cdot(1+\mathbf{d}_{n-1}^{2^{t_1}}+\cdots+\mathbf{d}_{\sigma_1-1}^{2^{t_1}})^{\mu_1\cdot 2^{s_1}-(\lambda_1+\omega_1)}\end{aligned}$$

modulo $(\mathbf{d}_0^{2^{t_{\ell+1}}},\ldots,\mathbf{d}_{n-1}^{2^{t_{\ell+1}}})$. In the bitter end this analysis leads to the formula for the product

$$\begin{aligned}&\mathbf{Q}_0^{2^{t_{\ell+1}}-2^{t_0}}\cdots\mathbf{Q}_{\sigma_0-2}^{2^{t_{\ell+1}}-2^{t_0}}\cdot\mathbf{Q}_{\sigma_0-1}^{2^{t_{\ell+1}}-\wp_0 2^{t_0}}\\&\quad\cdots\\&\quad\cdot\mathbf{Q}_{\sigma_{i-1}}^{2^{t_{\ell+1}}-2^{t_i}}\cdots\mathbf{Q}_{\sigma_i-2}^{2^{t_{\ell+1}}-2^{t_i}}\cdot\mathbf{Q}_{\sigma_i-1}^{2^{t_{\ell+1}}-\wp_i 2^{t_i}}\\&\quad\cdots\\&\quad\cdot\mathbf{Q}_{\sigma_{\ell-1}}^{2^{t_{\ell+1}}-2^{t_\ell}}\cdots\mathbf{Q}_{\sigma_\ell-2}^{2^{t_{\ell+1}}-2^{t_\ell}}\cdot\mathbf{Q}_{\sigma_\ell-1}^{2^{t_{\ell+1}}-\wp_\ell 2^{t_\ell}}\end{aligned}$$

modulo $(\mathbf{d}_0^{2^{t_{\ell+1}}}, \ldots, \mathbf{d}_{n-1}^{2^{t_{\ell+1}}})$, where $s_i = t_{\ell+1} - t_i$ for $i = 0, \ldots, \ell$, that follows:

$$\begin{aligned}&\mathbf{d}_0^{2^{t_{\ell+1}}-2^{t_0}}\cdots\mathbf{d}_{\sigma_0-2}^{2^{t_{\ell+1}}-2^{t_0}}\cdot\mathbf{d}_{\sigma_0-1}^{2^{t_{\ell+1}}-\wp_0 2^{t_0}}\\&\quad\cdots\\&\cdot\mathbf{d}_{\sigma_{i-1}}^{2^{t_{\ell+1}}-2^{t_i}}\cdots\mathbf{d}_{\sigma_i-2}^{2^{t_{\ell+1}}-2^{t_i}}\cdot\mathbf{d}_{\sigma_i-1}^{2^{t_{\ell+1}}-\wp_i 2^{t_i}}\\&\quad\cdots\\&\cdot\mathbf{d}_{\sigma_{\ell-1}}^{2^{t_{\ell+1}}-2^{t_\ell}}\cdots\mathbf{d}_{\sigma_\ell-2}^{2^{t_{\ell+1}}-2^{t_\ell}}\cdot\mathbf{d}_{\sigma_\ell-1}^{2^{t_{\ell+1}}-\wp_\ell 2^{t_\ell}}\\&\cdot\left(1+\mathbf{d}_{n-1}^{2^{t_0}}+\cdots+\mathbf{d}_{\sigma_0-1}^{2^{t_0}}\right)^{\mu_0\cdot 2^{s_0}-(\lambda_0+\wp_0)}\\&\quad\cdots\\&\quad\cdot\left(1+\mathbf{d}_{n-1}^{2^{t_i}}+\cdots+\mathbf{d}_{\sigma_i-1}^{2^{t_i}}\right)^{\mu_i\cdot 2^{s_i}-(\lambda_i+\wp_i)}\\&\quad\cdots\\&\quad\cdot\left(1+\mathbf{d}_{n-1}^{2^{t_\ell}}+\cdots+\mathbf{d}_{\sigma_\ell-1}^{2^{t_\ell}}\right)^{\mu_\ell\cdot 2^{s_\ell}-(\lambda_\ell+\wp_\ell)}.\end{aligned}$$

This in turn gives by means of the $K \subset L$ paradigm and Theorem III.3.5 the following formula for the conjuate Wu classes of an $\mathscr{A}^*$-invariant ideal generated by powers of Dickson polynomials.

THEOREM V.4.3: *Let $n, a_0, \ldots, a_{n-1} \in \mathbb{N}$ and suppose the ideal*

$$\mathfrak{d}(a_0, \ldots, a_{n-1}) = \left(\mathbf{d}_{n,0}^{a_0}, \mathbf{d}_{n,1}^{a_1}, \ldots, \mathbf{d}_{n,n-1}^{a_{n-1}}\right) \subset \mathbb{F}_2[z_1, \ldots, z_n]$$

is $\mathscr{A}^*$-invariant. Let $t_0 < \cdots < t_\ell$ be the distinct values of the 2-adic valuations of the exponents $a_0, \ldots, a_{n-1}$ and choose $t_{\ell+1}$ so that $2^{t_{\ell+1}} \geq a_{n-1}$. Then

$$\chi\mathrm{Wu}\big(\mathfrak{d}(a_0, \ldots, a_{n-1})\big) = \prod_{i=0}^{\ell}\left(1 + \mathbf{d}_{n-1}^{2^{t_i}} + \cdots + \mathbf{d}_{\sigma_i-1}^{2^{t_i}}\right)^{\mu_i\cdot 2^{s_i}-(\lambda_i+\wp_i)},$$

where $s_i = t_{\ell+1} - t_i$ for $i = 0, \ldots, \ell$. This formula is independent of $t_{\ell+1}$. □

REMARK: Note carefully that the entire computation up to the final step took place in quotients of $\mathbf{D}(n)$. One might therefore wonder why Theorem V.4.3 is stated only for the ideal $\mathfrak{d}(a_0, \ldots, a_{n-1})$ in $\mathbb{F}_2[z_1, \ldots, z_n]$ and not also for the ideal with the same generators in $\mathbf{D}(n)$. The answer is both important and subtle. Namely, the ideals $\mathfrak{d}(2^t, \ldots, 2^t)$ in $\mathbb{F}_2[z_1, \ldots, z_n]$ have trivial Wu classes, however, regarded as ideals in $\mathbf{D}(n)$ this is no longer the case: for example the ideal $(\mathbf{d}_{2,0}^2, \mathbf{d}_{2,1}^2) = (\mathbf{d}_{2,0}, \mathbf{d}_{2,1})^{[2]}$ regarded as an ideal in $\mathbf{D}(2)$ does not have trivial Wu classes. Indeed, as Example 1 in Section VI.6 shows

$$\mathrm{Wu}_2((\mathbf{d}_{2,0}^2, \mathbf{d}_{2,1}^2)) = \mathbf{d}_{2,1} \neq 0 \in \mathbf{D}(2)/(\mathbf{d}_{2,0}^2, \mathbf{d}_{2,1}^2).$$

Therefore when applying the $K \subset L$ paradigm and some variant of Theorem III.3.5, such as Proposition III.3.1, one *cannot* assume that the Frobenius powers of the maximal ideal in $\mathbf{D}(n)$ have trivial Wu classes: the Wu classes of $\overline{\mathbf{D}(n)}^{[q^m]}$, regarded as an ideal in $\mathbf{D}(n)$, being nontrivial enter into the final result as per Theorem III.3.5.

To pass from information about the Wu classes of the ideal $\mathfrak{d}(a_0, \ldots, a_{n-1})$ regarded as an ideal in $\mathbb{F}_2[z_1, \ldots, z_n]$ to information about the Wu classes of the ideal with the same generators in $\mathbf{D}(n)$ (and vice versa) requires a correction term. This correction term is what S. A. Mitchell and R. E. Stong [61] called the Wu classes of the adjoint representation of an unstable polynomial algebra over the Steenrod algebra. In essence it is the conjugate Wu class of $\overline{\mathbf{D}(n)}^{[q^m]}$ as $m \rightsquigarrow \infty$. See the discussion of change of rings in Sections VI.5 and VI.7 as well as the discussion in Section VI.8 of the *Hit Problem* for the Dickson algebra, and [27]. We intend to return to this circle of ideas in a future investigation.

The final goal of this section is to determine which of the $\mathscr{A}^*$-invariant ideals $\mathfrak{d}(a_0, \ldots, a_{n-1})$ have trivial Wu classes. We continue to employ the notations introduced preceding Theorem V.4.3. Our strategy is to fix the number $\ell + 1$ of distinct 2-adic valuations among $a_0, \ldots, a_{n-1}$ as well as their values $t_0 < \cdots < t_\ell$, and multiplicities $\mu_0, \ldots, \mu_\ell$. This allows us to arrange the sequence $a_0, \ldots, a_{n-1}$ into a rectangular array

$$\underbrace{\overbrace{2^{t_0}, \ldots, 2^{t_0}}^{\lambda_0}, 2^{t_0} \cdot \omega_0}_{\mu_0} = a_{\sigma_0}$$
$$\vdots$$
$$\underbrace{\overbrace{2^{t_i}, \ldots, 2^{t_i}}^{\lambda_i}, 2^{t_i} \cdot \omega_i}_{\mu_i} = a_{\sigma_i}$$
$$\vdots$$
$$\underbrace{\overbrace{2^{t_\ell}, \ldots, 2^{t_\ell}}^{\lambda_\ell}, 2^{t_\ell} \cdot \omega_\ell}_{\mu_\ell} = a_{\sigma_\ell}$$

where each line of the array corresponds to one of the $\ell + 1$ distinct 2-adic valuations $t_0 < \cdots < t_\ell$, the integers $\omega_0, \ldots, \omega_\ell$ are odd, and $2^{t_{i+1}} > 2^{t_i} \cdot \omega_i$, for $i = 0, \ldots, \ell - 1$. We refer to the lines as **blocks** and to

$\ell, t_0, \ldots, t_\ell, \mu_0, \ldots, \mu_\ell$ as the **block parameters** of the exponent sequence $a_0, \ldots, a_{n-1}$. For a fixed set of block parameters the exponent sequences indexing an $\mathscr{A}^*$-invariant ideal $\mathfrak{d}(a_0, \ldots, a_{n-1})$ correspond to the different values of $\mathscr{a}_0, \ldots, \mathscr{a}_\ell$. The formula of Theorem V.4.3 for conjugate Wu classes is therefore a function of $\mathscr{a}_0, \ldots, \mathscr{a}_\ell$. We propose to analyze the product representation of the conjugate Wu classes[8]

$$\chi\mathrm{Wu}(a_0, \ldots, a_{n-1}) = \prod_{i=0}^{\ell} \big(1 + \mathbf{d}_{n,n-1} + \cdots + \mathbf{d}_{n,\sigma_i-1}\big)^{\mu_i \cdot 2^{t_{\ell+1}} - (\lambda_i + \mathscr{a}_i)\cdot 2^{t_i}}$$

modulo the ideal $\mathfrak{d}(a_0, \ldots, a_{n-1})$ to determine precise conditions for the vanishing of the Wu classes of $\mathfrak{d}(a_0, \ldots, a_{n-1})$.

To this end, note that the term

$$\big(1 + \mathbf{d}_{n,n-1} + \cdots + \mathbf{d}_{n,\sigma_0-1}\big)^{\mu_0 \cdot 2^{t_{\ell+1}} - (\lambda_0 + \mathscr{a}_0)\cdot 2^{t_0}}$$

is the only term of the product involving the forms $\mathbf{d}_{n,\sigma_1-2}, \ldots, \mathbf{d}_{n,\sigma_0-1}$. Therefore the triviality of the Wu classes of the ideal $\mathfrak{d}(a_0, \ldots, a_{n-1})$ requires that the terms of the expansion involving these forms must all vanish or, what is the same thing, that

$$\begin{aligned}&\big(1 + \mathbf{d}_{n,n-1} + \cdots + \mathbf{d}_{n,\sigma_0-1}\big)^{\mu_0 \cdot 2^{t_{\ell+1}} - (\lambda_0 + \mathscr{a}_0)\cdot 2^{t_0}}\\ &\quad\equiv \big(1 + \mathbf{d}_{n,n-1} + \cdots + \mathbf{d}_{n,\sigma_1-1}\big)^{\mu_0 \cdot 2^{t_{\ell+1}} - (\lambda_0 + \mathscr{a}_0)\cdot 2^{t_0}} \bmod \mathfrak{d}(a_0, \ldots, a_{n-1}).\end{aligned}$$

If this is the case, the product

$$\begin{aligned}&\big(1 + \mathbf{d}_{n,n-1} + \cdots + \mathbf{d}_{n,\sigma_1-1}\big)^{\mu_1 \cdot 2^{t_{\ell+1}} - (\lambda_1 + \mathscr{a}_1)\cdot 2^{t_1}} \cdot \big(1 + \mathbf{d}_{n,n-1} + \cdots + \mathbf{d}_{n,\sigma_1-1}\big)^{\mu_0 \cdot 2^{t_\ell} - (\lambda_0 + \mathscr{a}_0)\cdot 2^{t_0}}\\ &\quad= \big(1 + \mathbf{d}_{n,n-1} + \cdots + \mathbf{d}_{n,\sigma_1-1}\big)^{\sigma_1 \cdot 2^{t_\ell} - (\lambda_1 + \mathscr{a}_1)\cdot 2^{t_1} - (\lambda_0 + \mathscr{a}_0)\cdot 2^{t_0}}\end{aligned}$$

(remember $\sigma_1 = \mu_1 + \mu_0$) becomes the only term in the resulting formula for $\chi\mathrm{Wu}(a_0, \ldots, a_{n-1})$ that involves the forms $\mathbf{d}_{n,\sigma_2-2}, \ldots, \mathbf{d}_{\sigma_1-1}$, and so on. This means that the sought for conditions on $\mathscr{a}_0, \ldots, \mathscr{a}_\ell$ can be recovered recursively once we have determined integers $\varphi_0, \ldots, \varphi_\ell$ such that

$$\begin{aligned}&\big(1 + \mathbf{d}_{n,n-1} + \cdots + \mathbf{d}_{n,\sigma_i-1}\big)^{\varphi_i}\\ &\quad\equiv \big(1 + \mathbf{d}_{n,n-1} + \cdots + \mathbf{d}_{n,\sigma_{i+1}-1}\big)^{\varphi_i} \bmod \mathfrak{d}(a_0, \ldots, a_{n-1}),\end{aligned}$$

for $i = 0, \ldots, \ell - 1$, and

$$\big(1 + \mathbf{d}_{n,\sigma_\ell-1}\big)^{\varphi_\ell} \equiv 1 \bmod \mathfrak{d}(a_0, \ldots, a_{n-1}).$$

We do this next.

---

[8] We have rewritten the formula given by Theorem V.4.3 to eliminate $s_0, \ldots, s_\ell$.

Expand $\left(1 + \mathbf{d}_{n,\,n-1} + \cdots + \mathbf{d}_{n,\,\sigma_i - 1}\right)^{\varphi_i}$ in powers of $\mathbf{d}_{n,\,\sigma_i-1}$. The terms of the expansion are

$$\binom{\varphi_i}{\alpha}\left(1 + \mathbf{d}_{n,\,n-1} + \cdots + \mathbf{d}_{n,\,\sigma_i}\right)^{\varphi_i - \alpha} \cdot \mathbf{d}^{\alpha}_{n,\,\sigma_i-1} \qquad 0 \le \alpha \le \varphi_i.$$

Since $\left(1 + \mathbf{d}_{n,\,n-1} + \cdots + \mathbf{d}_{n,\,\sigma_i}\right)^{\varphi_i - \alpha}$ is a monic polynomial we see that for $\alpha \neq 0$ this term belongs to the ideal $\mathfrak{d}(a_0, \ldots, a_{n-1})$ if and only if

$$\binom{\varphi_i}{\alpha} \equiv 0 \bmod 2$$

or

$$\mathbf{d}^{\alpha}_{n,\,\sigma_i-1} \in \mathfrak{d}(a_0, \ldots, a_{n-1}).$$

Since $a_{\sigma_i-1} = 2^{t_i} \cdot \vartheta_i$ the second condition holds provided $\alpha \ge 2^{t_i} \cdot \vartheta_i$. The significance of the first condition is revealed by the following lemma.

LEMMA V.4.4: *Suppose $a,\, b \in \mathbb{N}$ and define $\omega \in \mathbb{N}_0$ by $2^{\omega-1} < b \le 2^{\omega}$. Then*

$$\binom{a}{c} \equiv 0 \bmod 2$$

for all $0 < c < b$ if and only if 2^{ω} divides a.

PROOF: By Lucas' formula, see e.g. [97] Chapter I Lemma 2.6,

$$\binom{a}{c} \equiv 0 \bmod 2 \quad \forall\, 0 < c < b$$

if and only if

$$\binom{a}{2^d} \equiv 0 \bmod 2 \quad \forall\, 0 < 2^d < b,$$

and therefore the least power of 2 occurring in the dyadic expansion of a is greater than or equal to 2^{ω}. □

DEFINITION: *For $k \in \mathbb{N}$ define $\omega(k)$ by the requirement $2^{\omega(k)-1} < k \le 2^{\omega(k)}$.*

NOTATION: *Set $\omega_i = \omega(\vartheta_i)$, for $i = 0, \ldots, \ell$.*

From Lemma V.4.4 we conclude that

$$\binom{\varphi_i}{\alpha} \equiv 0 \bmod 2 \quad \forall\, 0 < \alpha < 2^{t_i} \cdot \vartheta_i$$

if and only if $2^{t_i + \omega_i}$ divides φ_i. Therefore we have shown the following.

LEMMA V.4.5: *With the preceding notations we have*

$$\begin{aligned}&\left(1 + \mathbf{d}_{n,\,n-1} + \cdots + \mathbf{d}_{n,\,\sigma_i-1}\right)^{\varphi_i}\\ &\quad \equiv \left(1 + \mathbf{d}_{n,\,n-1} + \cdots + \mathbf{d}_{n,\,\sigma_i}\right)^{\varphi_i} \bmod \mathfrak{d}(a_0, \ldots, a_{n-1})\end{aligned}$$

if and only if $2^{t_i+\omega_i}$ *divides* φ_i. □

The expansion of $\left(1 + \mathbf{d}_{n,n-1} + \cdots + \mathbf{d}_{n,\sigma_i}\right)^{\varphi_i}$ in powers of $\mathbf{d}_{n,\sigma_i}$ contains the terms

$$\binom{\varphi_i}{\alpha}\left(1 + \mathbf{d}_{n,n-1} + \cdots + \mathbf{d}_{n,\sigma_i+1}\right)^{\varphi_i-\alpha}\mathbf{d}^{\alpha}_{n,\sigma_i} \quad 0 \le \alpha \le \varphi_i.$$

Again, $\left(1 + \mathbf{d}_{n,n-1} + \cdots + \mathbf{d}_{n,\sigma_i+1}\right)^{\varphi_i-\alpha}$ is a monic polynomial so reasoning as before we see that these terms belong to $\mathfrak{d}(a_0, \ldots, a_{n-1})$ only if $2^{t_{i+1}}$ divides φ_i. If $2^{t_{i+1}}$ divides φ_i then

$$\begin{aligned}
&\left(1 + \mathbf{d}_{n,n-1} + \cdots + \mathbf{d}_{n,\sigma_i}\right)^{\varphi_i}\\
&\quad= \left(1 + \mathbf{d}^{2^{t_{i+1}}}_{n,n-1} + \cdots + \mathbf{d}^{2^{t_{i+1}}}_{n,\sigma_{i+1}-1} + \mathbf{d}^{2^{t_{i+1}}}_{n,\sigma_{i+1}-2} + \cdots + \mathbf{d}^{2^{t_{i+1}}}_{n,\sigma_i}\right)^{\varphi_i/2^{t_{i+1}}}\\
&\quad= \left(1 + \mathbf{d}^{2^{t_{i+1}}}_{n,n-1} + \cdots + \mathbf{d}^{2^{t_{i+1}}}_{n,\sigma_{i+1}-1}\right)^{\varphi_i/2^{t_{i+1}}}\\
&\quad\equiv \left(1 + \mathbf{d}_{n,n-1} + \cdots + \mathbf{d}_{n,\sigma_{i+1}-1}\right)^{\varphi_i} \bmod \mathfrak{d}(a_0, \ldots, a_{n-1})
\end{aligned}$$

since

$$\mathbf{d}^{2^{t_{i+1}}}_{n,\sigma_{i+1}-2}, \ldots, \mathbf{d}^{2^{t_{i+1}}}_{n,\sigma_i} \in \mathfrak{d}(a_0, \ldots, a_{n-1}).$$

Hence combining this with Lemma V.4.5 and repeating the above argument for $\mathbf{d}_{n,\sigma_i+1}, \ldots, \mathbf{d}_{n,\sigma_{i+1}-2}$ leads to the following conclusion.

LEMMA V.4.6: *With the preceding notations one has for* $0 \le i \le \ell - 1$

$$\begin{aligned}
&\left(1 + \mathbf{d}_{n,n-1} + \cdots + \mathbf{d}_{n,\sigma_i-1}\right)^{\varphi_i}\\
&\quad\equiv \left(1 + \mathbf{d}_{n,n-1} + \cdots + \mathbf{d}_{n,\sigma_{i+1}-1}\right)^{\varphi_i} \bmod \mathfrak{d}(a_0, \ldots, a_{n-1})
\end{aligned}$$

if and only if φ_i *is divisible by* $2^{t_{i+1}}$. *For* $i = \ell$

$$\left(1 + \mathbf{d}_{n,\sigma_\ell-1}\right)^{\varphi_\ell} \equiv 1 \bmod \mathfrak{d}(a_0, \ldots, a_{n-1})$$

if and only if $2^{t_\ell+\omega_\ell}$ *divides* φ_ℓ.

PROOF: For $i = 0, \ldots, \ell - 1$ the $\mathscr{A}^*$-invariance of $\mathfrak{d}(a_0, \ldots, a_{n-1})$ implies that $t_{i+1} \ge t_i + \omega_i$ and therefore the preceding discussion has disposed of these cases. For $i = \ell$ the binomial theorem gives

$$\left(1 + \mathbf{d}_{n,\sigma_\ell-1}\right)^{\varphi_\ell} = \sum_{\alpha=0}^{\varphi_\ell}\binom{\varphi_\ell}{\alpha}\mathbf{d}^{\alpha}_{n,\sigma_\ell-1},$$

so $\left(1 + \mathbf{d}_{n,\sigma_\ell-1}\right)^{\varphi_\ell} \equiv 1$ modulo $\mathfrak{d}(a_0, \ldots, a_{n-1})$ if and only if

$$\binom{\varphi_\ell}{\alpha} \equiv 0 \bmod 2$$

for $0 < \alpha < a_{\sigma_\ell-1} = 2^{t_\ell} \cdot \mathscr{O}_\ell$ and the result follows from Lemma V.4.4. □

THEOREM V.4.7: *Let $\mathfrak{d}(a_0, \ldots, a_{n-1}) \subset \mathbb{F}_2[z_1, \ldots, z_n]$ be an $\mathscr{A}^*$-invariant ideal with block parameters $\ell, t_0, \ldots, t_\ell, \mu_0, \ldots, \mu_\ell$. Then with the above notations $\mathfrak{d}(a_0, \ldots, a_{n-1})$ has trivial Wu classes if and only if*

$$\sum_{j=0}^{i} 2^{t_j} \cdot (\lambda_j + \omega_j)$$

is divisible by $2^{t_{i+1}}$ for $0 \leq i < \ell$, and

$$\sum_{j=0}^{\ell} 2^{t_j} \cdot (\lambda_j + \omega_j)$$

is divisible by $2^{t_\ell + \omega_\ell}$.

PROOF: Choose $t_{\ell+1}$ so that the ideal $\mathfrak{d}(2^{t_{\ell+1}}, \ldots, 2^{t_{\ell+1}})$ is contained in $\mathfrak{d}(a_0, \ldots, a_{n-1})$. Then using the preceding notations we see that $2^{t_{\ell+1}} \geq 2^{t_\ell} \cdot \omega_\ell > \cdots > 2^{t_0} \cdot \omega_0$. Set $\theta_i = 2^{t_{\ell+1}} \cdot \mu_i - (\lambda_i + \omega_i)2^{t_i}$ for $i = 0, \ldots, \ell$. Then Theorem V.4.3 says

$$\chi\mathrm{Wu}(a_0, \ldots, a_{n-1}) = \prod_{i=0}^{\ell} \left(1 + \mathbf{d}_{n,n-1} + \cdots + \mathbf{d}_{n,\sigma_i - 1}\right)^{\theta_i}.$$

Define $\varphi_0, \ldots, \varphi_\ell$ by

$$\varphi_i = \theta_i + \cdots + \theta_0 \quad 0 \leq i \leq \ell.$$

Then

$$\varphi_i = 2^{t_{\ell+1}} \cdot \sigma_i - \sum_{j=0}^{i} 2^{t_j} \cdot (\lambda_j + \omega_j)$$

so the hypotheses are equivalent to: φ_i is divisible by $2^{t_{i+1}}$ for $i = 0, \ldots, \ell - 1$, and φ_ℓ is divisible by $2^{t_\ell + \omega_\ell}$.

Since $\varphi_0 = \theta_0$ is divisible by 2^{t_1} we may apply Lemma V.4.6 to conclude that

$$\chi\mathrm{Wu}(a_0, \ldots, a_{n-1}) = \prod_{i=2}^{\ell} \left(1 + \mathbf{d}_{n,n-1} + \cdots + \mathbf{d}_{n,\sigma_i - 1}\right)^{\theta_i} \cdot \left(1 + \mathbf{d}_{n,n-1} + \cdots + \mathbf{d}_{n,\sigma_1 - 1}\right)^{\varphi_1}.$$

Proceeding inductively we may assume that

$$\chi\mathrm{Wu}(a_0, \ldots, a_{n-1}) = \prod_{i=j+1}^{\ell} \left(1 + \mathbf{d}_{n,n-1} + \cdots + \mathbf{d}_{n,\sigma_i - 1}\right)^{\theta_i} \cdot \left(1 + \mathbf{d}_{n,n-1} + \cdots + \mathbf{d}_{n,\sigma_j - 1}\right)^{\varphi_j}.$$

Again Lemma V.4.6 applies since φ_j is divisible by $2^{t_{j+1}}$, so by induction we obtain

$$\chi\mathrm{Wu}(a_0, \ldots, a_{n-1}) = \left(1 + \mathbf{d}_{n,\, \sigma_{\ell-1}}\right)^{\varphi_\ell}$$

from which the result follows by one last application of Lemma V.4.6 since φ_ℓ is divisible by $2^{t_\ell + \omega_\ell}$. □

In the extreme case where $\ell = 0$, i.e., if all the exponents $a_0, \ldots, a_{n-1}$ have the same 2-adic valuation, then the condition of Theorem V.4.7 is particularly simple.

COROLLARY V.4.8: *If $t, \varpi \in \mathbb{N}_0$, and ϖ is odd, then the ideal $(\mathbf{d}_{n,0}^{2^t}, \ldots, \mathbf{d}_{n,n-2}^{2^t}, \mathbf{d}_{n,n-1}^{2^t \cdot \varpi})$ in $\mathbb{F}_2[z_1, \ldots, z_n]$ is $\mathscr{A}^*$-invariant. It has trivial Wu classes if and only if $2^{\omega(\varpi)}$ divides $n - 1 + \varpi$.*

PROOF: The $\mathscr{A}^*$-invariance is a consequence of Proposition V.4.1. For the rest, apply Theorem V.4.7 with $\ell = 0$. We have only one block with block parameters $t_0 = t$, $\lambda_0 = n - 1$, and $\varpi_0 = \varpi$ yielding the desired conclusion. □

V.5 Powers of mod 2 Stiefel–Whitney classes in 3 variables

Recall that $\tilde{\tau} : \Sigma_{n+1} \hookrightarrow \mathrm{GL}(n, \mathbb{F}_2)$ denotes the restriction of the tautological representation of Σ_{n+1} on $\mathbb{F}_2^{n+1}$ to the subspace of vectors with coordinate sum zero. The ring of invariants $\mathbb{F}[z_1, \ldots, z_n]^{\Sigma_{n+1}}$ is a polynomial algebra $\mathbb{F}[w_2, \ldots, w_{n+1}]$ whose generators we call **Stiefel–Whitney classes**, following topological terminology (cf. Section IV.4). By Proposition IV.4.2 the corresponding ring of coinvariants $\mathbb{F}[z_1, \ldots, z_n]_{\Sigma_{n+1}}$ has trivial Wu classes. By Proposition III.6.1 and Corollary III.6.5 this means that for $s \in \mathbb{N}_0$ the ideals $(w_{n+1}^{2^s} \cdots w_2^{2^s}) \subset \mathbb{F}_2[z_1, \ldots, z_n]$ are $\mathscr{A}^*$-invariant and the corresponding Poincaré duality quotient algebras $\mathbb{F}_2[z_1, \ldots, z_n]/(w_{n+1}^{2^s} \cdots w_2^{2^s})$ have trivial Wu classes.[9]

In this section we study the ideals $(w_4^a, w_3^b, w_2^c) \subset \mathbb{F}_2[x, y, z]$, $a, b, c \in \mathbb{N}$, determine which of these are $\mathscr{A}^*$-invariant, which have trivial Wu classes, and apply this to the *Hit Problem* for $\mathbb{F}_2[x, y, z]$. We introduce the notation $\omega(a, b, c)$ for the ideal[10] (w_4^a, w_3^b, w_2^c) in $\mathbb{F}_2[x, y, z]$. In the study of the Wu classes of these ideals we often need to write exponents in the form of a

[9] This does *not* imply that the quotient algebra $\mathbb{F}_2[w_{n+1}, \ldots, w_2]/(w_{n+1}^{2^s}, \ldots, w_2^{2^s})$ also has trivial Wu classes. See Sections VI.6 and VI.8, as well as [61] and [27].

[10] As with the ideals $\mathfrak{d}(a_0, \ldots, a_{n-1})$ generated by powers of Dickson polynomials we write the generators of the ideals $\omega(a, b, c)$ in decreasing order by degree.

power of 2 times an odd number. To facilitate this we employ the notation introduced in Section V.4, $n = 2^{\nu(n)} \cdot \wp(n)$, $\forall\, n \in \mathbb{N}$ where $\wp(n)$ is odd.

We begin by recalling the following formulae for how the Steenrod operations act on the Stiefel–Whitney classes in $\mathbb{F}_2[x, y, z]$. These are special cases of Wu's formula (cf. [1]):

$$\begin{aligned}\mathbf{Sq}(w_4) &= w_4 + w_2w_4 + w_3w_4 + w_4^2\\ \mathbf{Sq}(w_3) &= w_3 + w_2w_3 + w_3^2\\ \mathbf{Sq}(w_2) &= w_2 + w_3 + w_2^2.\end{aligned}$$

Note that w_4 and w_3 are Thom classes in $\mathbb{F}_2[x, y, z]$, so to check whether an ideal $\omega(a, b, c) = (w_4^a, w_3^b, w_2^c) \subset \mathbb{F}_2[x, y, z]$ is $\mathscr{A}^*$-invariant entails examining only the behavior of the Steenrod squares on w_2^c. One aid to do so is a special case of Wu's formula:

$$\begin{aligned}\mathrm{Sq}^{2^{\nu(c)}}(w_2^c) &= \mathrm{Sq}^{2^{\nu(c)}}(w_2^{2^{\nu(c)}\cdot\wp(c)}) = \left(\mathrm{Sq}^1(w_2^{\wp(c)})\right)^{2^{\nu(c)}}\\ &= \left(\wp(c)w_2^{\wp(c)-1}w_3\right)^{2^{\nu(c)}} = w_2^{(\wp(c)-1)2^{\nu(c)}}w_3^{2^{\nu(c)}} = w_2^{c-2^{\nu(c)}}w_3^{2^{\nu(c)}}.\end{aligned}$$

PROPOSITION V.5.1: *If $a, b, c \in \mathbb{N}$ then the ideal $\omega(a, b, c) \subset \mathbb{F}_2[x, y, z]$ is $\mathscr{A}^*$-invariant if and only if $b \le 2^{\nu(c)}$, so $\nu(b) \le \nu(c)$.*

PROOF: If the ideal $\omega(a, b, c)$ is to be $\mathscr{A}^*$-invariant then

$$\mathrm{Sq}^{2^{\nu(c)}}(w_2^c) = w_2^{c-2^{\nu(c)}}w_3^{2^{\nu(c)}}$$

has to lie in $\omega(a, b, c)$. Since $c = \wp(c)\cdot 2^{\nu(c)} > c - 2^{\nu(c)}$ this can only happen if $w_3^{2^{\nu(c)}} \in \omega(a, b, c)$, which means $2^{\nu(c)} \ge b$.

It remains to show that $\omega(a, b, c)$ is $\mathscr{A}^*$-invariant provided $b \le 2^{\nu(c)}$. To prove this we apply the $K \subset L$ paradigm.

Consider the ideal $\omega(a, 1, \wp(c))$. Note that

$$\begin{aligned}\mathbf{Sq}(w_2^{\wp(c)}) &= \left(\mathbf{Sq}(w_2)\right)^{\wp(c)} = (w_2 + w_3 + w_2^2)^{\wp(c)}\\ &\equiv w_2^{\wp(c)}(1 + w_2)^{\wp(c)} \bmod \omega(a, 1, \wp(c)) \ \text{ since } w_3 \in \omega(a, 1, \wp(c))\\ &\equiv 0 \bmod \omega(a, 1, \wp(c)) \ \text{ since } w_2^{\wp(c)} \in \omega(a, 1, \wp(c)).\end{aligned}$$

Since w_4 and w_3 are Thom classes this shows that the ideal $\omega(a, 1, \wp(c))$ is $\mathscr{A}^*$-invariant for any $a \in \mathbb{N}$. By Proposition III.6.1 so are the ideals $\omega(a \cdot 2^{\nu(c)}, 2^{\nu(c)}, \wp(c)\cdot 2^{\nu(c)}) = \omega(a \cdot 2^{\nu(c)}, 2^{\nu(c)}, c)$. Next note that

$$\omega(a, b, c) = \left(\omega(a \cdot 2^{\nu(c)}, 2^{\nu(c)}, c) : w_4^{a(2^{\nu(c)}-1)}w_3^{2^{\nu(c)}-b}\right).$$

The principal ideal $(w_4^{a(2^{\nu(c)}-1}w_3^{2^{\nu(c)}-b})$ is $\mathscr{A}^*$-invariant since both w_4 and w_3 are Thom classes. Therefore $\omega(a, b, c)$ is $\mathscr{A}^*$-invariant by Lemma III.1.3. $\square$

Note that $\mathscr{A}^*$-invariance of $\omega(a, b, c)$ imposes no condition on a, but, by contrast, it forces $b = 1$ if c is odd.

LEMMA V.5.2: *Let $a, b, c \in \mathbb{N}$. Let the ideal $\omega(a, b, c) \subset \mathbb{F}_2[x, y, z]$ be $\mathscr{A}^*$-invariant. Choose $s \in \mathbb{N}$ such that $2^s > \max\{a, b, c\}$. Then*

$$\chi\mathrm{Wu}(\omega(a, b, c)) \equiv (1 + w_2 + w_3 + w_4)^{2^s-a}(1 + w_2 + w_3)^{2^s-b}(1 + w_2)^{2^s-c}$$

modulo $\omega(a, b, c)$. If a is a power of 2 this simplifies to

$$\chi\mathrm{Wu}(\omega(a, b, c)) \equiv (1 + w_2 + w_3)^{2^{s+1}-2^{\nu(a)}-b}(1 + w_2)^{2^s-c}$$

modulo $\omega(a, b, c)$.

PROOF: We apply the $K \subset L$ paradigm (Theorem III.3.5). Denote by K the ideal $\omega(2^s, 2^s, 2^s) = (w_4^{2^s}, w_3^{2^s}, w_2^{2^s})$ and set $L = \omega(a, b, c) = (w_4^a, w_3^b, w_2^c)$. Then $(K : L) = (h) + K$, where for h we may choose the element $w_4^{2^s-a} w_3^{2^s-b} w_2^{2^s-c}$. The ideal $\omega(2^s, 2^s, 2^s)$ has trivial Wu classes so if we write $\mathbf{Sq}(h)$ in the form $h \cdot \overline{w}$ then $\chi\mathrm{Wu}(L) = \overline{w}$. Using Wu's formulae we find for $\mathbf{Sq}(h)$

$$w_4^{2^s-a}(1 + w_2 + w_3 + w_4)^{2^s-a} w_3^{2^s-b}(1 + w_2 + w_3)^{2^s-b}(w_2(1 + w_2) + w_3)^{2^s-c}.$$

By Proposition V.5.1 $2^{\nu(c)} \geq b$ so $w_3^{2^s-b} w_3^{2^{\nu(c)}} \equiv 0$ modulo K. Since $2^{\nu(c)} \mid 2^s - c$ this means every term in the expansion of $(w_2(1 + w_2) + w_3)^{2^s-c}$ that contains a nonzero power of w_3 is annihilated by $w_3^{2^s-b}$ modulo K. Therefore we can eliminate w_3 from the last term above and arrive at the following expression

$$\mathbf{Sq}(h) \underset{\mathrm{mod} K}{\equiv} h(1 + w_2 + w_3 + w_4)^{2^s-a}(1 + w_2 + w_3)^{2^s-b}(1 + w_2)^{2^s-c}$$

for $\mathbf{Sq}(h)$ modulo K. By Theorem III.3.5 this implies the first formula from which the second formula for the case where a is a power of 2 follows easily. □

Thus to determine which ideals $\omega(a, b, c)$ have trivial Wu classes we pick s such that $2^s > \max\{a, b, c\}$ and must determine the conditions on a, b, c such that

$$(1 + w_2 + w_3 + w_4)^{2^s-a}(1 + w_2 + w_3)^{2^s-b}(1 + w_2)^{2^s-c} \in \omega(a, b, c).$$

LEMMA V.5.3: *Let $a, b, c \in \mathbb{N}$. If the ideal $\omega(a, b, c) \subset \mathbb{F}_2[x, y, z]$ is $\mathscr{A}^*$-invariant and has trivial Wu classes then a is a power of 2.*

PROOF: This follows from the first formula in Lemma V.5.2. □

Lemma V.5.3 provides us with a first condition on a for an $\mathscr{A}^*$-invariant ideal $\omega(a, b, c)$ to have trivial Wu classes, namely a is a power of 2, i.e.,

$a = 2^{\nu(a)}$. What remains to be done is determine the conditions on b, c, and $\nu(a)$ such that

$$(1 + w_2 + w_3)^{2^{s+1} - 2^{\nu(a)} - b}(1 + w_2)^{2^s - c} \in \omega(a, b, c).$$

So far all we have is the condition $b \leq 2^{\nu(c)}$ imposed by the assumption that $\omega(a, b, c)$ be $\mathscr{A}^*$-invariant (see Proposition V.5.1).

PROPOSITION V.5.4: *Let $a, b, c \in \mathbb{N}$. If the ideal $\omega(a, b, c) \subset \mathbb{F}_2[x, y, z]$ is $\mathscr{A}^*$-invariant and $\nu(a) < \nu(b)$ then* $\mathrm{Wu}(\omega(a, b, c)) \neq 1$.

PROOF: The second formula of Lemma V.5.2 shows that the total conjugate Wu class $\chi\mathrm{Wu}(\omega(a, b, c))$ contains the isolated term $w_3^{2^{\nu(a)}}$. Since $2^{\nu(a)} < 2^{\nu(b)} \leq 2^{\nu(b)} \cdot \mathcal{O}(b) = b$ this term does not belong to $\omega(a, b, c)$. □

So, if the ideal $\omega(a, b, c)$ is $\mathscr{A}^*$-invariant and has trivial Wu classes, then $\nu(b) \leq \min\{\nu(a), \nu(c)\}$. Let us summarize what we know up to this point by combining Propositions V.5.1 and V.5.4 with Lemma V.5.3.

COROLLARY V.5.5: *Let $a, b, c \in \mathbb{N}$. If the ideal $\omega(a, b, c) \subset \mathbb{F}_2[x, y, z]$ is $\mathscr{A}^*$-invariant and has trivial Wu classes then*

(i) a is a power of 2,

(ii) $b \leq 2^{\nu(c)}$, so if c is odd then $b = 1$, and

(iii) $\nu(b) \leq \nu(a)$.

Together (ii) and (iii) imply $\nu(b) \leq \min\{\nu(a), \nu(c)\}$. □

The case of equal 2-adic valuations can be easily disposed of with this and the second formula for the conjugate Wu classes of an $\mathscr{A}^*$-invariant ideal $\omega(a, b, c)$ derived in Lemma V.5.2.

PROPOSITION V.5.6: *Let $a, b, c \in \mathbb{N}$ with $\nu(a) = \nu(b) = \nu(c)$. If the ideal $\omega(a, b, c) \subset \mathbb{F}_2[x, y, z]$ is $\mathscr{A}^*$-invariant and has trivial Wu classes, then $1 = \mathcal{O}(a) = \mathcal{O}(b) = \mathcal{O}(c)$. So the ideal $\omega(a, b, c)$ is a Frobenius power of the ideal $\omega(1, 1, 1)$.*

PROOF: Since $\nu(a) = \nu(b) = \nu(c)$ we may suppose in view of Theorem III.6.4 that a, b, and c are odd. By Corollary V.5.5 $a = 1 = b$ so we need to show that $c = 1$. If we substitute $a = 1 = b$ into the second formula of Lemma V.5.2 we obtain after simplification that

$$\chi\mathrm{Wu}(\omega(a, b, c)) = (1 + w_2)^{2^{s+1} - 2} \cdot (1 + w_2)^{2^s - c} = (1 + w_2)^{2^{s+1} + 2^s - 2 - c}.$$

Since $2^{s+1} + 2^s - 2 - c$ is odd the expansion of the right hand side of this equation contains the term w_2, which belongs to $\omega(a, b, c)$ if and only if $c = 1$. □

It remains to consider the $\mathscr{A}^*$-invariant ideals $\omega(a, b, c)$ with exponents of unequal 2-adic valuations.

LEMMA V.5.7: *Suppose that $t, r \in \mathbb{N}$, $t > r$, and $2^t > 2^r \cdot m$ where $m \in \mathbb{N}$ is odd. Then $2^t \geq 2^r \cdot (m+1)$.*

PROOF: If $t > r$ then 2^{t-r} is even. Since $m < 2^{t-r}$ and m is odd it follows $m + 1 \leq 2^{t-r}$. Multiplying both sides of this equality by 2^r yields the result. □

PROPOSITION V.5.8: *Let $a, b, c \in \mathbb{N}$ with $\nu(a) = \nu(b) < \nu(c)$. If the ideal $\omega(a, b, c) \subset \mathbb{F}_2[x, y, z]$ is $\mathscr{A}^*$-invariant then it has trivial Wu classes if and only if*

$$\alpha(b) = 2^{\nu(c)-\nu(b)} - 1,$$

and

$$\alpha(c) = 2^{\sigma} - 1,$$

for some $\sigma \in \mathbb{N}$.

PROOF: If we substitute $\nu(a) = \nu(b)$ in the second formula of Lemma V.5.2 then we obtain

$$\chi\mathrm{Wu}(\omega(a, b, c)) = (1 + w_2 + w_3)^{2^{s+1} - 2^{\nu(b)}\cdot(\alpha(b)+1)} \cdot (1 + w_2)^{2^s - 2^{\nu(c)}\cdot\alpha(c)}.$$

If we write $\alpha(b) + 1$ in the form $2^\gamma \cdot \delta$, with $\gamma, \delta \in \mathbb{N}$ and δ odd, then the preceding expression may be rewritten as

$$\left(1 + w_2^{2^{\nu(b)+\gamma}} + w_3^{2^{\nu(b)+\gamma}}\right)^{2^{s+1-\nu(b)-\gamma} - \delta} \cdot (1 + w_2)^{2^s - 2^{\nu(c)}\cdot\alpha(c)}.$$

Expanding this as a polynomial in w_3 we see that we have an isolated term $w_3^{2^{\nu(b)+\gamma}}$ since $2^{s+1-\nu(b)-\gamma} - \delta$ is odd. This term belongs to $\omega(a, b, c)$ if and only if $2^{\nu(b)+\gamma} \geq b = 2^{\nu(b)}\alpha(b)$, equivalently $2^\gamma \geq \alpha(b)$. Since $\alpha(b) + 1 = 2^\gamma \cdot \delta$ with δ odd, this leads to $\delta = 1$, so $\alpha(b) = 2^\gamma - 1$.

From this it follows that

$$\begin{aligned}(1 + w_2 + w_3)^{2^{s+1} - 2^{\nu(b)}\cdot(\alpha(b)+1)} &= (1 + w_2 + w_3)^{2^{s+1} - 2^{\nu(b)+\gamma}} \\ &= \left(1 + w_2^{2^{\nu(b)+\gamma}} + w_3^{2^{\nu(b)+\gamma}}\right)^{2^{s+1-\nu(b)-\gamma} - 1} \\ &= \left(1 + w_2^{2^{\nu(b)+\gamma}}\right)^{2^{s+1-\nu(b)-\gamma} - 1} \\ &= (1 + w_2)^{2^{s+1} - 2^{\nu(b)+\gamma}}\end{aligned}$$

in $\mathbb{F}_2[x, y, z]/\omega(a, b, c)$, since $b < 2^{\nu(b)+\gamma}$ implies $w_3^{2^{\nu(b)+\gamma}} \in \omega(a, b, c)$. Hence,

$$(\because) \qquad \chi\mathrm{Wu}(\omega(a, b, c)) = (1 + w_2)^{2^{s+1} + 2^s - 2^{\nu(b)+\gamma} - 2^{\nu(c)}\cdot\alpha(c)}.$$

Since $\mathbb{F}_2[x, y, z]/\omega(a, b, c)$ was assumed to have trivial Wu classes this expression must lie in $\omega(a, b, c)$.

We consider three cases based on the relation between $\nu(b) + \gamma$ and $\nu(c)$.

CASE: $\nu(b) + \gamma < \nu(c)$. Then

$$(1 + w_2)^{2^{s+1}+2^s-2^{\nu(b)+\gamma}-2^{\nu(c)}\cdot\wp(c)} = (1 + w_2^{2^{\nu(b)+\gamma}})^{\lambda},$$

where $\lambda = 2^{s+1-(\nu(b)+\gamma)} + 2^{s-(\nu(b)+\gamma)} - 1 - 2^{\nu(c)-(\nu(b)+\gamma)} \cdot \wp(c)$ is odd. Therefore the term $w_2^{2^{\nu(b)+\gamma}}$ occurs in the expansion of (✣). It does not belong to $\omega(a, b, c)$ so $\chi\mathrm{Wu}(\omega(a, b, c)) \neq 1$. Hence this case does not occur.

CASE: $\nu(b) + \gamma = \nu(c)$. Then we need to deal with the expansion of

$$(1 + w_2)^{2^{s+1}+2^s-2^{\nu(c)}\cdot(\wp(c)+1)}.$$

We write $\wp(c) + 1 = 2^{\sigma} \cdot \tau$ with $\sigma, \tau \in \mathbb{N}$ and τ odd. The expansion contains the term $w_2^{2^{\nu(c)+\sigma}}$. For this term to belong to $\omega(a, b, c)$ we need $2^{\nu(c)+\sigma} \geq c = 2^{\nu(c)} \cdot \wp(c)$. So $2^{\sigma} \geq \wp(c)$, and since $\wp(c) + 1 = 2^{\sigma} \cdot \tau$ where τ is odd, we must have $\tau = 1$. Therefore $\wp(c) = 2^{\sigma} - 1$. Since $\gamma = \nu(c) - \nu(b)$ and $\wp(b) = 2^{\gamma} - 1$ we conclude

$$\wp(b) = 2^{\nu(c)-\nu(b)} - 1$$

and

$$\wp(c) = 2^{\sigma} - 1$$

as claimed. The fact that $\omega(a, b, c)$ has trivial Wu classes under these conditions follows by direct computation from the formulae given by Lemma V.5.2.

CASE: $\nu(b) + \gamma > \nu(c)$. This case cannot occur as we next show. To see this note that $b \leq 2^{\nu(c)}$ by Proposition V.5.1 and $\nu(b) < \nu(c)$ by hypothesis. Therefore from Lemma V.5.7 we have that $2^{\nu(c)} \geq 2^{\nu(b)} \cdot (\wp(b) + 1) = 2^{\nu(b)+\gamma}$ which is a contradiction to the fact that $\nu(b) + \gamma > \nu(c)$ in this case. □

It remains to consider the $\mathscr{A}^*$-invariant ideals $\omega(a, b, c)$ with $\nu(a) > \nu(b)$.

PROPOSITION V.5.9: *Let $a, b, c \in \mathbb{N}$ with $\nu(a) > \nu(b)$. Suppose that the ideal $\omega(a, b, c) \subset \mathbb{F}_2[x, y, z]$ is $\mathscr{A}^*$-invariant. It has trivial Wu classes if and only if*

(i) $\nu(b) = \nu(c)$,

(ii) $\wp(b) = 1$,

and either

(iii-a) $\wp(c) = 2^{\nu(a)-\nu(b)} \cdot (2^{\zeta} - 1) - 1$,

for some $\zeta \in \mathbb{N}$, or

(iii-b) $$\mathfrak{o}(c) = 2^{\sigma} - 1,$$
for some $\sigma \in \mathbb{N}$, *with* $\sigma < \nu(a) - \nu(c)$.

Proof: In the situation $\nu(a) > \nu(b)$ the formula for $\chi\mathrm{Wu}(\omega(a, b, c))$ (see the first formula in Lemma V.5.2) contains the term $w_3^{2^{\nu(b)}}$ in its expansion. This belongs to $\omega(a, b, c)$ if and only if $2^{\nu(b} \geq b = 2^{\nu(b)} \cdot \mathfrak{o}(b)$, so $\mathfrak{o}(b) = 1$. Therefore $b = 2^{\nu(b)}$ is a power of 2. The formula for the total conjugate Wu class therefore may be simplified to yield

$$\chi\mathrm{Wu}(a, b, c) = (1 + w_2)^{2^{s+1} + 2^s - 2^{\nu(a)} - 2^{\nu(b)} - c}.$$

Next use that $\nu(b) \leq \min\{\nu(a), \nu(c)\}$ to rewrite this further in the form

$$(1 + w_2^{2^{\nu(b)}})^{\lambda}$$

where the exponent λ is

$$\lambda = 3 \cdot 2^{s - \nu(b)} - 2^{\nu(a) - \nu(b)} - 2^{\nu(c) - \nu(b)} \mathfrak{o}(c) - 1.$$

The term $3 \cdot 2^s - 2^{\nu(b)}$ is even since $s \geq \max\{a, b, c\}$. We claim $\nu(c) = \nu(b)$. The term $2^{\nu(a) - \nu(b)}$ is even since $\nu(a) > \nu(b)$ by hypothesis. If $\nu(c)$ were strictly greater than $\nu(b)$ then the integer $2^{\nu(c) - \nu(b)} \mathfrak{o}(c)$ would also be even. This would mean that λ would be odd and that therefore the term $w_2^{2^{\nu(b)}}$ would appear in the expansion. However, $w_2^{2^{\nu(b)}}$ does not belong to $\omega(a, b, c)$ since $2^{\nu(b)} < 2^{\nu(c)} \leq 2^{\nu(c)} \mathfrak{o}(c) = c$, yielding a contradiction to the assumption that $\omega(a, b, c)$ has trivial Wu classes. Therefore $\nu(c) \leq \nu(b)$ and since $\nu(b) \leq \min\{\nu(a), \nu(c)\}$ by Corollary V.5.5, we must have $\nu(c) = \nu(b)$.

To summarize: if the $\mathscr{A}^*$-invariant ideal $\omega(a, b, c)$ has trivial Wu classes then under the hypothesis $\nu(a) > \nu(b)$ the following conditions hold:

$$\mathfrak{o}(a) = 1 \text{ so } a = 2^{\nu(a)},$$
$$\mathfrak{o}(b) = 1 \text{ so } b = 2^{\nu(b)},$$

and

$$\nu(b) = \nu(c).$$

If we put this into the second formula of Lemma V.5.2 for the conjugate Wu classes of $\omega(a, b, c)$ and use that $w_3^{2^{\nu(b)}}, w_3^{2^{\nu(c)}} \in \omega(a, b, c)$ we then obtain

$$\begin{aligned} \chi\mathrm{Wu}(\omega(a, b, c)) &= (1 + w_2 + w_3)^{2^{s+1} - 2^{\nu(a)} - 2^{\nu(b)}} (1 + w_2)^{2^s - c} \\ &= (1 + w_2)^{2^{s+1} - 2^{\nu(a)} - 2^{\nu(b)}} (1 + w_2)^{2^s - c}. \end{aligned}$$

Since $\nu(b) = \nu(c)$ and $\mathfrak{o}(b) = 1$ we obtain

$$\chi\mathrm{Wu}(\omega(a, b, c)) = (1 + w_2)^{2^{s+1} + 2^s - 2^{\nu(a)} - 2^{\nu(c)}(\mathfrak{o}(c) + 1)}.$$

Write $\mathfrak{o}(c) + 1 = 2^{\sigma} \cdot \tau$ with $\sigma, \tau \in \mathbb{N}$ and τ odd. The formula for the total conjugate Wu class of $\omega(a, b, c)$ contains the terms of the expansion of

$$(\maltese) \qquad (1 + w_2)^{2^{s+1}+2^s-2^{\nu(a)}-2^{\nu(c)+\sigma}\cdot\tau}.$$

We consider three cases based on the relation of σ to $\nu(a) - \nu(c)$.

CASE: $\sigma > \nu(a) - \nu(c)$. Then formula $(\maltese)$ may be rewritten in the form

$$\chi \mathrm{Wu}(\omega(a, b, c)) = (1 + w_2^{2^{\nu(a)}})^{2^{s+1-\nu(a)}+2^{s-\nu(a)}-2^{\nu(c)-\nu(a)+\sigma}\cdot\tau-1}.$$

The integers $2^{s+1-\nu(a)}$, $2^{s-\nu(a)}$ and $2^{\sigma-(\nu(a)-\nu(c))}$ are all even, so the exponent $2^{s+1-\nu(a)} + 2^{s-\nu(a)} - 2^{\nu(c)-\nu(a)+\sigma} \cdot \tau - 1$ is odd. This means that $w_2^{2^{\nu(a)}}$ appears in the expansion of $(\maltese)$. Since we have asumed that $\omega(a, b, c)$ has trivial Wu classes this implies

$$2^{\nu(a)} \geq c = 2^{\nu(c)} \alpha(c)$$

giving

$$2^{\nu(a)-\nu(c)} + 1 \geq \alpha(c) + 1 = 2^{\sigma} \cdot \tau \geq 2^{\sigma} \geq 2^{\nu(a)-\nu(c)+1}$$

which is a contradiction since $\nu(a) > \nu(b) = \nu(c)$. Hence this case does not occur.

CASE: $\sigma = \nu(a) - \nu(c)$. The formula $(\maltese)$ becomes

$$(1 + w_2)^{2^{s+1}+2^s-2^{\nu(a)}(\tau+1)}.$$

If we write $\tau + 1 = 2^{\zeta} \cdot \varkappa$ where $\zeta, \varkappa \in \mathbb{N}$ and $\varkappa$ is odd then this simplifies to

$$(1 + w_2)^{2^{s+1}+2^s-2^{\nu(a)+\zeta}\cdot\varkappa}.$$

The expansion of this expression contains the term $w_2^{2^{\nu(a)+\zeta}}$, which belongs to $\omega(a, b, c)$ if and only if $2^{\nu(a)+\zeta} \geq c = 2^{\nu(c)} \cdot \alpha(c)$. Since $\sigma = \nu(a) - \nu(c)$ this is equivalent to $2^{\sigma+\zeta} \geq \alpha(c)$. If we substitute $\alpha(c) + 1 = 2^{\sigma} \cdot \tau$ we get $2^{\sigma+\zeta} + 1 \geq 2^{\sigma} \cdot \tau$ so dividing by 2^{σ} gives $2^{\zeta} + \frac{1}{2^{\sigma}} \geq \tau$. Since $\sigma, \zeta > 0$ and τ is odd this implies $\tau \leq 2^{\zeta} - 1$ yielding $2^{\zeta} \cdot \varkappa = \tau + 1 \leq 2^{\zeta}$ and hence $\varkappa = 1$. From this it follows that $\tau = 2^{\zeta} - 1$ and substituting into the equalities

$$\alpha(c) + 1 = 2^{\sigma} \cdot \tau$$

and

$$\nu(a) - \nu(c) = \sigma$$

and solving for $\alpha(c)$ gives

$$\alpha(c) = 2^{\nu(a)-\nu(c)} \cdot (2^{\zeta} - 1) - 1$$

which is condition (iii-a).

CASE: $\sigma < \nu(a) - \nu(c)$. The term $w_2^{2^{\nu(c)+\sigma}}$ appears in the expansion of $(\maltese)$. This monomial belongs to $\omega(a, b, c)$ if and only if $2^{\nu(c)+\sigma} \geq c$, so we must have $2^\sigma \geq \wp(c)$. Since $\wp(c) + 1 = 2^\sigma \cdot \tau$ with τ odd, this implies that $\tau = 1$. Hence $\wp(c) = 2^\sigma - 1$ which yields condition (iii-b).

Conversely, substituting either of the conditions (iii-a) or (iii-b) into the second formula of Lemma V.5.2 yields that the conjugate Wu classes of $\omega(a, b, c)$ are trivial. □

This completes the determination of the $\mathscr{A}^*$-invariant ideals $\omega(a, b, c)$ in $\mathbb{F}_2[x, y, z]$ with trivial Wu classes. Putting together Propositions V.5.6, V.5.8, and V.5.9 we arrive at the following complete result.

THEOREM V.5.10: *The ideals $\omega(a, b, c)$ such that $\mathbb{F}_2[x, y, z]/\omega(a, b, c)$ have trivial Wu classes are the Frobenius powers of the following four basic types:*

(i) (w_4, w_3, w_2), *or*

(ii) $(w_4, w_3^{2^\alpha - 1}, w_2^{2^\alpha(2^\beta - 1)})$ *for some* $\alpha, \beta \in \mathbb{N}$, *or*

(iii) $(w_4^{2^\alpha}, w_3, w_2^{2^\alpha(2^\beta-1)-1})$ *for some* $\alpha, \beta \in \mathbb{N}$, *or*

(iv) $(w_4^{2^\alpha}, w_3, w_2^{2^\beta - 1})$ *for some* $\alpha, \beta \in \mathbb{N}$, *with* $\beta < \alpha$. □

EXAMPLE 1: Consider the ideal $\omega(1, 1, 2) = (w_4, w_3, w_2^2)$ in $\mathbb{F}_2[x, y, z]$. It is $\mathscr{A}^*$-invariant and has trivial Wu classes by Theorem V.5.10: it is of type (ii) with $\alpha = 1$ and $\beta = 1$. The corresponding Poincaré duality quotient has formal dimension 8, and a monomial representing the fundamental class can be found by making use of the following facts.

(1) $\big(\omega(2, 2, 2) : \omega(1, 1, 2)\big) = (w_4 w_3) + \omega(2, 2, 2)$.

(2) A fundamental class for $\mathbb{F}_2[x, y, z]/\omega(1, 1, 1)$ is given by the monomial xy^2z^3 (see Section II.5 Example 3).

(3) A fundamental class for $\mathbb{F}_2[x, y, z]/\omega(2, 2, 2)$ is given by the monomial $xyz(xy^2z^3)^2$ (use Theorem II.6.6).

(4) A short computation shows that a Poincaré dual for $w_4 w_3$ in $\mathbb{F}_2[x, y, z]/\omega(2, 2, 2)$ is given by xy^2z^5.

(5) By Corollary I.2.3 xy^2z^5 is a monomial representing a fundamental class for $\mathbb{F}_2[x, y, z]/\omega(1, 1, 2)$.

This provides a natural explanation for the $\mathscr{A}^*$-indecomposable monomial xy^2z^5 which we first discussed in Section III.7 Example 2: it is a representative for the fundamental class of $\mathbb{F}_2[x, y, z]/(w_4, w_3, w_2^2)$ which is an unstable $\mathscr{A}^*$-Poincaré duality algebra with trivial Wu classes.

We close this section with a few comments on the relation between Theorem V.5.10 and the *Hit Problem* for $\mathbb{F}_2[x, y, z]$. In the manuscripts on this problem known to us the emphasis has been on finding conditions on monomials $x^\lambda y^\mu z^\nu$ that assure they are $\mathscr{A}^*$-decomposable, or $\mathscr{A}^*$-indecomposable. This may however not be the most efficacious way to approach the problem. Other viewpoints are possible, and we presented one such here. Namely, to investigate the $\mathscr{A}^*$-invariant $\mathfrak{m}$-primary irreducible ideals I in $\mathbb{F}_2[z_1, \ldots, z_n]$ such that the quotient algebra $\mathbb{F}[z_1, \ldots, z_n]/I$ has trivial Wu classes and to organize these so a few basic ideals and constructions yield all examples.

It is of course possible to obtain results about $\mathscr{A}^*$-indecomposable monomials from this point of view. If $a, b, c \in \mathbb{N}$ then $\mathbb{F}_2[w_4, w_3, w_2]$ is a free module over the subalgebra $\mathbb{F}_2[w_4^a, w_3^b, w_2^c]$ with basis the monomials $w_4^\alpha w_3^\beta w_2^\gamma$ such that $0 \le \alpha \le a-1$, $0 \le \beta \le b-1$, and $0 \le \gamma \le c-1$. A fundamental class for $\mathbb{F}_2[x, y, z]/(w_4, w_3, w_2)$ is the monomial xy^2z^3 (see e.g. [44]). If we combine these two facts we see that $xy^2z^3w_4^{a-1}w_3^{b-1}w_2^{c-1}$ is a fundamental class for $\mathbb{F}_2[x, y, z]/\omega(a, b, c)$. Therefore from Proposition III.5.3 and Theorem V.5.10 we obtain the following.

COROLLARY V.5.11: *Let $a, b, c \in \mathbb{N}$. The form $xy^2z^3w_4^{a-1}w_3^{b-1}w_2^{c-1} \in \mathbb{F}_2[x, y, z]$ is $\mathscr{A}^*$-indecomposable if it satisfies one of the following conditions*

(i) $\nu(a) = \nu(b) = \nu(c)$, *and* $\alpha(a) = \alpha(b) = \alpha(c) = 1$, *or*

(ii) $\nu(a) = \nu(b) < \nu(c)$, $\alpha(b) = 2^{\nu(c)-\nu(b)} - 1$, *and* $\alpha(c) = 2^\sigma - 1$ *for some* $\sigma \in \mathbb{N}$, *or*

(iii) $\nu(a) > \nu(b) = \nu(c)$, $\alpha(b) = 1$, *and* $\alpha(c) = 2^{\nu(a)-\nu(c)}(2^\zeta - 1) - 1$ *for some* $\zeta \in \mathbb{N}$, *or*

(iv) $\nu(a) > \nu(b) = \nu(c)$, $\alpha(b) = 1$ *and* $\alpha(c) = 2^\sigma - 1$ *for some* $\sigma \in \mathbb{N}$ *with* $\sigma < \nu(a) - \nu(c)$. □

Part VI
Macaulay's inverse systems and applications

MACAULAY 's theory of irreducible ideals based on Hopf algebra duality was discussed in Section II.2. This was particularly attractive in connection with the motivating problems from invariant theory, our study of Poincaré duality quotients of $\mathbb{F}[V]$, and the applications involving Frobenius powers and Steenrod operations. It seems however ill adapted to discussing a number of naturality properties, which we choose to formulate as change of rings results in the spirit of Cartan–Eilenberg [15]. For this we return to the source, [49] Part IV, of Macaulay's theory of $\overline{\mathbb{F}[V]}$-primary irreducible ideals and present a version of it using *inverse polynomials*. This is basically the way[1] Macaulay presented his theory in one of his last papers [50]. We use it to explain the computational tool of *catalecticant matrices* for computing ancestor ideals as this version of Macaulay's theory seems better adapted for this purpose than the formulation using Hopf algebra duality. This tool was important in building up a stock of basic examples. Using it we supply the missing details for a number of assertions made in the text, e.g., in Example 1 of Section III.7.

We show how to introduce a Steenrod algebra action into this formulation and use it to determine the formula generalizing the equality $\mathrm{Wu}(I^{[q]}) = \mathrm{Wu}(I)^q \in \mathbb{F}_q[V]/I^{[q]}$ for an $\overline{\mathbb{F}_q[V]}$-primary irreducible ideal in $\mathbb{F}_q[V]$ (see Theorem III.6.4) to arbitrary polynomial algebras over the Steenrod algebra. We also give a new proof of a result of S. A. Mitchell and R. E. Stong generalizing the triviality of Wu classes of rings of coinvariants (see Corollary IV.2.3).

The two variations of Macaulay's theory viz., the Hopf algebra duality ver-

[1] We emphasize that we are only referring to the theory as it applies to *homogeneous* ideals, what Macaulay calls H-ideals.

sion of Part II, and the inverse polynomial version here in Part VI, by no means exhaust all the possibilities of reformulation! See [21] Exercise 21.7 for an even more remarkable interpretation in the characteristic zero case. Using local cohomology it is also possible to generalize Macaulay's Double Duality Theorem to other algebras than polynomial algebras.

VI.1 Macaulay's inverse principal systems

We devote this section to a discussion of Macaulay's theory of irreducible ideals in the language of inverse polynomials. Fix a ground field $\mathbb{F}$. Let $P = \mathbb{F}[X_1, \ldots, X_n]$ be a graded connected polynomial algebra over $\mathbb{F}$ in the formal variables $X_1, \ldots, X_n$. We do *not* demand that $X_1, \ldots, X_n$ have degree 1: they may have any strictly positive degrees whatever. We denote these degrees[2] by $d_i = \deg(X_i) \in \mathbb{N}$ for $i = 1, \ldots, n$.

We let $P^{-1} = \mathbb{F}[X_1^{-1}, \ldots, X_n^{-1}]$ be the graded algebra in the formal variables $X_1^{-1}, \ldots, X_n^{-1}$ with degrees $\deg(X_i^{-1}) = -\deg(X_i)$ for $i = 1, \ldots, n$. This is the algebra of **inverse polynomials**. Note that P^{-1} is a negatively graded connected $\mathbb{F}$-algebra, i.e., the homogeneous components of strictly positive degrees are zero and the unit $1 \in P^{-1}$ is a basis for the homogeneous component of degree 0. For $F \in \mathbb{N}_0^n$ we denote by X^{-F} the inverse monomial $X_1^{-f_1} \cdots X_n^{-f_n} \in \mathbb{F}[X_1^{-1}, \ldots, X_n^{-1}]$. There is a pairing $P_i^{-1} \times P_j \longrightarrow \mathbb{F}$, denoted by $<- \mid ->$, which is defined on monomials by

$$<X^{-F} \mid X^E> = \begin{cases} 1 & \text{if } E = F \in \mathbb{N}_0^n \\ 0 & \text{otherwise.} \end{cases}$$

By using this pairing in place of the pairing obtained from Hopf algebra duality $<- \mid -> : \Gamma(u_1, \ldots, u_n) \times \mathbb{F}[z_1, \ldots, z_n] \longrightarrow \mathbb{F}$ employed in Section II.2 we may associate to an inverse polynomial $\theta \in P^{-1}$ an ideal $I(\theta) \subset P$ by interpreting θ as the linear form $<\theta \mid ->$ on P_d, where $d = -\deg(\theta)$, and taking the ideal $\mathfrak{A}(\ker(\theta))$. The proof of the following reformulation of Theorem II.2.1 is completely analagous to the proof of that theorem: it is due to F. S. Macaulay.

THEOREM VI.1.1 (F. S. Macaulay): *Let $\mathbb{F}$ be a field and $\mathbb{F}[X_1, \ldots, X_n]$ a graded polynomial algebra over $\mathbb{F}$ with generators $X_1, \ldots, X_n$ of strictly positive degrees. The assignment $\theta \rightsquigarrow I(\theta)$ sets up a correspondence between nonzero inverse polynomials $\theta \in \mathbb{F}[X_1^{-1}, \ldots, X_n^{-1}]$ of degree $-d$ and $\mathfrak{m}$-primary irreducible ideals $I(\theta) \subset \mathbb{F}[X_1, \ldots, X_n]$ such that the corresponding Poincaré duality quotient algebra $\mathbb{F}[X_1, \ldots, X_n]/I(\theta)$ has formal*

---

[2] We have chosen to denote the variables with capital letters X, Y, Z etc., hopefully to prevent confusion with the case where all the variables have degree 1.

dimension d. Two nonzero elements θ', $\theta'' \in \mathbb{F}[X_1^{-1}, \ldots, X_n^{-1}]$ determine the same ideal I if and only if they are nonzero multiples of each other; equivalently, they have the same kernel. □

The pairing $<- \mid ->$ between inverse polynomials and polynomials may be extended to an $\mathbb{F}[X_1, \ldots, X_n]$-module structure on $\mathbb{F}[X_1^{-1}, \ldots, X_n^{-1}]$ by setting

$$X^E \cap X^{-F} = \begin{cases} X^{-(F-E)} & \text{if } F - E \in \mathbb{N}_0^n \\ 0 & \text{otherwise.} \end{cases}$$

With this module structure $\mathbb{F}[X_1^{-1}, \ldots, X_n^{-1}]$ becomes an injective hull of $\mathbb{F}$ as graded $\mathbb{F}[X_1, \ldots, X_n]$-module (see e.g. [23] or [70]).

Give $\mathbb{F}[X_1, \ldots, X_n]$ the primitively generated Hopf algebra structure and let $\Gamma(U_1, \ldots, U_n)$ be the dual Hopf algebra. Then, the trick of turning the superscript grading index of a monomial in $\mathbb{F}[X_1^{-1}, \ldots, X_n^{-1}]$ into the subscript grading index of an element of $\Gamma(U_1, \ldots, U_n)$ by changing its sign, namely by mapping X^{-F} to γ_F, shows that $\Gamma(U_1, \ldots, U_n)$ and $\mathbb{F}[X_1^{-1}, \ldots, X_n^{-1}]$ are isomorphic as $\mathbb{F}[X_1, \ldots, X_n]$-modules.. This follows from the fact that in both cases the action of $\mathbb{F}[X_1, \ldots, X_n]$ is determined by how monomials act on monomials. The divided power monomials and inverse monomials are each indexed by the same index set, viz., the elements of $\mathbb{N}_0^n$, and the action of $X^E \in \mathbb{F}[X_1, \ldots, X_n]$ on either $\gamma_F \in \Gamma(U_1, \ldots, U_n)$ or $X^{-F} \in \mathbb{F}[X_1^{-1}, \ldots, X_n^{-1}]$ is given by the contraction pairing (see Sections II.2 and VI.2).

Here is the module version of Macaulay's Double Annihilator Theorem in the language of inverse polynomials. It needs no proof in view of the preceding discussion.

THEOREM VI.1.2 (F. S. Macaulay): *There is a bijective correspondence between nonzero cyclic $\mathbb{F}[X_1, \ldots, X_n]$-submodules of $\mathbb{F}[X_1^{-1}, \ldots, X_n^{-1}]$ and proper $\mathfrak{m}$-primary irreducible ideals in $\mathbb{F}[X_1, \ldots, X_n]$. It is given by associating to a nonzero cyclic submodule $M(\theta) \subset \mathbb{F}[X_1^{-1}, \ldots, X_n^{-1}]$ its annihilator ideal $I(\theta) = \mathrm{Ann}_{\mathbb{F}[X_1, \ldots, X_n]}(M(\theta)) \subset \mathbb{F}[X_1, \ldots, X_n]$, and to a proper $\mathfrak{m}$-primary irreducible ideal $I \subset \mathbb{F}[X_1, \ldots, X_n]$ the submodule $\mathrm{Ann}_{\mathbb{F}[X_1^{-1}, \ldots, X_n^{-1}]}(I)$ of elements in $\mathbb{F}[X_1^{-1}, \ldots, X_n^{-1}]$ annihilated by I.* □

We adapt the terminology of Section II.2 to this new context and refer to the $\mathbb{F}[X_1, \ldots, X_n]$-submodule of $\mathbb{F}[X_1^{-1}, \ldots, X_n^{-1}]$ associated to an $\mathfrak{m}$-primary irreducible ideal I also as **Macaulay's inverse system** of I and denote it by I^{-1}. We call a generator of I^{-1} a **Macaulay inverse** for I and denote it by θ_I. It follows from Theorems VI.1.1 and VI.1.2 that Macaulay's theory of irreducible ideals in the language of Hopf algebras and *dual* principal systems is equivalent to using *inverse* principal systems and the language of inverse polynomials. The choice between dual or inverse principal systems is a matter of taste and the context.

REMARK: After some cosmetic changes[3] to the statements of theorems such as Theorem II.5.1 and Theorem II.6.6 they remain valid for irreducible ideals in $\mathbb{F}[X_1, \ldots, X_n]$ that are primary for the maximal ideal even if the generators $X_1, \ldots, X_n$ do not all have degree 1. The proofs use the inverse system and a Macaulay inverse to an $\mathfrak{m}$-primary irreducible ideal in much the same way as the proofs in Part II use the dual system and a Macaulay dual.

VI.2 Catalecticant Matrices and Ancestor Ideals

The topic of this section is a procedure for computing ancestor ideals based on the action of the polynomial algebra $\mathbb{F}[z_1, \ldots, z_n]$ on the inverse polynomial algebra $\mathbb{F}[z_1^{-1}, \ldots, z_n^{-1}]$. It allows us to compute ideal generators for the ideal defined by a nonzero inverse polynomial using linear algebra. To simplify the discussion we restrict ourselves to the case where all the variables $z_1, \ldots, z_n$ have degree 1.

CONVENTION: *Both the algebras $\mathbb{F}[z_1, \ldots, z_n]$ and $\mathbb{F}[z_1^{-1}, \ldots, z_n^{-1}]$ have bases consisting of monomials in the variables. We agree for this section to index these monomials by the elements $E \in \mathbb{N}_0^n$.*

Thus $z^E = z_1^{e_1} \cdots z_n^{e_n}$ belongs to $\mathbb{F}[z_1, \ldots, z_n]$ and $z^{-E} = z_1^{-e_1} \cdots z_n^{-e_n}$ belongs to $\mathbb{F}[z_1^{-1}, \ldots, z_n^{-1}]$.

Let $i, j, d \in \mathbb{N}_0$, with $i + j = d$, and let $0 \neq \theta \in \mathbb{F}[z_1^{-1}, \ldots, z_n^{-1}]$ be a nonzero inverse polynomial. We use θ to define a bilinear pairing

$$\mathit{cat}_\theta(i, j) : \mathbb{F}[z_1, \ldots, z_n]_i \times \mathbb{F}[z_1, \ldots, z_n]_j \longrightarrow \mathbb{F}$$

by

$$\mathit{cat}_\theta(i, j)(f, h) = \langle \theta \mid f \cdot h \rangle = (f \cdot h) \cap \theta$$

(remember $i + j = d$). The matrix of this pairing with respect to the basis pair $\{z^I \mid |I| = i\}$ for $\mathbb{F}[z_1, \ldots, z_n]_i$ and $\{z^J \mid |J| = j\}$ for $\mathbb{F}[z_1, \ldots, z_n]_j$ is called the (i, j)-th **catalecticant matrix**[4] **of** θ and is denoted by $\mathbf{Cat}_\theta(i, j)$. Therefore the (I, J)-th entry of the matrix $\mathbf{Cat}_\theta(i, j)$ is $\theta(z^I \cdot z^J) = \theta(z^{I+J}) = \theta(z^J \cdot z^I)$, where we are regarding θ as a linear form on $\mathbb{F}[z_1, \ldots, z_n]_d$,

[3] Replace *dual* systems and generators with *inverse* systems and generators. We see no need to restate these in this new, or old, depending on your viewpoint, context.

[4] To talk of a matrix we need to choose an ordering on the monomials $\{z^E \mid E \in \mathbb{N}_0^n\}$. We assume that this has been done once and for all.

and hence $\mathbf{Cat}_\theta(i, j) = \mathbf{Cat}_\theta(j, i)^{\mathrm{tr}}$. This means the catalecticant matrices $\mathbf{Cat}_\theta(i, j)$ for all i and j summing up to d are determined by those where $i \le d/2$.

The nonzero inverse polynomial θ of degree $-d$ defines a codimension 1 subspace $W = \ker(\theta)$ in $\mathbb{F}[z_1, \dots, z_n]_d$. The ideal $I(\theta)$ inverse to θ is nothing but $\mathfrak{A}(W)$, where $\mathfrak{A}(W)$ is the big ancestor ideal of W (see Section I.3). Thus to find generators for the ideal $I(\theta)$ it is more than enough to find additive bases for the homogeneous components $\mathfrak{a}(W)_j$ for $j = 0, \dots, d$ of the little ancestor ideal $\mathfrak{a}(W)$. This amounts to solving the systems of linear equations

$$(\cdot\!\!\cdot\!\!\cdot) \qquad \mathbf{Cat}_\theta(d-j, j) \cdot \begin{bmatrix} z^{E_1} \\ \vdots \\ z^{E_m} \end{bmatrix} = 0,$$

where $m = \binom{n-1+j}{j} = \dim_{\mathbb{F}}(\mathbb{F}[z_1, \dots, z_n]_j)$ and $z^{E_1}, \dots, z^{E_m}$ are the elements of a monomial basis for the forms of degree j in $\mathbb{F}[z_1, \dots, z_n]$. We call (⁘) the **catalecticant equations**. The dimension of the solution space to the equation (⁘) is the number of columns of $\mathbf{Cat}_\theta(d-j, j)$ minus $\mathrm{rk}(\mathbf{Cat}_\theta(d-j, j))$, where $\mathrm{rk}(\mathbf{Cat}_\theta(d-j, j))$ denotes the rank of the matrix $\mathbf{Cat}_\theta(d-j, j)$; viz., $m - \mathrm{rk}(\mathbf{Cat}_\theta(d-j, j))$. So the dimension of the quotient algebra $\mathbb{F}[z_1, \dots, z_n]/I(\theta)$ in degree j is $\mathrm{rk}(\mathbf{Cat}_\theta(d-j, j))$ and the Poincaré polynomial of the quotient algebra $\mathbb{F}[z_1, \dots, z_n]/I(\theta)$ is therefore

$$P(\mathbb{F}[z_1, \dots, z_n]/I(\theta), t) = \sum_{j=0}^{d} \mathrm{rk}(\mathbf{Cat}_\theta(d-j, j)) \cdot t^j.$$

Since $\mathbf{Cat}_\theta(d-j, j)$ and $\mathbf{Cat}_\theta(j, d-j)$ are transposes of each other this is a palindromic polynomial just as it should be.

Solving the catalecticant equations is just linear algebra, and amounts to putting the catalecticant matrices into a suitable standard form, such as row-echelon form. Here is a simple example to illustrate how this works in practice.

EXAMPLE 1: Consider the inverse polynomial $x^{-4} + y^{-4} + z^{-4}$ in $\mathbb{F}[x^{-1}, y^{-1}, z^{-1}]$. There are 5 catalecticant matrices $\mathbf{Cat}_\theta(4-j, j)$ for $j = 0, \dots, 4$, but really we need only compute two of them, $\mathbf{Cat}_\theta(3, 1)$ and $\mathbf{Cat}_\theta(2, 2)$. This is straightforward and we collect the results in Table VI.2.1.

Both of the matrices $\mathbf{Cat}_\theta(3, 1)$ and $\mathbf{Cat}_\theta(2, 2)$ have rank 3, so we know that the Poincaré series of the quotient algebra $\mathbb{F}[x, y, z]/I(\theta)$ is the palindromic biquadratic polynomial

$$P(\mathbb{F}[x, y, z]/I(\theta), t) = 1 + 3t + 3t^2 + 3t^3 + t^4.$$

There are no nonzero elements in the little ancestor ideal of $\ker(\theta)$ in degree 1. In degree 2 the little ancestor ideal has as basis the monomials xy, xz, yz. In degree 3 it has as basis the monomials $x^a y^b z^c$ where (a, b, c) satisfies $a + b + c = 3$ but is not one of $(3, 0, 0)$, $(0, 3, 0)$, or $(0, 0, 3)$. All these cubic forms already lie in the ideal generated by the quadratic forms in $I(\theta)$. In degree 4 the little ancestor ideal has as vector space basis the two binomials $y^4 - x^4$ and $y^4 - z^4$ together with all monomials $x^a y^b z^c$ with $a + b + c = 4$ where (a, b, c) is not one of the index triples $(4, 0, 0)$, $(0, 4, 0)$, or $(0, 0, 4)$. All these monomials also lie in the ideal generated by the quadratic forms in $I(\theta)$, so we conclude

$\mathbf{Cat}_\theta(3, 1)$	x	y	z
x^3	1	0	0
x^2y	0	0	0
xy^2	0	0	0
xz	0	0	0
xz^2	0	0	0
y^3	0	1	0
y^2z	0	0	0
z^3	0	0	1
y^2y	0	0	0
xyz	0	0	0

$\mathbf{Cat}_\theta(2, 2)$	x^2	xy	xz	y^2	yz	z^2
x^2	1	0	0	0	0	0
xy	0	0	0	0	0	0
xz	0	0	0	0	0	0
y^2	0	0	0	1	0	0
yz	0	0	0	0	0	0
z^2	0	0	0	0	0	1

TABLE VI.2.1: $\mathbf{Cat}_\theta(3, 1)$ and $\mathbf{Cat}_\theta(2, 2)$

$$I(\theta) = (xy, xz, yz, y^4 - x^4, y^4 - z^4) \subset \mathbb{F}[x, y, z].$$

The monomials xy, xz, yz, $y^4 - x^4$, $y^4 - z^4$ are a minimal ideal basis for $I(\theta)$. The quotient algebra $\mathbb{F}[x, y, z]/I(\theta)$ is the connected sum of three copies of the algebra $\mathbb{F}[u]/(u^4)$. This follows from the fact that the inverse polynomial $\theta = x^{-4} + y^{-4} + z^{-4}$ is a sum of forms in distinct sets of variables (see the discussion of the connected sum operation in Section I.5). The inverse polynomial θ is also invariant under the action of the symmetric group Σ_3 by permutation of the variables. This invariance is not reflected in the generators of $I(\theta)$, rather in the fact that the ideal $I(\theta)$ is stable under the action of Σ_3, i.e., the elements of Σ_3 map $I(\theta)$ into itself.

One easily checks that the rank sequence of the catalecticant matrices defined by the inverse polynomial $\psi = y^4 - x^2y^2 \in \mathbb{F}[x^{-1}, y^{-1}, z^{-1}]$ is also 1, 3, 3, 3, 1, so the Poincaré series of the quotient algebra $\mathbb{F}[x, y, z]/I(\psi)$ is also $1+3t+3t^2+3t^3+t^4$. However, the quotient algebras $\mathbb{F}[x, y, z]/I(\theta)$ and $\mathbb{F}[x, y, z]/I(\psi)$ are not isomorphic since there is no linear change of coordinates taking θ to ψ.

REMARK: If we regard the coefficients of $\theta \in \mathbb{F}[z_1^{-1}, \ldots, z_n^{-1}]_d$ as variables in their own right then the catalecticant matrices are functions of these coefficients. The ranks of these matrices are polynomial functions of these coefficients regarded as variables since the rank of a matrix may be computed by examining the vanishing of the determinants of its square submatrices. Therefore, specifying a sequence of ranks, specifies an algebraic set in the projective space of the vector space of dimension $\binom{n+d-1}{d}$ (n.b. this is the dimension of the vector space of forms of degree $-d$ in n inverse polynomial variables of degree -1). This geometric viewpoint is at the heart of [34].[5] It is an open problem to determine the conditions on such a sequence that assure this set is nonempty (but see [95]). Although all the examples occurring in this manuscript have the property that the rank sequences are nondecreasing up to the middle dimension, this is not always the case. In [95] R. P. Stanley constructs (loc. cit. Example 3) an example where this fails in rank 13; it holds in all examples of rank at most 3 (loc. cit. Theorem 4.2). There is clearly room for more work here.

The next example provides the computations with catalecticant[6] matrices promised in Example 1 in Section III.7.

ILLUSTRATION VI.2.2: $\mathbf{Cat}_{\nabla}(6, 1)$, $\mathbf{Cat}_{\nabla}(5, 2)$, and $\mathbf{Cat}_{\nabla}(4, 3)$

EXAMPLE 2: Consider the inverse polynomial ∇ in $\mathbb{F}_2[x^{-1}, y^{-1}, z^{-1}]$ supported on the set $\mathcal{D}$ consisting of the monomials

$$xyz^5,\ xy^5z,\ x^5yz,\ xy^2z^4,\ x^2yz^4,\ xy^4z^2,$$
$$x^2y^4z,\ x^4yz^2,\ x^4y^2z,\ x^2y^2z^3,\ x^2y^3z^2,\ x^3y^2z^2.$$

---

5 This book also contains a short discussion (see page XVI) of the origin of the term catalecticant. Though the discussion in [24] pages 104–105 is considerably more amusing.

6 Since catalecticant is very tricky to pronounce properly, and for one of us to spell correctly, these matrices were referred to between us as *the cats*. Hence the graphic to aid the imagination.

We emphasize that this example was not found ad hoc: $\mathscr{D}$ is the $\mathscr{A}^*$-equivalence class of the $\mathscr{A}^*$-indecomposable monomial $xyz^5 \in \mathbb{F}_2[x, y, z]$, which is one of the lowest degree less obvious $\mathscr{A}^*$-indecomposables. We let $\mathfrak{A}(\mathscr{D})$ be the corresponding irreducible ideal in $\mathbb{F}_2[x, y, z]$.

The catalecticant matrices relevant to the computation of the generators for the ideal $\mathfrak{A}(\mathscr{D})$ are the three matrices $\mathbf{Cat}_\nabla(6, 1)$, $\mathbf{Cat}_\nabla(5, 2)$, and $\mathbf{Cat}_\nabla(4, 3)$. We list these in the two Diagrams VI.2.3 and VI.2.4. The first entry in each column and each row is a triple of exponents (a, b, c) for a monomial $x^a y^b z^c$. The entry at the intersection of the row headed by (a_1, b_1, c_1) with the column headed by (a_2, b_2, c_2) is the value of the linear form defined by ∇ on $\mathbb{F}_2[x, y, z]_7$ on the product of the monomials $x^{a_1}y^{b_1}z^{c_1}$ and $x^{a_2}y^{b_2}z^{c_2}$, i.e., on the monomial $x^{a_1+a_2}y^{b_1+b_2}z^{c_1+c_2}$.

$\mathbf{Cat}_\nabla(6, 1)$	1, 0, 0	0, 1, 0	0, 0, 1
6, 0, 0	0	0	0
5, 1, 0	0	0	1
5, 0, 1	0	1	0
4, 2, 0	0	0	1
4, 1, 1	1	1	1
4, 0, 2	0	1	0
3, 3, 0	0	0	0
3, 2, 1	1	0	1
3, 1, 2	1	1	0
3, 0, 3	0	0	0
2, 4, 0	0	0	1
2, 3, 1	0	1	1
2, 2, 2	1	1	1
2, 1, 3	0	1	1
2, 0, 4	0	1	0
1, 5, 0	0	0	1
1, 4, 1	1	1	1
1, 3, 2	1	1	0
1, 2, 3	1	0	1
1, 1, 4	1	1	1
1, 0, 5	0	1	0
0, 6, 0	0	0	0
0, 5, 1	1	0	0
0, 4, 2	1	0	0
0, 3, 3	0	0	0
0, 2, 4	1	0	0
0, 1, 5	1	0	0
0, 0, 6	0	0	0

$\mathbf{Cat}_\nabla(5, 2)$	2, 0, 0	1, 1, 0	1, 0, 1	0, 2, 0	0, 1, 1	0, 0, 2
5, 0, 0	0	0	0	0	1	0
4, 1, 0	0	0	1	0	1	1
4, 0, 1	0	1	0	1	1	0
3, 2, 0	0	0	1	0	0	1
3, 1, 1	1	1	1	0	1	0
3, 0, 2	0	1	0	1	0	0
2, 3, 0	0	0	0	0	1	1
2, 2, 1	1	0	1	1	1	1
2, 1, 2	1	1	0	1	1	1
2, 0, 3	0	0	0	1	1	0
1, 4, 0	0	0	1	0	1	1
1, 3, 1	0	1	1	1	1	0
1, 2, 2	1	1	1	1	0	1
1, 1, 3	0	1	1	0	1	1
1, 0, 4	0	1	0	1	1	0
0, 5, 0	0	0	1	0	0	0
0, 4, 1	1	1	1	0	0	0
0, 3, 2	1	1	0	0	0	0
0, 2, 3	1	0	1	0	0	0
0, 1, 4	1	1	1	0	0	0
0, 0, 5	0	1	0	0	0	0

DIAGRAM VI.2.3: $\mathbf{Cat}_\nabla(6, 1)$ and $\mathbf{Cat}_\nabla(5, 2)$

$\mathbf{Cat}_{\nabla}(4, 3)$	3, 0, 0	2, 1, 0	2, 0, 1	1, 2, 0	1, 1, 1	1, 0, 2	0, 3, 0	0, 2, 1	0, 1, 2	0, 0, 3
4, 0, 0	0	0	0	0	1	0	0	1	1	0
3, 1, 0	0	0	1	0	1	1	0	0	1	0
3, 0, 1	0	1	0	1	1	0	0	1	0	0
2, 2, 0	0	0	1	0	0	1	0	1	1	1
2, 1, 1	1	1	1	0	1	0	1	1	1	1
2, 0, 2	0	1	0	1	0	0	1	1	1	0
1, 3, 0	0	0	0	0	1	1	0	1	1	0
1, 2, 1	1	0	1	1	1	1	1	1	0	1
1, 1, 2	1	1	0	1	1	1	1	0	1	1
1, 0, 3	0	0	0	1	1	0	0	1	1	0
0, 4, 0	0	0	1	0	1	1	0	0	0	0
0, 3, 1	0	1	1	1	1	0	0	0	0	0
0, 2, 2	1	1	1	1	0	1	0	0	0	0
0, 3, 1	0	1	1	0	1	1	0	0	0	0
0, 0, 4	0	1	0	1	1	0	0	0	0	0

DIAGRAM VI.2.4: $\mathbf{Cat}_{\nabla}(4, 3)$

A certain amount of manipulation is required to work out the ranks of these matrices. It is pretty easy to see that $\mathbf{Cat}_{\nabla}(6, 1)$ has rank 3. If one brings the other two matrices into upper triangular form then one sees their ranks are 6 and 10. From this it follows that the Poincaré series of the Poincaré duality quotient algebra $\mathbb{F}_2[x, y, z]/\mathfrak{A}(\mathscr{D})$ is given by the polynomial

$$P(\mathbb{F}_2[x, y, z]/\mathfrak{A}(\mathscr{D}), t) = 1 + 3t + 6t^2 + 10t^3 + 10t^4 + 6t^5 + 3t^6 + t^7.$$

There are no nonzero elements in $\mathfrak{A}(\mathscr{D})$ of degree less than 4 as the dimensions of $\mathbb{F}_2[x, y, z]_i$ are 1, 3, 6, 10 for $i = 0, 1, 2, 3$. Since the dimension of $\mathbb{F}_2[x, y, z]_4$ as an $\mathbb{F}_2$-vector space is 15 we also see that the ideal $\mathfrak{A}(\mathscr{D})$ must have 5 generators of degree 4. These may be determined by finding linear combinations of the rows that add up to zero, and reinterpreting in terms of monomials. For example it is not difficult to see that the rows labeled by the exponent sequences

3, 1, 0   3, 0, 1   1, 0, 3   0, 1, 3

add up to zero. This says that the form

$$x^3y + x^3z + xz^3 + yz^3$$

belongs to $\mathfrak{A}(\mathscr{D})$. Continuing in this way one finds that the five polynomials

$$f_1 = x^3y + x^3z + xz^3 + yz^3$$
$$f_2 = xy^3 + y^3z + xz^3 + yz^3$$
$$f_3 = xy^2z + xyz^2 + xy^3 + x^3y + xz^3 + x^3z$$
$$f_4 = x^2yz + xy^2z + x^3z + xz^3 + y^3z + yz^3$$
$$f_5 = x^4 + y^4 + z^4 + x^2y^2 + x^2z^2 + y^2z^2 + x^2yz + xy^2z + xyz^2 + x^3y + x^3z$$

belong to the ideal $\mathfrak{A}(\mathscr{D})$ in $\mathbb{F}_2[x, y, z]$. Poincaré series considerations show that no further generators are needed and $f_1, \ldots, f_5$ form a minimal set of ideal generators for $\mathfrak{A}(\mathscr{D})$.

This ideal provides an example to show that complete intersections do not generate the semigroup of Poincaré duality quotients of $\mathbb{F}_2[V]$ under connected sum as V ranges over the finite dimensional $\mathbb{F}_2$-vector spaces. To justify this remark we introduce a lemma.

LEMMA VI.2.1: *If $H = \mathbb{F}[V]/I$ is a Poincaré duality quotient of $\mathbb{F}[V]$ that can be written as a nontrivial connected sum, then I contains a nonzero quadratic form.*

PROOF: Write $H = H' \# H''$ where

$$H' = \mathbb{F}[z_1', \ldots, z_{m'}']/I'$$
$$H'' = \mathbb{F}[z_1'', \ldots, z_{m''}'']/I'',$$

$m', m'' > 0$, and neither I' nor I'' contains any linear forms. Then I contains the nonzero quadratic form $z_{m'}' z_{m''}''$. □

So to show that the complete intersections do not generate the semigroup of Poincaré duality quotients of $\mathbb{F}_2[V]$ under the connected sum operation one needs an irreducible $\overline{\mathbb{F}[V]}$-primary ideal containing no quadratic forms. Since $\mathfrak{A}(\mathscr{D})$ contains no nonzero quadratic forms the corresponding algebra $\mathbb{F}_2[x, y, z]/\mathfrak{A}(\mathscr{D})$ cannot be written as a nontrivial connected sum of Poincaré duality quotients. It is clearly not a complete intersection.

VI.3 Regular ideals

In this section we study **regular ideals**, by which we mean ideals generated by a regular sequence. We examine the behavior of $\Xi = (I : -)$ on the regular over ideals of a regular ideal I. If we restrict attention to the regular over ideals with the same grade[7] as I we can establish an analog of Theorem I.2.1 which gives an explicit computation[8] of a transition element a for I over J, i.e., an element a with $(I : J) = (a) + I$ and $J = (I : a)$. This applies to the special case where I and J are parameter ideals in a Cohen–Macaulay algebra, since in such an algebra parameter ideals are generated by regular sequences. Here is one formulation of what we need (see e.g. [105] Satz 3). The proof is a variant of that of [87] Theorem 6.5.1.

PROPOSITION VI.3.1: *Let $I \subseteq J$ be ideals in a commutative graded connected algebra A over the field $\mathbb{F}$. Suppose that $I = (u_1, \ldots, u_r)$ and $J = (w_1, \ldots, w_r)$ where $u_1, \ldots, u_r$ and $w_1, \ldots, w_r$ are regular sequences*

[7] The grade of an ideal is the length of the longest regular sequence it contains (see e.g. [11] Section 1.2).

[8] This would appear to have been known to F. S. Macaulay ([49] Section 73), but it is hard to say due to the enormous difference between his terminology and our own.

in A. Write

$$\begin{bmatrix} u_1 \\ \vdots \\ u_r \end{bmatrix} = \begin{bmatrix} a_{1,1} & \cdots & a_{1,r} \\ \vdots & \ddots & \vdots \\ a_{r,1} & \cdots & a_{r,r} \end{bmatrix} \cdot \begin{bmatrix} w_1 \\ \vdots \\ w_r \end{bmatrix}.$$

where $[a_{i,j}]$ *is an* $r \times r$ *matrix with entries in A. If* $a = \det[a_{i,j}]$ *then* $(I : J) = (a) + I$ *and* $J = (I : a)$.

PROOF: If we pass down from A to A/I then what needs to be proved are $\mathrm{Ann}_{A/I}(J/I) = (a)$ and $\mathrm{Ann}_{A/I}(a) = J/I$. To this end introduce the subalgebra of A generated by $u_1, \ldots, u_r$, which we call R, and the subalgebra generated by $w_1, \ldots, w_r$, which we call S. Both R and S are polynomial algebras, viz.,

$$R = \mathbb{F}[u_1, \ldots, u_r]$$
$$S = \mathbb{F}[w_1, \ldots, w_r],$$

since $u_1, \ldots, u_r$ and $w_1, \ldots, w_r$ are regular sequences ([87] Proposition 6.2.1), and A is free as a module over both R and S ([87] Corollary 6.2.8).

If we regard A as a module over $R \otimes S$ then standard change of rings arguments ([15] Proposition 4.1.1) yield isomorphisms

$$\mathrm{Tor}_*^S(\mathbb{F}, A/I) \cong \mathrm{Tor}_*^{R\otimes S}(\mathbb{F}, A) \cong \mathrm{Tor}_*^R(\mathbb{F}, A/J).$$

Since $u_1, \ldots, u_r \in I \subseteq J$ the R-module structure on A/J is trivial, i.e., factors through the augmentation map $R \longrightarrow \mathbb{F}$. Therefore by using a Koszul complex (see e.g. [87] Section 6.2) to resolve $\mathbb{F}$ as an R-module we obtain

$$\mathrm{Tor}_*^R(\mathbb{F}, A/J) \cong \mathrm{Tor}_*^R(\mathbb{F}, \mathbb{F}) \otimes A/J \cong E(su_1, \ldots, su_r) \otimes A/J,$$

where $E(su_1, \ldots, su_n)$ denotes a bigraded exterior algebra on the variables[9] $su_1, \ldots, su_n$ of bidegrees $(1, \deg(u_1)), \ldots, (1, \deg(u_n))$. On the other hand using a Koszul resolution of $\mathbb{F}$ as an S-module, we find that

$$\mathrm{Tor}_*^S(\mathbb{F}, A/I) \cong H_*(E(sw_1, \ldots, sw_r) \otimes A/I; \partial),$$

where the derivation ∂ is characterized by

$$\begin{aligned} \partial(sw_i \otimes 1) &= 1 \otimes w_i & i &= 1, \ldots, r \\ \partial(1 \otimes c) &= 0 & \forall\, c &\in A/I. \end{aligned}$$

This allows us to compute $\mathrm{Tor}_r^S(\mathbb{F}, A/I)$ as follows: a chain of homological degree r in this complex is of the form $sw_1 \wedge \cdots \wedge sw_r \otimes c$ for some $c \in A/I$. The boundary of such a chain is

$$\partial(sw_1 \wedge \cdots \wedge sw_r \otimes c) = \sum_{i=1}^{r} (-1)^i sw_1 \wedge \cdots \wedge \widehat{sw_i} \wedge \cdots \wedge sw_r \otimes (w_i \cdot c),$$

[9] This notation comes from algebraic topology and is derived from the suspension homomorphism (see e.g. [84]).

with the usual convention that the term under the $\widehat{}$ is omitted, and $\wedge$ denotes the product in the exterior algebra. So the chain $sw_1 \wedge \cdots \wedge sw_r \otimes c$ of homological degree r is a cycle if and only if $w_1c = \cdots w_rc = 0$ in A/I, i.e., c annihilates $J/I \subset A/I$. Since there are no nonzero boundaries of degree r we obtain

$$\mathrm{Tor}_r^S(\mathbb{F}, A/I) \cong (sw_1 \wedge \cdots \wedge sw_r) \cdot \mathrm{Ann}_{A/I}(J/I).$$

Under the change of rings isomorphism

$$\mathrm{Tor}_*^R(\mathbb{F}, A/J) \longrightarrow \mathrm{Tor}_*^S(\mathbb{F}, A/I)$$

the element $su_i \in \mathrm{Tor}_1^R(\mathbb{F}, A/J)$ is mapped to $\sum_{i=1}^r a_{i,j} sw_j \in \mathrm{Tor}_1^S(\mathbb{F}, A/I)$, and hence $su_1 \wedge \cdots \wedge su_r$ is mapped to $(sw_1 \wedge \cdots \wedge sw_r) \cdot a$ by the definition of the determinant. Therefore $\mathrm{Ann}_{A/I}(J/I) = (a)$ is the principal ideal generated by a in A/I.

The second assertion, that $\mathrm{Ann}_{A/I}(a) = J/I$, follows from Lemma I.1.1. □

NOTATION: *If $A \hookrightarrow B$ is an extension of algebras and $I \subset A$ an ideal then the extension of I to B, viz., $B \cdot I$ will be denoted by I^{ex}.*

Proposition VI.3.1 implies that a transition element for a pair of regular ideals of the same grade is in a certain sense independent of the containing algebra. Specifically we have the following.

COROLLARY VI.3.2: *Suppose that A is a commutative graded connected algebra over the field $\mathbb{F}$ and $K \subseteq L \subset A$ regular ideals of the same grade. Let $K = (u_1, \ldots, u_r)$ and $L = (w_1, \ldots, w_r)$, where $u_1, \ldots, u_r$ and $w_1, \ldots, w_r$ are regular sequences in A. If $B \supseteq A$ is an extension algebra of A in which $u_1, \ldots, u_r$ and $w_1, \ldots, w_r$ remain regular sequences then a transition element for L over K in A is also a transition element for K^{ex} over L^{ex} in B.*

PROOF: Write

$$\begin{bmatrix} u_1 \\ \vdots \\ u_r \end{bmatrix} = \begin{bmatrix} a_{1,1} & \cdots & a_{1,r} \\ \vdots & \ddots & \vdots \\ a_{r,1} & \cdots & a_{r,r} \end{bmatrix} \cdot \begin{bmatrix} w_1 \\ \vdots \\ w_r \end{bmatrix}$$

where $[a_{i,j}]$ is an $r \times r$ matrix with entries in A. Note that it does not matter whether we compute $\det[a_{i,j}]$ in A or B. Any transition element for L over K in A is equivalent modulo K to this determinant. □

The condition of Corollary VI.3.2 is fulfilled if for example A and B are Cohen–Macaulay and $A \hookrightarrow B$ is a finite extension. Proposition VI.3.1 also has the following neat little consequence, which is no doubt well known.[10]

COROLLARY VI.3.3: *Suppose that* $a_1, \ldots, a_n$ *is a regular sequence in a commutative graded connected algebra A over the field* $\mathbb{F}$. *If* $a \in A$ *satisfies* $a \cdot (a_1 \cdots a_n)^s \in (a_1^{s+1}, \ldots, a_n^{s+1})$ *then* $a \in (a_1, \ldots, a_n)$.

PROOF: There is the inclusion $(a_1^{s+1}, \ldots, a_n^{s+1}) \subset (a_1, \ldots, a_n)$ and moreover $(a_1, \ldots, a_n) = \big((a_1^{s+1}, \ldots, a_n^{s+1}) : (a_1 \cdots a_n)^s\big)$. By hypothesis $a \in \big((a_1^{s+1}, \ldots, a_n^{s+1}) : (a_1 \cdots a_n)^s\big)$ so the result follows. □

In a Cohen–Macaulay algebra parameter ideals are minimally generated by regular sequences, so we have the following special case of Proposition VI.3.1.

COROLLARY VI.3.4: *Let* $I \subseteq J$ *be parameter ideals in a commutative graded connected Cohen–Macaulay algebra A over the field* $\mathbb{F}$. *Choose minimal ideal bases* $u_1, \ldots, u_r$ *for I and* $w_1, \ldots, w_r$ *for J and write*

$$\begin{bmatrix} u_1 \\ \vdots \\ u_r \end{bmatrix} = \begin{bmatrix} a_{1,1} & \cdots & a_{1,r} \\ \vdots & \ddots & \vdots \\ a_{r,1} & \cdots & a_{r,r} \end{bmatrix} \cdot \begin{bmatrix} w_1 \\ \vdots \\ w_r \end{bmatrix},$$

where $[\,a_{i,j}\,]$ *is an* $r \times r$ *matrix with entries in A. If* $a = \det[\,a_{i,j}\,]$ *then* $(I : J) = (a) + I$ *and* $J = (I : a)$. □

We draw one final consequence from Proposition VI.3.1 that will be of use in our treatment of the *lying over* for $\mathfrak{m}$-primary irreducible ideals in Gorenstein algebras. It is a variant of Corollary VI.3.2. To wit, a transition element for one parameter ideal over another is independent of the containing algebra provided the algebra is Cohen–Macaulay. This is because all Proposition VI.3.2 requires is that the regularity property be preserved on passage to a larger algebra, and parameter ideals in a Cohen–Macaulay algebra are regular.

COROLLARY VI.3.5: *Let* $\mathbb{F}$ *be a field, and* $A \overset{\varphi}{\hookrightarrow} B$ *be a finite extension of commutative graded connected Cohen–Macaulay algebras over* $\mathbb{F}$, *and* $P \subset Q$ *parameter ideals in A. Then there is an element* $a \in A$ *such that*

$$(P : Q) = (a) + P$$

and in addition

$$(P^{\mathrm{ex}} : Q^{\mathrm{ex}}) = (a) + P^{\mathrm{ex}}.$$

[10] At least to those who know it well.

PROOF: The ideals P and P^{ex} are generated by the same regular sequence, as are Q and Q^{ex}, so the result is a special case of Corolllary VI.3.4. □

VI.4 Lying over for irreducible ideals

If $A \hookrightarrow B$ is a finite ring extension of Noetherian rings and $\mathfrak{p} \subset A$ is a prime ideal, then the classical Lying Over Theorem (see e.g. [4] Theorem 3.1.16) says that there is a prime ideal $\mathfrak{P} \subset B$ with $\mathfrak{P} \cap A = \mathfrak{p}$. In this case one says **$\mathfrak{P}$ lies over $\mathfrak{p}$**. If A and B are commutative connected graded Gorenstein algebras we show how to exploit Theorem I.2.1 to prove in favorable cases an analog of this classical result for irreducible ideals. We begin with the case where the algebras have Krull dimension zero, i.e., where A and B are Poincaré duality algebras.

PROPOSITION VI.4.1: *If $H' \subset H''$ is an inclusion of Poincaré duality algebras, and $I' \subset H'$ an irreducible ideal, then there exists an irreducible ideal $I'' \subset H''$ such that $I'' \cap H' = I'$.*

PROOF: By Corollary I.2.2 $\text{Ann}_{H'}(I') = (h')$ for some element $h' \in H'$. If we set $I'' = \text{Ann}_{H''}(h') \subset H''$ a second application of Corollary I.2.2 shows I'' is irreducible. The verification that $I'' \cap H' = I'$ is routine. □

EXAMPLE 1: If $H' \subset H''$ is an inclusion of Poincaré duality algebras and $I' \subset H'$ is an irreducible ideal, then the extended ideal $I'^{\text{ex}} \subset H''$ need not be irreducible. To see this let $H'' = \mathbb{F}[x, y]/(x^2, y^2)$. Then the subalgebra H' generated by the image of xy in the quotient is a Poincaré duality subalgebra. The ideal I' in H' generated by the image of xy is irreducible, since the quotient algebra is $\mathbb{F}$. However the extended ideal $I'^{\text{ex}} \subset H''$ is not irreducible as $\mathbb{F}[x, y]/(x^2, xy, y^2)$ is not a Poincaré duality algebra. Topologists should recognize these algebras as $H'' = H^*(S^1 \times S^1; \mathbb{F})$, $H' = H^*(S^2; \mathbb{F})$, where the inclusion $H' \hookrightarrow H''$ is induced by collapsing onto the top cell of the torus. The quotient H''/I'^{ex} is the cohomology of the wedge of two circles which is not a manifold.

If we attempt to extend Proposition VI.4.1 from Poincaré duality algebras (i.e., graded Gorenstein algebras of Krull dimension zero) to finite extensions $A \hookrightarrow B$ of Gorenstein algebras of positive Krull dimension by passage to the quotients by a system of parameters we encounter a problem. Namely, if $Q \subset A$ is a parameter ideal and $Q^{\text{ex}} \subset B$ the extension of Q to B, then the induced map $A/Q \longrightarrow B/Q^{\text{ex}}$ need not be a monomorphism. It is a monomorphism if and only if $Q^{\text{ex}} \cap A = Q$. The following example illustrates this can fail.

EXAMPLE 2: Let A be the subalgebra of $\mathbb{F}[x]$ generated by x^2 and x^3 and $B = \mathbb{F}[x]$. The ideal $Q = (x^2) \subset A$ is a parameter ideal in A, and $A/Q \cong \mathbb{F}[x^3]/(x^6)$ is a Poincaré duality algebra, so Q is irreducible. The extended ideal Q^{ex} in B contains the element x^3 which belongs to A but not to Q. Hence $A \cap Q^{\text{ex}} \neq Q$.

Note that the algebra A is not integrally closed. If $\mathbb{F}$ has characteristic zero and A is an integrally closed integral domain, this type of behavior cannot occur (see [30] Section 2 G). If the characteristic of $\mathbb{F}$ is $p \neq 0$ then it can occur. Perhaps the most striking example originates in invariant theory. Recall that the ring of invariants of a finite group is always integrally closed. If $p < n$ then the ideal $(e_1, \ldots, e_n)$ generated by the elementary symmetric polynomials $e_1, \ldots, e_n$ in the ring of invariants of the alternating group $\mathbb{F}[z_1, \ldots, z_n]^{A_n}$ if first extended up to $\mathbb{F}[z_1, \ldots, z_n]$ and then restricted back to $\mathbb{F}[z_1, \ldots, z_n]^{A_n}$ becomes the entire Hilbert ideal $\mathfrak{h}(A_n) = (e_1, \ldots, e_n, \nabla_n)$ (see e.g. [25] Section 11, [83], and [93].) The ideal $\mathfrak{h}(A_n)$ properly contains $(e_1, \ldots, e_n)$.

To overcome the difficulty posed by examples such as Example 2 we require an additional assumption on the extension $A \hookrightarrow B$ (see e.g. [31] Proposition 10).

LEMMA VI.4.2: *Let $\varphi : A \hookrightarrow B$ be a finite extension of commutative graded connected algebras over the field $\mathbb{F}$ and $I \subset A$ an ideal. If φ splits as a map of A-modules then $I^{\text{ex}} \cap A = I$.*

PROOF: Choose a set of generators for the ideal I and let C be the subalgebra of A that they generate. Choose an A-module homomorphism $\sigma : B \longrightarrow A$ splitting φ. It is also a splitting of C-modules. The commutative diagram, in which $\tilde{\varphi}$ is induced by φ,

$$\begin{array}{ccc} A/I & \xrightarrow{\tilde{\varphi}} & B/I^{\text{ex}} \\ \cong \downarrow & & \downarrow \cong \\ \mathbb{F} \otimes_C A & \xrightarrow{\mathbb{F}\otimes\varphi} & \mathbb{F} \otimes_C B. \\ \uparrow & \xleftarrow{\mathbb{F}\otimes\sigma} & \end{array}$$

shows that $A/I \longrightarrow B/I^{\text{ex}}$ is a monomorphism, and since $(A \cap I^{\text{ex}})/I$ is the kernel of this map, the result follows. □

Here is the first of the *lying over* results for irreducible ideals.

PROPOSITION VI.4.3: *Let $\varphi : A \hookrightarrow B$ be a finite extension of commutative graded connected Gorenstein algebras over the field $\mathbb{F}$. Assume that φ splits as a map of A-modules and $I \subset A$ is an $\overline{A}$-primary irreducible ideal. Then there exists a $\overline{B}$-primary irreducible ideal lying over I.*

PROOF: Choose a system of parameters $a_1, \ldots, a_d$ for A that is contained in I. Since $B \supseteq A$ is a finite extension $a_1, \ldots, a_d$ are also a system of parameters for B. Since A and B are Gorenstein the quotients A/Q and B/Q^{ex} are Poincaré duality algebras. By Lemma VI.4.2 $Q^{\text{ex}} \cap A = Q$ so the map $A/Q \longrightarrow B/Q^{\text{ex}}$ is a monomorphism. The ideal $I/Q \subset A/Q$ is irreducible, so by Proposition VI.4.1 there is an irreducible ideal in B/Q^{ex} lying over it. If $J \subset B$ is the preimage of this ideal under the quotient map $B \longrightarrow B/Q^{\text{ex}}$ then $J \subset B$ is $\overline{B}$-primary, irreducible, and $A \cap J = I$. □

The restriction imposed in Proposition VI.4.3 on a finite extension of commutative graded connected Gorenstein algebras $\varphi : A \hookrightarrow B$ over the field $\mathbb{F}$ is not severe. The map φ splits as a map of A-modules in several interesting cases. First, if A is a polynomial algebra, say $A = \mathbb{F}[a_1, \ldots, a_d]$, then $a_1, \ldots, a_d$ are a system of parameters for B. Since B is Cohen–Macaulay they form a regular sequence, and B becomes a free A-module, so φ splits as a map of A-modules. Further examples are provided by invariant theory. If $\rho : G \hookrightarrow \text{SL}(n, \mathbb{F})$, $|G| \in \mathbb{F}^{\times}$, and $H < G$ is a subgroup, then $A = \mathbb{F}[V]^G$ and $B = \mathbb{F}[V]^H$ are Gorenstein by a theorem of K. Watanabe [103] (see also [68] Theorem 5.7.6). The relative transfer $\text{Tr}^G_H : \mathbb{F}[V]^H \longrightarrow \mathbb{F}[V]^G$ is an $\mathbb{F}[V]^G$-module homomorphism, and after normalizing it by dividing by $|G : H|$ it becomes a splitting to the inclusion $\mathbb{F}[V]^G \hookrightarrow \mathbb{F}[V]^H$ (see e.g. [87] Section 2.4). In the modular case the inclusion need not split and these results may fail. Again examples are provided by the invariants of the tautological permutation representation of the alternating group A_n in the modular case ([25] Section 11, [83], and [93]). There is no irreducible ideal in $\mathbb{F}[z_1, \ldots, z_n]$ whose intersection with $\mathbb{F}[z_1, \ldots, z_n]^{A_n}$ is the ideal $(e_1, \ldots, e_n) \subset \mathbb{F}[z_1, \ldots, z_n]^{A_n}$ generated by the elementary symmetric polynomials. In $\mathbb{F}[z_1, \ldots, z_n]^{A_n}$ this ideal is regular, hence irreducible, since it is a parameter ideal and $\mathbb{F}[z_1, \ldots, z_n]^{A_n}$ is a complete intersection, so Cohen–Macaulay.

For a finite extension $\varphi : A \hookrightarrow B$ of a pair of commutative graded connected Gorenstein algebras and $I \subset A$ an $\overline{A}$-primary irreducible ideal, if φ splits as a map of A-modules, then the proof of Proposition VI.4.3 constructs for each parameter ideal $Q \subseteq I$ a $\overline{B}$-primary irreducible ideal $J \subset B$, which lies over I. Here is how to make this construction explicit.

CONSTRUCTION VI.4.4: *Let $\varphi : A \hookrightarrow B$ be a finite extension of commu-*

tative graded connected Gorenstein algebras and $I \subset A$ an $\overline{A}$-primary irreducible ideal. Assume φ splits as a map of A-modules. Let $Q \subseteq I$ be a parameter ideal for A. By Theorem I.2.1 $(Q : I) = (a_Q) + Q$ for some element $a_Q \in A$. To obtain an ideal $J_Q \subset B$ lying over I set $J_Q = (Q^{ex} : a_Q) \subset B$. By construction J_Q is $\overline{B}$-primary and Proposition VI.4.3 assures us that J_Q lies over I. The ideal $Q^{ex} \subset B$ is a parameter ideal in B, so irreducible and $\overline{B}$-primary since B is Gorenstein. The ideal J_Q is a $\overline{B}$-primary over ideal of Q^{ex} and $(Q^{ex} : J_Q) = (a_Q) + Q^{ex}$, so J_Q is irreducible by Theorem I.2.1.

The ideals arising from Construction VI.4.4 play a central role in the sequel. For this reason we introduce the following terminology.

DEFINITION: *Let $\varphi : A \hookrightarrow B$ be a finite extension of commutative graded connected Gorenstein algebras and assume that φ splits as a map of A-modules. If $I \subset A$ an $\overline{A}$-primary irreducible ideal and $Q \subseteq I$ a parameter ideal for A contained in I, then the ideal J_Q obtained from Construction VI.4.4 is called a* **fit** *extension of I to B.*

Not every $\overline{B}$-primary irreducible ideal lying over I need be fit. Here is a simple example to illustrate this.

EXAMPLE 3: Suppose that $A = \mathbb{F}[x^2] \subset \mathbb{F}[x] = B$ and $I = \overline{A}$ is the maximal ideal of A. The ideal $J = (x) \subset B$ lies over I, is $\overline{B}$-primary and irreducible. However it is not of the form $(a) + Q^{ex}$ for any parameter ideal Q of A and element $a \in A$ since $(a) + Q^{ex}$ cannot contain any nonzero element of degree 1.

If $\mathbb{F}[Y_1, \ldots, Y_n] \hookrightarrow \mathbb{F}[X_1, \ldots, X_n]$ is a finite extension of polynomial algebras then the $\mathfrak{m}$-primary regular ideals in $\mathbb{F}[Y_1, \ldots, Y_n]$ are irreducible. The extension of such an ideal to $\mathbb{F}[X_1, \ldots, X_n]$ is again regular and $\mathfrak{m}$-primary, and Corollary VI.3.2 shows that the ordinary extension of such an ideal is in fact fit.

We next show that fit extensions are unique.

PROPOSITION VI.4.5: *Suppose $\varphi : A \hookrightarrow B$ is a finite extension of commutative graded connected Gorenstein algebras and $I \subset A$ an $\overline{A}$-primary irreducible ideal. Suppose that φ splits by a map of A-modules. Then I has a unique* fit *extension.*

PROOF: Let $Q', Q'' \subset I$ be parameter ideals and $a_{Q'}, a_{Q''}$ transition elements for I over J' and J'' respectively. Let $J_{Q'}, J_{Q''} \subset B$ be the fit extensions of I defined by $J_{Q'} = ((Q')^{ex} : a_{Q'})$ and $J_{Q''} = ((Q'')^{ex} : a_{Q''})$.

Consider first the special case where $Q' \subseteq Q''$. Then we may write $Q'' =$

$(Q' : a)$ for some element $a \in A$ using Theorem I.2.1. Therefore

$$(Q' : a_{Q'}) = I = (Q'' : a_{Q''}) = ((Q' : a) : a_{Q''}) = (Q' : a a_{Q''}).$$

So without loss of generality we may suppose that $a_{Q'} = a a_{Q''}$. Then

$$\begin{aligned} J_{Q'} &= ((Q')^{\mathrm{ex}} : a_{Q'}) = ((Q')^{\mathrm{ex}} : a a_{Q''}) \\ &= (((Q')^{\mathrm{ex}} : a) : a_{Q''}) = ((Q' : a)^{\mathrm{ex}} : a_{Q''}) = ((Q'')^{\mathrm{ex}} : a_{Q''}) = J_{Q''}. \end{aligned}$$

In the general case where Q' and Q'' are arbitrary parameter ideals contained in I we may choose a third parameter ideal Q that is contained in both of them (e.g., $Q = (a_1^k, \ldots, a_r^k)$ for k large, where $a_1, \ldots, a_r \in A$ is any system of parameters). By what we just established $J_{Q'} = J_Q = J_{Q''}$ and the fit extension is therefore unique. □

DEFINITION: *If $\varphi : A \hookrightarrow B$ is an A-split finite extension of commutative graded connected Gorenstein algebras and $I \subset A$ an $\overline{A}$-primary irreducible ideal then the unique* fit *extension of I to B is denoted by I^{fit}.*

For an A-split finite extension $A \hookrightarrow B$ of commutative graded connected Gorenstein algebras the fit extension I^{fit} of an $\overline{A}$-primary irreducible ideal $I \subset A$ is a particular choice of $\overline{B}$-primary irreducible of B lying over I. If B is a free A-module and $I \subset A$ is an $\overline{A}$-primary regular ideal then so is $I^{\mathrm{ex}} \subset B$ and moreover Corollary VI.3.2 shows that $I^{\mathrm{ex}} = I^{\mathrm{fit}}$. So we have the following result.

COROLLARY VI.4.6: *If $\varphi : A \hookrightarrow B$ is an A-split finite extension of commutative graded connected Gorenstein algebras and $I \subset A$ an $\overline{A}$-primary regular ideal then the unique* fit *extension of I to B is the ideal I^{ex}.* □

Before discussing fit extensions in more detail we record some lemmas that will be useful.

LEMMA VI.4.7: *Suppose A is a commutative graded connected algebra and $K \subseteq I \subseteq L \subset A$ are $\overline{A}$-primary irreducible ideals. If*

$$(K : I) = (u) + K$$

and

$$(I : L) = (w) + I$$

then

$$(K : L) = (uw) + K.$$

PROOF: By Theorem I.2.1 we have $L = (I : w)$ so using basic properties of $(- : K)$ from Section I.2 (to do so we may harmlessly replace the principal ideal (w) by the over ideal $(w) + K$ of K without changing the outcome of the result) we have

$$(K:L) = (K:(I:w)) = (K:(I:((w)+K))) = ((w)+K)\cdot(K:I)$$
$$= ((w)+K)\cdot((u)+K) = (uw)+K$$

as required. □

LEMMA VI.4.8: *Let $A \hookrightarrow B$ be an A-split finite extension of commutative graded connected Gorenstein algebras and let $I \subseteq J \subset A$ be $\overline{A}$-primary irreducible ideals. Suppose that $(I \underset{A}{:} J) = (u) + I$ for some $u \in A$. Then*

$$(I^{\text{fit}} \underset{B}{:} J^{\text{fit}}) = (u) + I^{\text{fit}}.$$

PROOF: Choose a parameter ideal $Q \subseteq I$ and a transition element $a_I \in A$ for I over Q, i.e., $(Q:I) = (a_I) + Q$. By Lemma VI.4.7 $(Q:J) = (ua_I) + Q$. Therefore

$$J^{\text{fit}} = (Q^{\text{ex}} : ua_I) = ((Q^{\text{ex}} : a_I) : u) = (I^{\text{fit}} : u).$$

So the result follows from Theorem I.2.1. □

LEMMA VI.4.9: *Suppose $\varphi : A \hookrightarrow B$ is an A-split finite extension of commutative graded connected Gorenstein algebras. The passage from an $\overline{A}$-primary irreducible ideal $I \subset A$ to $I^{\text{fit}} \subset B$ preserves inclusions, i.e., if $I' \subseteq I'' \subset A$ are $\overline{A}$-primary irreducible ideals then $(I')^{\text{fit}} \subseteq (I'')^{\text{fit}}$.*

PROOF: To see this choose a parameter ideal $Q \subseteq I'$ and note that by Lemma VI.4.7 we may also choose transition elements a'_Q and a''_Q for I' and I'' respectively such that $a''_Q = aa'_Q$ where $(I' : I'') = (a) + I'$. Then $(I'')^{\text{fit}} = (Q^{\text{ex}} : a''_Q) = (Q^{\text{ex}} : a'_Q a) = ((Q^{\text{ex}} : a'_Q) : a) = ((I')^{\text{fit}} : a) \supseteq (I')^{\text{fit}}$. □

Since the fit extension I^{fit} of I lies over I the map $A/I \longrightarrow B/I^{\text{fit}}$ induced by the inclusion $A \hookrightarrow B$ is also monic. The cokernel of the map $A/I \hookrightarrow B/I^{\text{fit}}$ is the algebra $(B/I^{\text{fit}})/\!/(A/I) \cong B/(\overline{A}^{\text{ex}} + I^{\text{fit}})$. The case $\overline{A}^{\text{ex}} = \overline{A}^{\text{fit}}$ is particularly attractive, since then $\overline{A}^{\text{ex}} = \overline{A}^{\text{fit}} \supseteq I^{\text{fit}}$ so this quotient algebra $(B/I^{\text{fit}})/\!/(A/I)$ is the Poincaré duality algebra $B/\!/A$, and we obtain a coexact sequence of Poincaré duality algebras

$$\mathbb{F} \longrightarrow A/I \longrightarrow B/I^{\text{fit}} \longrightarrow B/\!/A \longrightarrow \mathbb{F}$$

which is split. Let us analyze this situation a bit more carefully.

Suppose $\varphi : A \hookrightarrow B$ is an A-split finite extension of commutative graded connected Gorenstein algebras and $I \subset A$ an $\overline{A}$-primary irreducible ideal with $I^{\text{fit}} = I^{\text{ex}}$. By Lemma VI.4.8 there is an element $u \in A$ such that

$$(I \underset{A}{:} \overline{A}) = (u) + I$$

and

$$(I^{\text{fit}} \underset{B}{:} \overline{A}^{\text{fit}}) = (u) + I^{\text{fit}}.$$

Passing down to the quotient algebra A/I gives $\mathrm{Ann}_{A/I}(\overline{A}/I) = (u)$ so, since $\overline{A}/I = \overline{A/I}$, we conclude from the definition of Poincaré duality that u represents a fundamental class of A/I. Passing down to B/I^{fit} tells us that $\mathrm{Ann}_{B/I^{\text{fit}}}(\overline{A}^{\text{fit}}/I^{\text{fit}}) = (u)$. Since $\overline{A}^{\text{fit}}/I^{\text{fit}}$ is the kernel of the natural map $B/I^{\text{fit}} \longrightarrow (B/I^{\text{fit}})/(\overline{A}^{\text{fit}}/I^{\text{fit}}) \cong B/\!/A$ Corollary I.2.3 tells us that a Poincaré dual $u^\vee$ to u in B/I^{fit} represents a fundamental class of $B/\!/A$. Therefore

$$\text{f-dim}(B/I^{\text{fit}}) = \deg(uu^\vee) = \deg(u) + \deg(u^\vee) = \text{f-dim}(A/I) + \text{f-dim}(B/\!/A).$$

Hence we have proven the following result.

LEMMA VI.4.10: *Suppose $\varphi : A \hookrightarrow B$ is an A-split finite extension of commutative graded connected Gorenstein algebras and that $I \subset A$ is an $\overline{A}$-primary irreducible ideal. If $I^{\text{fit}} = I^{\text{ex}}$ then*

$$\text{f-dim}(B/I^{\text{fit}}) = \text{f-dim}(A/I) + \text{f-dim}(B/\!/A). \quad \square$$

Not every short coexact sequence of Poincaré duality algebras

$$\mathbb{F} \longrightarrow H' \xrightarrow{\varphi'} H \longrightarrow H'' \xrightarrow{\varphi''} \mathbb{F}$$

satisfies the additivity formula

$$\text{f-dim}(H) = \text{f-dim}(H') + \text{f-dim}(H'')$$

of the preceding lemma: not even if the map φ' is split as map of H'-modules. Here is an example to illustrate this.

EXAMPLE 4: Let $A = \mathbb{F}[X]$ with X of degree 3 and $B = \mathbb{F}[z]$ with z of degree 1. Define $\varphi(X) = z^3$ and let $I = (X^2) \subset A$ and $J = (z^4) \subset B$. Both the ideals I and J are $\mathfrak{m}$-primary and irreducible. The inclusion $\varphi : A \hookrightarrow B$ splits as a map of A-modules because B is a free A-module with basis $1, z, z^2$. A splitting map is given by sending $1 \in \mathbb{F}[z] = B$ to $1 \in \mathbb{F}[X] = A$ and both z and z^2 to zero. Note that J lies over I, so the sequence of Poincaré duality quotient algebras[11]

$$\mathbb{F} \longrightarrow \mathbb{F}[X]/(X^2) \xrightarrow{\alpha} \mathbb{F}[z]/(z^4) \longrightarrow \mathbb{F}[z]/(z^3) \longrightarrow \mathbb{F},$$

is coexact and splits as a sequence of $\mathbb{F}[X]/(X^2)$-modules. However, it is not of the form considered in Lemma VI.4.10 since $J \neq I^{\text{fit}} = I^{\text{ex}}$, and

$$3 = \text{f-dim}(\mathbb{F}[z]/(z^4)) \neq \text{f-dim}(\mathbb{F}[X]/(X^2)) + \text{f-dim}(\mathbb{F}[z]/(z^3)) = 3 + 2 = 5,$$

---

[11] This sequence corresponds to the cohomology exact sequence of the cofibration

$$\mathbb{CP}(2) \hookrightarrow \mathbb{CP}(3) \longrightarrow S^6$$

after regrading.

so the additivity formula of Lemma VI.4.10 is not satisfied.

Short coexact sequences of Poincaré duality algebras with the additivity property for formal dimensions have a special property expressed in the next lemma.

LEMMA VI.4.11: *Suppose that*

$$\mathbb{F} \longrightarrow H' \xrightarrow{\psi'} H \xrightarrow{\psi''} H'' \longrightarrow \mathbb{F}$$

is a coexact sequence of Poincaré duality algebras. If

$$\text{f-dim}(H') + \text{f-dim}(H'') = \text{f-dim}(H)$$

then H is a free H'-module.

PROOF: The ideal $(\overline{H}')^{\text{ex}} = \ker(\psi'') \subset H$ is irreducible and $\overline{H}$-primary by Lemma I.1.3. By Corollary I.2.2 there is a $u \in H$ such that $(u) = \text{Ann}_H((\overline{H}')^{\text{ex}})$ and $(\overline{H}')^{\text{ex}} = \text{Ann}_H(u)$. Since $H'' = H/\text{Ann}_H(u)$ Corollary I.2.3 tells us that $\text{f-dim}(H'') = \text{f-dim}(H) - \deg(u)$. But $\text{f-dim}(H'') = \text{f-dim}(H) - \text{f-dim}(H')$ by hypothesis so it follows that $\deg(u) = \text{f-dim}(H')$. If $[H'] \in H'$ is a fundamental class then $[H']$ annihilates $\overline{H}'$ in H' so $\psi'([H'])$ annihilates $(\overline{H}')^{\text{ex}}$ in H and therefore $\psi'([H']) \in (u)$. As u and $\psi'([H'])$ have the same degree and $\psi([H'])$ is a multiple of u they must be proportional, so they generate the same principal ideal in H. Therefore $\text{Ann}_H((\overline{H}')^{\text{ex}}) = (\psi([H'])$.

We may regard $H'' = \mathbb{F} \otimes_{H'} H$ as the vector space of H'-indecomposable elements in H. Therefore we may construct an epimorphism from the free H' module with the same indecomposables, viz., $\varphi : H' \otimes H'' \longrightarrow H$, onto H such that the induced map on indecomposables is the identity.

We claim that this map φ is an isomorphism. To prove this it is enough to show it is a monomorphism which we do by downward induction on the degree. It is certainly the case in degrees strictly larger than degree $d = \text{f-dim}(H) = \text{f-dim}(H') + \text{f-dim}(H'')$ as both modules are zero in such degrees. In degree d it follows from the fact that φ is an epimorphism, and $\dim_{\mathbb{F}}((H' \otimes H'')_d) = 1 = \dim_{\mathbb{F}}(H_d)$.

So suppose that we have shown that φ is monic in all degrees strictly greater than k for some $k < d$. Let

$$0 \neq \sum h'_j \otimes h''_j \in H' \otimes H''$$

be a nonzero element of degree k written in an irredundant form: namely no h'_j is a nonzero scalar multiple of an h'_i for $i \neq j$ and likewise for the elements $\{h''_j\}$. We consider two cases.

CASE 1: There is an i such that $\deg(h_i') < \text{f-dim}(H')$. Using Poincaré duality we may find an element $(h_i')^\vee \in H'$ such that

$$h_j'(h_i')^\vee = \begin{cases} [H'] & \text{if } i = j \\ 0 & \text{otherwise.} \end{cases}$$

Then $\deg((h_i')^\vee) > 0$ and

$$(h_i')^\vee\left(\sum h_j' \otimes h_j''\right) = (h_i')^\vee h_i' \otimes h_i'' \neq 0.$$

Since the degree of this element is $> k$ we have

$$0 \neq \varphi((h_i')^\vee h_i' \otimes h_i'') = \varphi\left((h_i')^\vee\left(\sum h_j' \otimes h_j''\right)\right) = (h_i')^\vee \varphi\left(\sum h_j' \otimes h_j''\right)$$

so $\varphi(\sum h_j' \otimes h_j'') \neq 0$.

CASE 2: Suppose $\deg(h_i') = \text{f-dim}(H')$ for all i. In this case, since the representation is irredundant, we may write $[H'] \otimes h''$ for the element in question. Then

$$\varphi([H'] \otimes h'') = [H'] \cdot \varphi(1 \otimes h'').$$

By construction of φ the element $1 \otimes_{H'} \varphi(1 \otimes h'') = h''$ is nonzero in H''. Let $h''^\vee \in H$ be a lift of a Poincaré dual in H'' for it. Then, since $\ker(\psi'') = \text{Ann}_H((\overline{H}')^{\text{ex}}) = (\psi'([H'])$, Corollary I.2.5 implies

$$\varphi([H'] \cdot (1 \otimes h'')) \cdot h''^\vee = [H'] \cdot (\varphi(1 \otimes h'') \cdot h''^\vee) = [H] \neq 0$$

as $\varphi((1 \otimes h'')) \cdot h''^\vee)$ is a fundamental class for H''. Hence $\varphi([H'] \otimes h'') \neq 0$ in this case also. □

PROPOSITION VI.4.12: *Suppose $\varphi : A \hookrightarrow B$ is a finite extension of commutative graded connected Gorenstein algebras and $I \subset A$ an $\overline{A}$-primary irreducible ideal. Assume that φ is split as a map of A-modules and $\overline{A}^{\text{ex}} = \overline{A}^{\text{fit}}$. Then the sequence of Poincaré duality algebras*

$$\mathbb{F} \longrightarrow A/I \longrightarrow B/I^{\text{fit}} \longrightarrow B/\!/A \longrightarrow \mathbb{F}$$

is split coexact, B/I^{fit} is a free A/I-module, and

$$\text{f-dim}(B/I^{\text{fit}}) = \text{f-dim}(A/I) + \text{f-dim}(B/\!/A).$$

PROOF: The discussion preceding Corollary I.2.5 shows $\text{f-dim}(B/I^{\text{fit}}) = \text{f-dim}(A/I) + \text{f-dim}(B/\!/A)$ so the result follows from Lemmas VI.4.10 and VI.4.11. □

A number of other properties are preserved by the passage $I \rightsquigarrow I^{\text{fit}}$. For the sake of completeness we close this section with a short list of a few of these. Their proofs are no more difficult than those already presented in Lemmas VI.4.7 and VI.4.8 and are left to the reader.

(1) Since fitness preserves inclusions, both *going up* and *going down* hold for fit extensions.

(2) Fitness is transitive: i.e., if $A \overset{\varphi}{\hookrightarrow} B \overset{\psi}{\hookrightarrow} C$ are finite extensions of commutative graded connected Gorenstein algebras, where φ splits as a map of A-modules and ψ splits as a map of B-modules, then $(I^{\mathrm{fit}})^{\mathrm{fit}} = I^{\mathrm{fit}}$.

(3) If $A \hookrightarrow B$ is a finite extension of polynomial algebras and the ground field $\mathbb{F}$ has characteristic p, then fitness commutes with the Frobenius, i.e., for an $\overline{A}$-primary irreducible ideal $I \subset A$ we have $(I^{\mathrm{fit}})^{[p]} = (I^{[p]})^{\mathrm{fit}}$. That the ideal $I^{[p]}$ is $\overline{A}$-primary and irreducible is a consequence of Sharp's theorem, Theorem II.6.5, which requires the ambient algebra to be a polynomial algebra.

VI.5 Change of rings and inverse polynomials

In this section and the next we explain another[12] reason why we have introduced Macaulay's theory of irreducible ideals using inverse polynomials. Namely, we plan by these means to establish a number of change of rings results for irreducible ideals using the *lying over* results of Sections VI.3 and VI.4.

The elements of the inverse polynomial algebra $\mathbb{F}[X_1^{-1}, \ldots, X_n^{-1}]_{-k}$ of degree $-k$ define linear maps $\mathbb{F}[X_1, \ldots, X_n]_* \longrightarrow \mathbb{F}[X_1, \ldots, X_n]_{*-k}$ in the following way: to the inverse monomial X^{-E}, with exponent sequence $E \in \mathbb{N}_0$, we assign the linear map $- \cap X^{-E}$ defined by

$$X^F \cap X^{-E} = \begin{cases} X^{F-E} & \text{if } F - E \in \mathbb{N}_0^n \\ 0 & \text{otherwise.} \end{cases}$$

We call this the **cocontraction pairing**. For an arbitrary inverse polynomial θ of degree $-k$ the linear map $- \cap \theta$ is defined in the obvious way by linear extension. If we restrict $- \cap X^{-E}$ to polynomials of degree $|E|$, and identify $\mathbb{F}$ with the homogeneous component of $\mathbb{F}[X_1, \ldots, X_n]$ of degree 0, then the cocontraction pairing defines a linear form denoted by ζ_E. The linear form defined on $\mathbb{F}[X_1, \ldots, X_n]_k$ corresponding to an arbitrary inverse polynomial θ of degree $-k$ is denoted by ζ_θ. The correspondence $\theta \leftrightarrow \zeta_\theta$ allows us to identify $\mathbb{F}[X_1^{-1}, \ldots, X_n^{-1}]_{-k}$ with the space of linear forms on $\mathbb{F}[X_1, \ldots, X_n]_k$.

---

[12] The first reason was to introduce the computational tool of catalecticant matrices which we used for several computations in the text. Catalecticant matrices are discussed in Section VI.2 which contains several illustrative examples.

If $\varphi : \mathbb{F}[Y_1, \ldots, Y_n] \longrightarrow \mathbb{F}[X_1, \ldots, X_n]$ is a map of algebras passing to dual vector spaces we obtain an induced map

$$\varphi^* : \mathbb{F}[X_1^{-1}, \ldots, X_n^{-1}] \longrightarrow \mathbb{F}[Y_1^{-1}, \ldots, Y_n^{-1}]$$

defined by assigning to $\theta \in \mathbb{F}[X_1^{-1}, \ldots, X_n^{-1}]_k$ the linear form

$$\mathbb{F}[Y_1, \ldots, Y_n]_k \xrightarrow{\varphi_k} \mathbb{F}[X_1, \ldots, X_n]_k \xrightarrow{\zeta_\theta} \mathbb{F}.$$

The maps φ, φ^*, and the contraction pairing are related by the formula

$$\zeta_\theta(\varphi(f)) = \zeta_{\varphi^*(\theta)}(f)$$

or

$$\varphi(f) \cap \theta = f \cap \varphi^*(\theta)$$

which holds for all $f \in \mathbb{F}[Y_1, \ldots, Y_n]$ of degree k and $\theta \in \mathbb{F}[X_1^{-1}, \ldots, X_n^{-1}]$ of degree $-k$.

Suppose that $\iota : \mathbb{F}[Y_1, \ldots, Y_n] \subseteq \mathbb{F}[X_1, \ldots, X_n]$ is a finite extension of commutative graded connected polynomial algebras over the field $\mathbb{F}$. This means that $Y_1, \ldots, Y_n \in \mathbb{F}[X_1, \ldots, X_n]$ is a regular sequence of maximal length in $\mathbb{F}[X_1, \ldots, X_n]$. Therefore $\mathbb{F}[X_1, \ldots, X_n]/(Y_1, \ldots, Y_n)$ is a Poincaré duality algebra (see e.g. [86]). So the maximal ideal of $\mathbb{F}[Y_1, \ldots, Y_n]$ extends to an irreducible $\mathfrak{m}$-primary ideal $(Y_1, \ldots, Y_n)^{\text{ex}}$ in $\mathbb{F}[X_1, \ldots, X_n]$. Since $Y_1, \ldots, Y_n \in \mathbb{F}[X_1, \ldots, X_n]$ is a regular sequence $\mathbb{F}[X_1, \ldots, X_n]$ is a free module over $\mathbb{F}[Y_1, \ldots, Y_n]$ (see e.g. [87] Theorem 6.2.3) and therefore the inclusion of algebras $\mathbb{F}[Y_1, \ldots, Y_n] \subseteq \mathbb{F}[X_1, \ldots, X_n]$ is split by a map of $\mathbb{F}[Y_1, \ldots, Y_n]$-modules. Hence from Proposition VI.4.5 every $\mathfrak{m}$-primary irreducible ideal $I \subset \mathbb{F}[Y_1, \ldots, Y_n]$ has a unique fit extension I^{fit} to an irreducible $\mathfrak{m}$-primary ideal in the algebra $\mathbb{F}[X_1, \ldots, X_n]$. The inclusion $\iota : \mathbb{F}[Y_1, \ldots, Y_n] \subseteq \mathbb{F}[X_1, \ldots, X_n]$ induces as above an epimorphism $\iota^* : \mathbb{F}[X_1^{-1}, \ldots, X_n^{-1}] \longrightarrow \mathbb{F}[Y_1^{-1}, \ldots, Y_n^{-1}]$, which can be used to compare the inverse system of I with that of I^{fit}. This allows us to study how properties of the Poincaré duality quotient algebras $\mathbb{F}[Y_1, \ldots, Y_n]/I$ and $\mathbb{F}[X_1, \ldots, X_n]/I^{\text{fit}}$ relate to each other: particularly in the presence of a Steenrod algebra action we can study how their Wu classes relate.

PROPOSITION VI.5.1: *Let $\iota : \mathbb{F}[Y_1, \ldots, Y_n] \subseteq \mathbb{F}[X_1, \ldots, X_n]$ be a finite extension, $I \subset \mathbb{F}[Y_1, \ldots, Y_n]$ an $\mathfrak{m}$-primary irreducible ideal, and $I^{\text{fit}} \subset \mathbb{F}[X_1, \ldots, X_n]$ its* fit extension. *If $\theta_{I^{\text{fit}}} \in \mathbb{F}[X_1^{-1}, \ldots, X_n^{-1}]$ generates the inverse system of I^{fit}, then*

$$\iota^*(h \cap \theta_{I^{\text{fit}}}) \in \mathbb{F}[Y_1^{-1}, \ldots, Y_n^{-1}]$$

generates the inverse system of I, where $h \in \mathbb{F}[X_1, \ldots, X_n]$ represents a fundamental class of $\mathbb{F}[X_1, \ldots, X_n]/(\iota(Y_1), \ldots, \iota(Y_n))$.

PROOF: The elements $Y_1, \dots, Y_n \in \mathbb{F}[X_1, \dots, X_n]$ form a regular sequence since the extension $\mathbb{F}[Y_1, \dots, Y_n] \hookrightarrow \mathbb{F}[X_1, \dots, X_n]$ is finite. Therefore ι splits as a map of $\mathbb{F}[Y_1, \dots, Y_n]$-modules. Hence the sequence of Poincaré duality algebras

$$\mathbb{F} \longrightarrow \frac{\mathbb{F}[Y_1, \dots, Y_n]}{I} \longrightarrow \frac{\mathbb{F}[X_1, \dots, X_n]}{I^{\text{fit}}} \longrightarrow \frac{\mathbb{F}[X_1, \dots, X_n]}{(\iota(Y_1), \dots, \iota(Y_n))} \longrightarrow \mathbb{F}$$

is split coexact by Proposition VI.4.12 and

$$\text{f-dim}\left(\frac{\mathbb{F}[X_1, \dots, X_n]}{I^{\text{fit}}}\right) = \text{f-dim}\left(\frac{\mathbb{F}[Y_1, \dots, Y_n]}{I}\right) + \text{f-dim}\left(\frac{\mathbb{F}[X_1, \dots, X_n]}{(\iota(Y_1), \dots, \iota(Y_n))}\right).$$

Therefore by Corollary I.2.5, if $f \in \mathbb{F}[Y_1, \dots, Y_n]$ represents a fundamental class for $\mathbb{F}[Y_1, \dots, Y_n]/I$, then $\iota(f)h \in \mathbb{F}[X_1, \dots, X_n]$ represents a fundamental class for $\mathbb{F}[X_1, \dots, X_n]/I^{\text{fit}}$. Hence $\iota(f)h \cap \theta_{I^{\text{fit}}} \neq 0$. Without loss of generality we may suppose that $\iota(f)h \cap \theta_{I^{\text{fit}}} = 1$. Choose a generator θ_I for the inverse system of I such that $f \cap \theta_I = 1$. By Proposition VI.4.12

$$\deg(h \cap \theta_{I^{\text{fit}}}) = \deg(\theta_I),$$

and since

$$f \cap \iota^*(h \cap \theta_{I^{\text{fit}}}) = \iota(f) \cap \left(h \cap \theta_{I^{\text{fit}}}\right) = \iota(f)h \cap \theta_{I^{\text{fit}}} = 1,$$

the inverse polynomial $\iota^*(h \cap \theta_{I^{\text{fit}}}) \in \mathbb{F}[Y_1^{-1}, \dots, Y_n^{-1}]$ is nonzero. In fact $f \cap \iota^*(h \cap \theta_{I^{\text{fit}}}) = 1 = f \cap \theta_I$.

We claim that the diagram

$$\begin{array}{ccc} \mathbb{F}[Y_1, \dots, Y_n]_k & \xrightarrow{\iota} & \mathbb{F}[X_1, \dots, X_n]_k \\ \downarrow \zeta_{\theta_I} & & \downarrow \zeta_{h \cap \theta_{I^{\text{fit}}}} \\ \mathbb{F} & = & \mathbb{F} \end{array}$$

commutes where k is the degree of f. To see this extend $f = f_1$ to an $\mathbb{F}$-basis for $\mathbb{F}[Y_1, \dots, Y_n]_k$, say $f_1, f_2, \dots, f_m$, by choosing a basis $f_2, \dots, f_m$ for $\ker(\zeta_{\theta_I}) = I_k$. The elements $\iota(f_2)h, \dots, \iota(f_m)h \in \mathbb{F}[X_1, \dots, X_n]$ all lie in I^{ex}, and since $I^{\text{fit}} \supseteq I^{\text{ex}}$, they also lie in I^{fit}. Hence for $i = 2, \dots, m$

$$\begin{aligned} \zeta_{\iota^*(h \cap \theta_{I^{\text{fit}}})}(f_i) &= f_i \cap \iota^*(h \cap \theta_{I^{\text{fit}}}) = \iota(f_i) \cap (h \cap \theta_{I^{\text{fit}}}) = \iota(f_i)h \cap \theta_{I^{\text{fit}}} \\ &= 0 = \theta_I(f_i) = f_i \cap \theta_I = \zeta_{\theta_I}(f_i). \end{aligned}$$

Therefore $\iota^*(h \cap \theta_{I^{\text{fit}}})$ agrees with θ_I on all the basis elements $f_1, f_2, \dots, f_m$ and hence they define the same inverse polynomial. □

We refer to Proposition VI.5.1 as the **contravariant change of rings** formula[13] for Macaulay inverses. We apply it to translate and generalize our

[13] There is also a **covariant change of rings** formula which appears naturally if Macaulay's theory is formulated in terms of local cohomology.

formula for the Macaulay dual of a Frobenius power of an $\mathfrak{m}$-primary irreducible ideal in a polynomial algebra on generators of degree 1 (see Theorem II.6.6) to polynomial algebras on generators of arbitrary degrees.

Let $\mathbb{F}$ be a field of characteristic $p \neq 0$. Recall from Section II.6 the notation $\Phi(-)$ for the graded object $-$ regraded by setting

$$\Phi(-)_k = \begin{cases} -_{k/p} & \text{if } p \text{ divides } k \\ 0 & \text{otherwise.} \end{cases}$$

We write $\Phi(f) \in \Phi(\mathbb{F}[X_1, \ldots, X_n])_{dp}$ for the element corresponding to $f \in \mathbb{F}[X_1, \ldots, X_n]_d$. Note that $\Phi(\mathbb{F}[X_1, \ldots, X_n]) = \mathbb{F}[\Phi(X_1), \ldots, \Phi(X_n)]$. The map

$$\lambda : \Phi(\mathbb{F}[X_1, \ldots, X_n]) \longrightarrow \mathbb{F}[X_1, \ldots, X_n]$$

defined by

$$\lambda(\Phi(f)) = f^p \qquad \forall\ f \in \mathbb{F}[X_1, \ldots, X_n]$$

is a map of algebras (apart from a possible Frobenius automorphism of the ground field). In this way we obtain a finite extension

$$\Phi(\mathbb{F}[X_1, \ldots, X_n]) \overset{\lambda}{\hookrightarrow} \mathbb{F}[X_1, \ldots, X_n]$$

of polynomial algebras over $\mathbb{F}$.

Let $I \subset \mathbb{F}[X_1, \ldots, X_n]$ be an $\mathfrak{m}$-primary irreducible ideal. Then $\Phi(I) \subset \Phi(\mathbb{F}[X_1, \ldots, X_n])$ is also an $\mathfrak{m}$-primary irreducible ideal. Its extension to $\mathbb{F}[X_1, \ldots, X_n]$ is $I^{[p]}$. There are two basic facts pertaining to this situation.

Proposition VI.5.2: *Let $\mathbb{F}$ be a field of characteristic $p \neq 0$ and $I \subset \mathbb{F}[X_1, \ldots, X_n]$ an $\mathfrak{m}$-primary irreducible ideal. Then $\Phi(I)^{\text{ex}} = I^{[p]} = \Phi(I)^{\text{fit}}$.*

Proposition VI.5.3: *Let $\mathbb{F}$ be a field of characteristic $p \neq 0$ and $I \subset \mathbb{F}[X_1, \ldots, X_n]$ an $\mathfrak{m}$-primary irreducible ideal. If $\theta_I \in \mathbb{F}[X_1^{-1}, \ldots, X_n^{-1}]$ is a generator of the inverse system to I then $(X_1^{-1} \cdots X_n^{-1})^{p-1} \cdot (\theta_I)^p$ is a generator for the inverse system to $I^{[p]}$.*

Proof: The strategy of the proof is as follows: regard the inverse polynomial $(X_1^{-1} \cdots X_n^{-1})^{p-1} \cdot (\theta_I)^p$ as a linear form on $\mathbb{F}[X_1, \ldots, X_n]_{k(p-1)+pd}$, where $k = \deg(X_1) + \cdots + \deg(X_n)$ and $d = \text{f-dim}(\mathbb{F}[X_1, \ldots, X_n]/I)$. We will show that $I^{[p]}_{k(p-1)+pd} = \ker((X_1^{-1} \cdots X_n^{-1})^{p-1} \cdot (\theta_I)^p)$ from which the result follows by Corollary I.3.3.

Note that $\mathbb{F}[X_1, \ldots, X_n]/I^{[p]}$ is a Poincaré duality algebra of formal dimension $k(p-1) + pd$ (cf. Theorem II.6.6). Therefore $I^{[p]}_{k(p-1)+pd}$ is a codimension one subspace of $\mathbb{F}[X_1, \ldots, X_n]_{k(p-1)+pd}$. Since the inverse polynomial $(X_1 \cdots X_n^{-1})^{p-1} \cdot (\theta_I)^p$ is nonzero of degree $k(p-1) + pd$ its kernel is also a codimension one subspace of $\mathbb{F}[X_1, \ldots, X_n]_{k(p-1)+pd}$. Hence to establish the desired equality $I^{[p]}_{k(p-1)+pd} = \ker((X_1^{-1} \cdots X_n^{-1})^{p-1} \cdot (\theta_I)^p)$ it is enough to show that $(X_1^{-1} \cdots X_n^{-1})^{p-1} \cdot (\theta_I)^p$ vanishes on $I^{[p]}_{k(p-1)+pd}$.

So suppose that $f \in I^{[p]}_{k(p-1)+pd}$. Choose generators $f_1, \ldots, f_m$ for I. Then we may find $h_1, \ldots, h_m \in \mathbb{F}[X_1, \ldots, X_n]$ so that

$$f = \sum_{i=1}^{m} h_i f_i^p.$$

If we rewrite each element h_i as a sum of terms of the form $X^A \cdot H^p$ where $H \in \mathbb{F}[X_1, \ldots, X_n]$ and $A = (a_1, \ldots, a_n) \in \mathbb{N}_0^n$ is an exponent sequence with no entry larger than $p-1$, then we see it is enough to show that

$$(X^A \cdot (Hf_i)^p) \cap (X_1^{-1} \cdots X_n^{-1})^{p-1} \cdot (\theta_I)^p = 0.$$

Write $F = Hf_i$. If $a_1, \ldots, a_n = p-1$ then the preceding formula reduces to $F^p \cap \theta_I^p = (F \cap \theta_I)^p$, which in the module of inverse polynomials is zero since $F \in I_d$ and θ_I vanishes on I_d.

So suppose that $a_i < p-1$ for some index $1 \le i \le n$. Then $\deg(X^A) < k(p-1)$, so since $X^A F^p$ has degree $k(p-1) + pd$, it must be the case that $\deg(F^p) > pd$ so $\deg(F) > d = -\deg(\theta_I)$. Therefore $\deg(F \cap \theta_I) > 0$. Write $F \cap \theta_I$ as a linear combination of monomials, viz.,

$$F \cap \theta_I = \sum_{B \in \mathcal{B}} \tau_B X^B.$$

Then

$$F^p \cap \theta_I^p = \sum_{B \in \mathcal{B}} \tau_B^p X^{pB}$$

so each monomial X^{pB} in this representation contains at least one variable raised to a power at least equal to p. In the Laurent polynomial algebra $\mathbb{F}[X_1, \ldots, X_n, X_1^{-1}, \ldots, X_n^{-1}]$ we have

$$\begin{aligned}(X_1^{-1} \cdots X_n^{-1})^{p-1} \cdot \theta_I^p \cdot X^A \cdot F^p &= (X_1^{a_1-(p-1)} \cdots X_n^{a_n-(p-1)}) \cdot \theta_I^p \cdot F^p \\ &= \sum_{B \in \mathcal{B}} \tau_B^p X_1^{pb_1+a_1-(p-1)} \cdots X_n^{pb_n+a_n-(p-1)}.\end{aligned}$$

This is a linear combination of monomials each of which contains a variable raised to a strictly positive power. Each such monomial is zero in the module of inverse polynomials, and hence so is the entire sum. □

Proof of Proposition VI.5.2: The equality $I^{[p]} = (\Phi(I))^{\text{ex}}$ is elementary. Let $Q \subseteq I$ be a parameter ideal and h a transition element for I over Q, so

$$I = (Q : h).$$

By Proposition II.6.3

$$I^{[p]} = (Q^{[p]} : h^p).$$

Also $\Phi(Q) \subseteq \Phi(I)$ is a parameter ideal and

$$\Phi(I) = (\Phi(Q) \underset{\Phi}{:} \Phi(h)),$$

where $(- \underset{\Phi}{:} -)$ denotes the ideal of quotients taken in $\Phi(\mathbb{F}[X_1, \ldots, X_n])$. Applying λ to the last equation gives the equalities of graded sets

$$\lambda(\Phi(I)) = \lambda((\Phi(Q) \underset{\Phi}{:} \Phi(h))) = (\lambda(\Phi(Q)) : \lambda(\Phi(h))) = (\lambda(\Phi(Q)) : h^p),$$

where $\lambda(\Phi(Q))$ is the set of p-th powers of the elements in Q. Passing to the ideals generated by these sets we obtain $I^{[p]} = (Q^{[p]} : h^p)$ which shows that $I^{[p]}$ is the fit extension of $\Phi(I)$ as required. □

VI.6 Inverse systems and the Steenrod algebra

If the ground field is the Galois field $\mathbb{F}_q$ with $q = p^\nu$ elements, then we have seen in Part III that $\mathbb{F}_q[V]$ supports an additional structure: it is an unstable algebra over the Steenrod algebra. Such algebras are prevalent throughout algebraic topology and invariant theory over finite fields. For example the Dickson algebra $\mathbf{D}(n) = \mathbb{F}_q[V]^{\mathrm{GL}(n, \mathbb{F}_q)}$ is such an algebra. We have made many computations with the Poincaré duality quotients of $\mathbb{F}_q[z_1, \ldots, z_n]$ by ideals generated by powers of Dickson polynomials and applied these to the *Hit Problem* for $\mathbb{F}_q[z_1, \ldots, z_n]$. We would like to be able to use these same computations to study the *Hit Problem* for the corresponding quotient of the Dickson algebra $\mathbf{D}(n)$ itself by the ideal generated by the same polynomials. This is not as straightforward as it might seem. Already at this elementary level some surprising new phenomena appear. It is our intention to return to the many unexplored connections apparent here in an ulterior investigation.

EXAMPLE 1: Consider the Dickson algebra

$$\mathbf{D}(2) = \mathbb{F}_2[x, y]^{\mathrm{GL}(2, \mathbb{F}_2)} = \mathbb{F}_2[\mathbf{d}_{2,0}, \mathbf{d}_{2,1}]$$

and recall that the action of the mod 2 Steenrod algebra is given by

$$\begin{aligned} \mathbf{Sq}(\mathbf{d}_{2,1}) &= \mathbf{d}_{2,1} + \mathbf{d}_{2,0} + \mathbf{d}_{2,1}^2 \\ \mathbf{Sq}(\mathbf{d}_{2,0}) &= \mathbf{d}_{2,0}(1 + \mathbf{d}_{2,1} + \mathbf{d}_{2,0}). \end{aligned}$$

The maximal ideal $\overline{\mathbf{D}}(2) = (\mathbf{d}_{2,0}, \mathbf{d}_{2,1})$ in $\mathbf{D}(2)$ is certainly $\mathscr{A}^*$-invariant.

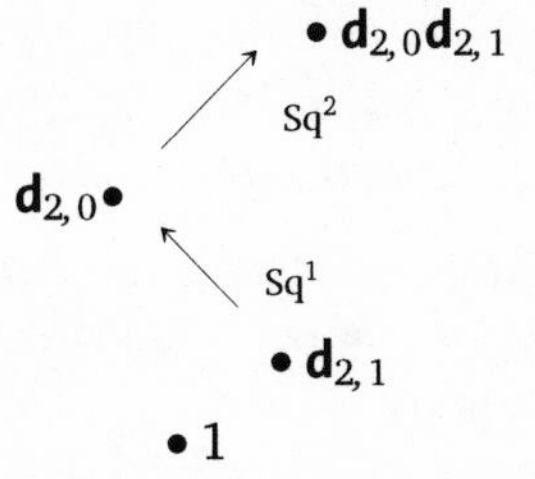

The algebra $\mathbf{D}(2)/(\mathbf{d}_{2,0}^2, \mathbf{d}_{2,1}^2)$

Since the quotient algebra $\mathbf{D}(2)/\overline{\mathbf{D}}(2)$ is just the ground field $\mathbb{F}_2$ it has trivial Wu classes. However, for the quotient algebra $\mathbf{D}(2)/(\mathbf{d}_{2,0}^2, \mathbf{d}_{2,1}^2)$ of $\mathbf{D}(2)$ by the first Frobenius power $\overline{\mathbf{D}}(2)^{[2]} = (\mathbf{d}_{2,0}^2, \mathbf{d}_{2,1}^2)$ of the maximal ideal one has that $\mathrm{Wu}_2((\mathbf{d}_{2,0}^2, \mathbf{d}_{2,1}^2)) = \mathbf{d}_{2,1}$, so the algebra $\mathbf{D}(2)/(\mathbf{d}_{2,0}, \mathbf{d}_{2,1})^{[2]}$ does not have trivial Wu classes. Indeed, as $\chi(\mathrm{Sq}^1) = \mathrm{Sq}^1$, $\chi(\mathrm{Sq}^2) = \mathrm{Sq}^2$, and $\mathrm{Sq}^3 = \mathrm{Sq}^1\mathrm{Sq}^2$ one has $\chi(\mathrm{Sq}^3) = \mathrm{Sq}^2\mathrm{Sq}^1$, so the graphic shows

$$\mathrm{Wu}(\mathbf{D}(2)) = 1 + \mathbf{d}_{2,1},$$

and

$$\chi\mathrm{Wu}(\mathbf{D}(2)) = 1 + \mathbf{d}_{2,1} + \mathbf{d}_{2,0}.$$

Therefore Corollary III.6.5 does not extend verbatim from $\mathbb{F}_q[V]$ to $\mathbf{D}(n)$. For another example of this sort see the example following Corollary VI.7.5.

This example also shows that S. A. Mitchell's Theorem (see [60] Appendix B or Corollary IV.2.3) does not generalize from $\mathscr{P}^*$-subalgebras of $\mathbb{F}_q[z_1, \ldots, z_n]$ to $\mathscr{P}^*$-subalgebras of an arbitrary unstable polynomial algebra. Namely $\mathbb{F}_2[\mathbf{d}_{2,0}^2, \mathbf{d}_{2,1}^2]$ is an $\mathscr{A}^*$-subalgebra of $\mathbf{D}(2)$, but the quotient $\mathbf{D}(2)/(\mathbf{d}_{2,0}^2, \mathbf{d}_{2,1}^2)$ does not have trivial Wu classes.

In order to study the phenomena brought to light by this example we extend Macaulay's theory of inverse systems to deal with $\mathscr{P}^*$-invariant $\mathfrak{m}$-primary irreducible ideals in an unstable polynomial algebra $\mathbb{F}_q[X_1, \ldots, X_n]$ where the variables need not have degree 1. To do so we first need to define an action of the Steenrod algebra on the inverse polynomials $\mathbb{F}_q[X_1^{-1}, \ldots, X_n^{-1}]$ compatible with the $\mathbb{F}[X_1, \ldots, X_n]$-module structure needed for the double annihilator correspondence. Here is one way to do so.

Let A be a commutative algebra over the Steenrod algebra which is an integral domain. Denote by $\boldsymbol{FF}(A)$ the graded field of fractions of A (see e.g. [87] Section 5.5). The key observation we need is that the Steenrod algebra acts on $\boldsymbol{FF}(A)$ extending the action on A in a unique way. This is a special case of a basic result of differential algebra [72] and as such has nothing to do with the instability condition. For the sake of completeness we record the simple proof from [106]. See [52] Proposition 1.14 for a somewhat different proof.

PROPOSITION VI.6.1: *Let A be a commutative algebra over the Steenrod algebra $\mathscr{P}^*$ of the Galois field $\mathbb{F}_q$. If $S \subset A$ is a multiplicatively closed* *subset of homogeneous elements, then there is a unique extension of the $\mathscr{P}^*$-algebra structure to the algebra $S^{-1}A$ obtained by formally inverting the elements of S. In particular if A is an integral domain the $\mathscr{P}^*$-algebra structure extends uniquely to the graded field of fractions* $\mathbf{FF}(A)$.

PROOF: Let $\mathscr{B}^*$ be the free graded associative algebra with unit over $\mathbb{F}_q$ generated by the symbols P^i for $i \in \mathbb{N}$ with $\deg(P^i) = i(q-1)$. We make $\mathscr{B}^*$ into a Hopf algebra via the coproduct ∇ given by

$$\nabla(P^k) = \sum_{i+j=k} P^i \otimes P^j,$$

with the convention that $P^0 = 1$. There is a natural map of Hopf algebras $\varphi : \mathscr{B}^* \longrightarrow \mathscr{P}^*$ given by $\varphi(P^i) = \mathcal{P}^i$ so by change of rings A becomes an algebra over $\mathscr{B}^*$. Let $A[[\xi]]$ be the formal power series algebra over A in the variable ξ of degree $1-q$. Define $\boldsymbol{P}_\xi : A \longrightarrow A[[\xi]]$ by

$$\boldsymbol{P}_\xi(a) = \sum_{k\geq 0} P^k(a)\xi^k \qquad \forall\, a \in A.$$

This is a homomorphism of $\mathbb{F}_q$-algebras. An extension of the $\mathscr{B}^*$-algebra structure on A to $S^{-1}A$ would of necessity satisfy $\boldsymbol{P}_\xi(b) \cdot \boldsymbol{P}_\xi(a/b) = \boldsymbol{P}_\xi(a)$ in $A[[\xi]]$ for any $a \in A$ and $b \in S$. If we regard this as an equation in the unknown quantity $\boldsymbol{P}_\xi(a/b)$ it can be solved recursively on the degree. So one obtains a formula that can be used to extend the definition of $\boldsymbol{P}_\xi$ to a ring homomorphism $\boldsymbol{P}_\xi : S^{-1}A \longrightarrow S^{-1}A[[\xi]]$. This gives us a $\mathscr{B}^*$-algebra structure on $S^{-1}A$.

Next we exploit the grading to show that this $\mathscr{B}^*$-algebra structure induces a $\mathscr{P}^*$-algebra structure on $S^{-1}A$. Let $\mathscr{I} \subset \mathscr{B}^*$ be the kernel of the quotient map $\varphi : \mathscr{B}^* \longrightarrow \mathscr{P}^*$. Then $\mathscr{I}$ annihilates A since A started out as a $\mathscr{P}^*$-algebra. If $Q \in \mathscr{B}^*$ acts as a derivation then one finds by the quotient rule for the $\mathscr{B}^*$-action that $Q(a/b) = \big(bQ(a) - aQ(b)\big)/b^2$. Hence the derivations in $\mathscr{I}$ annihilate $S^{-1}A$. Primitive elements in $\mathscr{B}^*$ act as derivations, so in particular the elements of $\mathscr{I}$ of minimal degree, being primitive, annihilate $S^{-1}A$. By induction on the degree, using that higher degree elements in $\mathscr{I}$ are primitive modulo lower degree elements we conclude that $\mathscr{I}$ annihilates $S^{-1}A$ and recover a $\mathscr{P}^*$-algebra structure on $S^{-1}A$. □

In the concrete case of $\mathbb{F}_q[z_1, \ldots, z_n, z_1^{-1}, \ldots, z_n^{-1}]$, with the variables $z_1, \ldots, z_n$ all of degree 1, one obtains a $\mathscr{P}^*$-algebra structure satisfying

$$\mathcal{P}\Big(\frac{1}{z_i}\Big) = \sum_{j=0}^{\infty}(-1)^j z_i^{j(q-1)-1} \qquad i = 1, \ldots, n.$$

Let $\mathbb{F}_q[X_1, \ldots, X_n]$ be an unstable polynomial algebra over the Steenrod algebra $\mathscr{P}^*$. Then Proposition VI.6.1 allows us to extend the action of $\mathscr{P}^*$

on $\mathbb{F}_q[X_1, \ldots, X_n]$ to $\mathbb{F}_q[X_1, \ldots, X_n, X_1^{-1}, \ldots, X_n^{-1}]$. The graded vector space $\mathfrak{X} \subset \mathbb{F}[X_1, \ldots, X_n, X_1^{-1}, \ldots, X_n^{-1}]$ spanned by all Laurent monomials X^E where $E = (e_1, \ldots, e_n) \neq (0, \ldots, 0)$ and such that at least one of $e_1, \ldots, e_n$ is positive is a $\mathscr{P}^*$-submodule. Dividing out this submodule we obtain a graded $\mathscr{P}^*$-module which we may identify with the module of inverse polynomials $\mathbb{F}[X_1^{-1}, \ldots, X_n^{-1}]$. The proof of Theorem II.2.2 for dual systems adapts to the context of inverse systems to yield that the inverse system $M(\theta_I) \subset \mathbb{F}_q[X_1^{-1}, \ldots, X_n^{-1}]$ of a $\mathscr{P}^*$-invariant $\mathfrak{m}$-primary irreducible ideal I in $\mathbb{F}_q[X_1, \ldots, X_n]$ is a $\mathscr{P}^*$-submodule, and conversely, a nonzero $\mathscr{P}^* \odot \mathbb{F}_q[X_1, \ldots, X_n]$-submodule of $\mathbb{F}_q[X_1^{-1}, \ldots, X_n^{-1}]$ that is cyclic as $\mathbb{F}_q[X_1, \ldots, X_n]$-module has an $\mathfrak{m}$-primary irreducible $\mathscr{P}^*$-invariant ideal as annihilator. In short, Theorem III.2.1 holds with $\Gamma(V)$ replaced by $\mathbb{F}_q[X_1^{-1}, \ldots, X_n^{-1}]$. Moreover, after making this same replacement, the argument used to prove Theorem III.3.4 applies, and so we also have the following theorem.

THEOREM VI.6.2: *Let $\mathbb{F}_q[X_1, \ldots, X_n]$ be an unstable algebra over the Steenrod algebra $\mathscr{P}^*$ and $I \subset \mathbb{F}_q[X_1, \ldots, X_n]$ a $\mathscr{P}^*$-invariant $\mathfrak{m}$-primary irreducible ideal. Let $\theta_I \in \mathbb{F}_q[X_1^{-1}, \ldots, X_n^{-1}]$ be a generator for the inverse principal system, then $\chi(\mathcal{P})(\theta_I) = \mathrm{Wu}(I) \cap \theta_I$ and $\mathcal{P}(\theta_I) = \chi\mathrm{Wu}(I) \cap \theta_I$.* □

We next examine the analog of Theorem III.6.4 in this wider context.

THEOREM VI.6.3: *Let $\mathbb{F}_q[X_1, \ldots, X_n]$ be an unstable algebra over the Steenrod algebra $\mathscr{P}^*$ of the Galois field $\mathbb{F}_q$ with $q = p^\nu$ elements, p a prime. Suppose that $I \subset \mathbb{F}_q[X_1, \ldots, X_n]$ is a $\mathscr{P}^*$-invariant $\mathfrak{m}$-primary irreducible ideal. Then*

$$\mathrm{Wu}(I^{[q]}) = \mathrm{Wu}(\mathfrak{m}^{[q]}) \cdot \mathrm{Wu}(I)^q \in \mathbb{F}_q[X_1, \ldots, X_n]/I^{[q]},$$

and

$$\chi\mathrm{Wu}(I^{[q]}) = \chi\mathrm{Wu}(\mathfrak{m}^{[q]}) \cdot \chi\mathrm{Wu}(I)^q \in \mathbb{F}_q[X_1, \ldots, X_n]/I^{[q]}.$$

PROOF: Let $\theta_I \in \mathbb{F}_q[X_1^{-1}, \ldots, X_n^{-1}]$ be a generator for the inverse system of I. By Proposition VI.5.3 the element $(X_1^{-1} \cdots X_n^{-1})^{q-1} \cdot (\theta_I)^q$ generates the inverse system of $I^{[q]}$. The element $(X_1^{-1} \cdots X_n^{-1})^{q-1}$ of $\mathbb{F}_q[X_1^{-1}, \ldots, X_n^{-1}]$ generates the inverse system of the ideal $\mathfrak{m}^{[q]} = (X_1^q, \ldots, X_n^q)$. Application of Theorem VI.6.2 yields

$$\begin{aligned}\chi(\mathcal{P})(\theta_{I^{[q]}}) &= \chi(\mathcal{P})\big((X_1^{-1} \cdots X_n^{-1})^{q-1} \cdot \theta_I^q\big)\\ &= \chi(\mathcal{P})\big((X_1^{-1} \cdots X_n^{-1})^{q-1}\big)\chi(\mathcal{P})(\theta_I^q)\\ &= \chi(\mathcal{P})\big((X_1^{-1} \cdots X_n^{-1})^{q-1}\big)\big(\chi(\mathcal{P})(\theta_I)\big)^q\\ &= \big(\mathrm{Wu}(\mathfrak{m}^{[q]}) \cap (X_1^{-1} \cdots X_n^{-1})^{q-1}\big) \cdot \big(\mathrm{Wu}(I)^q \cap \theta_I^q\big)\end{aligned}$$

$$= \left(\mathrm{Wu}(\mathfrak{m}^{[q]}) \cdot \mathrm{Wu}(I)^q\right) \cap \left((X_1^{-1} \cdots X_n^{-1})^{q-1} \theta_I^q\right)$$
$$= \left(\mathrm{Wu}(\mathfrak{m}^{[q]}) \cdot \mathrm{Wu}(I)^q\right) \cap \theta_{I^{[q]}}$$

from which the result follows by applying Theorem VI.6.2 one more time. The formula for the conjugate Wu clases is proven analagously. □

COROLLARY VI.6.4: *Let $\mathbb{F}_q[X_1, \ldots, X_n]$ be an unstable algebra over the Steenrod algebra $\mathscr{P}^*$ of the Galois field $\mathbb{F}_q$ with $q = p^\nu$ elements, p a prime. Then*

$$\mathrm{Wu}((X_1^{q^s}, \ldots, X_n^{q^s})) = \mathrm{Wu}((X_1^q, \ldots, X_n^q))^{q^{s-1}+\cdots+q+1}$$

and

$$\chi\mathrm{Wu}((X_1^{q^s}, \ldots, X_n^{q^s})) = \chi\mathrm{Wu}((X_1^q, \ldots, X_n^q))^{q^{s-1}+\cdots+q+1}$$

for all $s \geq 1$.

PROOF: It is enough to prove the formula for the Wu classes as the proof for the conjugate Wu classes is similar. The result is a tautology for $s = 1$ so we may proceed by induction on s. Suppose that the result has been established for all integers r strictly less than s, and $s > 1$. Apply Theorem VI.6.3 with $I = \mathfrak{m}^{[q^{s-1}]}$ to obtain from the induction hypothesis

$$\mathrm{Wu}(\mathfrak{m}^{[q^s]}) = \mathrm{Wu}(\mathfrak{m}^{[q]}) \cdot \mathrm{Wu}(\mathfrak{m}^{[q^{s-1}]})^q = \mathrm{Wu}(\mathfrak{m}^{[q]}) \cdot \left(\mathrm{Wu}(\mathfrak{m}^{[q]})^{q^{s-2}+\cdots+q+1}\right)^q$$
$$= \mathrm{Wu}(\mathfrak{m}^{[q]}) \cdot \mathrm{Wu}(\mathfrak{m}^{[q]})^{q^{s-1}+\cdots+q^2+q} = \mathrm{Wu}(\mathfrak{m}^{[q]})^{q^{s-1}+\cdots+q+1}$$

as required. □

Combining Theorem VI.6.3 with Corollary VI.6.4 we obtain by an induction argument the following result for higher Frobenius powers.

COROLLARY VI.6.5: *Let $\mathbb{F}_q[X_1, \ldots, X_n]$ be an unstable algebra over the Steenrod algebra $\mathscr{P}^*$ of the Galois field $\mathbb{F}_q$ with $q = p^\nu$ elements, p a prime. Suppose that $I \subset \mathbb{F}_q[X_1, \ldots, X_n]$ is a $\mathscr{P}^*$-invariant $\mathfrak{m}$-primary irreducible ideal. Then $\mathrm{Wu}(I^{[q^s]}) = \mathrm{Wu}(\mathfrak{m}^{[q^s]}) \cdot \mathrm{Wu}(I)^{q^s} \in \mathbb{F}_q[X_1, \ldots, X_n]/I^{[q^s]}$ and $\chi\mathrm{Wu}(I^{[q^s]}) = \chi\mathrm{Wu}(\mathfrak{m}^{[q^s]}) \cdot \chi\mathrm{Wu}(I)^{q^s} \in \mathbb{F}_q[X_1, \ldots, X_n]/I^{[q^s]}$.* □

These results make clear that in the study of the Wu classes of $\mathscr{P}^*$-Poincaré duality quotients of an unstable $\mathscr{P}^*$-algbebra over the Galois field $\mathbb{F}_q$ the first step is to determine the Wu classes $\mathrm{Wu}(A/\overline{A}^{[q^e]})$ of the quotients of A by Frobenius powers of the maximal ideal $\overline{A}$.

VI.7 Change of Rings and Wu Classes

Our goal in this section is to examine the behavior of the Wu classes for a $\mathscr{P}^*$-invariant ideal and its fit extension across a finite extension of polynomial algebras. The phenomenon that appears here was first examined by S. A. Mitchell and R. E. Stong in a different setting; see [61] Section 3. For simplicity we confine the discussion to the situation of a finite extension $\alpha : \mathbb{F}_q[X_1, \ldots, X_n] \hookrightarrow \mathbb{F}_q[z_1, \ldots, z_n]$ where $\mathbb{F}_q[X_1, \ldots, X_n]$ is a polynomial algebra with the structure of an unstable algebra over the Steenrod algebra, the variables $z_1, \ldots, z_n$ are of degree 1, and α is a map of algebras over the Steenrod algebra. In view of the main result of [2] and Theorem 5.3.4 of [66] this is a sort of universal situation since any unstable polynomial algebra $\mathbb{F}_q[X_1, \ldots, X_n]$ admits an embedding $\alpha : \mathbb{F}_q[X_1, \ldots, X_n] \hookrightarrow \mathbb{F}_q[z_1, \ldots, z_n]$ as unstable algebras making $\mathbb{F}_q[z_1, \ldots, z_n]$ into a finitely generated module over $\mathbb{F}_q[X_1, \ldots, X_n]$.

Fix an unstable polynomial algebra $\mathbb{F}_q[X_1, \ldots, X_n]$ over the Steenrod algebra of the Galois field $\mathbb{F}_q$. Suppose that $I \subset \mathbb{F}_q[X_1, \ldots, X_n]$ is an $\mathfrak{m}$-primary irreducible $\mathscr{P}^*$-invariant ideal and $I^{\text{fit}} \subset \mathbb{F}_q[z_1, \ldots, z_n]$ its fit extension to $\mathbb{F}_q[z_1, \ldots, z_n]$. Choose an integer s such that $(X_1^{q^s}, \ldots, X_n^{q^s}) \subseteq I$ and write

$$\big((X_1^{q^s}, \ldots, X_n^{q^s}) : I\big) = (h) + (X_1^{q^s}, \ldots, X_n^{q^s}),$$

where $h \in \mathbb{F}_q[X_1, \ldots, X_n]$. Then

$$\big((X_1^{q^s}, \ldots, X_n^{q^s})^{\text{fit}} : I^{\text{fit}}\big) = \big(\alpha(h)\big) + (X_1^{q^s}, \ldots, X_n^{q^s})^{\text{fit}}$$

as ideals in $\mathbb{F}_q[z_1, \ldots, z_n]$. By Theorem III.3.5 the element h becomes a Thom class in the quotient algebra $\mathbb{F}_q[X_1, \ldots, X_n]/(X_1^{q^s}, \ldots, X_n^{q^s})$ so we may write

$$\chi(\mathcal{P})(h) = w \cdot h \in \mathbb{F}_q[X_1, \ldots, X_n]/(X_1^{q^s}, \ldots, X_n^{q^s})$$

for some $w = 1 + w_1 + \cdots \in \mathbb{F}_q[X_1, \ldots, X_n]/(X_1^{q^s}, \ldots, X_n^{q^s})$. The induced map, namely

$$\alpha : \mathbb{F}_q[X_1, \ldots, X_n]/(X_1^{q^s}, \ldots, X_n^{q^s}) \mathbb{F}_q[z_1, \ldots, z_n]/(X_1^{q^s}, \ldots, X_n^{q^s})^{\text{fit}},$$

is a monomorphism so if we identify $\mathbb{F}_q[X_1, \ldots, X_n]/(X_1^{q^s}, \ldots, X_n^{q^s})$ with its image under α we also have

$$\chi(\mathcal{P})(h) = w \cdot h \in \mathbb{F}_q[z_1, \ldots, z_n]/(X_1^{q^s}, \ldots, X_n^{q^s})^{\text{fit}}.$$

From Theorem III.3.5 we obtain [14]

$$\mathrm{Wu}(I) = w \cdot \mathrm{Wu}\big((X_1^{q^s}, \ldots, X_n^{q^s})\big) \in \mathbb{F}_q[X_1, \ldots, X_n]/I$$

---

[14] Although this theorem is stated for the case of all variables being of degree 1 the proof carries over to the more general case needed here.

and

$$\mathrm{Wu}(I^{\mathrm{fit}}) = w \cdot \mathrm{Wu}\big((X_1^{q^s}, \ldots, X_n^{q^s})^{\mathrm{fit}}\big) \in \mathbb{F}_q[z_1, \ldots, z_n]/I^{\mathrm{fit}}.$$

The ideal $(X_1^{q^s}, \ldots, X_n^{q^s}) \subset \mathbb{F}_q[X_1, \ldots, X_n]$ is a regular ideal so its fit and ordinary extensions are the same and $\mathbb{F}[z_1, \ldots, z_n]$ is a free module over $\mathbb{F}_q[X_1, \ldots, X_n]$ (see Corollary VI.4.6). Hence

$$\mathrm{Wu}\big((X_1^{q^s}, \ldots, X_n^{q^s})^{\mathrm{fit}}\big) = \mathrm{Wu}\big((X_1^{q^s}, \ldots, X_n^{q^s})^{\mathrm{ex}}\big) = \mathrm{Wu}\left(\frac{\mathbb{F}_q[z_1, \ldots, z_n]}{(X_1^{q^s}, \ldots, X_n^{q^s})}\right).$$

Since $\mathbb{F}[X_1^{q^s}, \ldots, X_n^{q^s}]$ is a $\mathscr{P}^*$-subalgebra of $\mathbb{F}[z_1, \ldots, z_n]$ we may apply Mitchell's theorem (see Corollary IV.2.3) to conclude that

$$\mathrm{Wu}\big((X_1^{q^s}, \ldots, X_n^{q^s})^{\mathrm{fit}}\big) = 1 \in \mathbb{F}_q[z_1, \ldots, z_n]/(X_1^{q^s}, \ldots, X_n^{q^s})$$

and hence also in $\mathbb{F}_q[z_1, \ldots, z_n]/I^{\mathrm{fit}}$. This means $\alpha(w) = \mathrm{Wu}(I^{\mathrm{fit}})$, so we have proven the following result.

PROPOSITION VI.7.1: *Let $\alpha : \mathbb{F}_q[X_1, \ldots, X_n] \hookrightarrow \mathbb{F}_q[z_1, \ldots, z_n]$ be a finite extension of unstable polynomial algebras over the Steenrod algebra, where $z_1, \ldots, z_n$ all have degree 1. If $I \subset \mathbb{F}_q[X_1, \ldots, X_n]$ is an $\mathfrak{m}$-primary irreducible $\mathscr{P}^*$-invariant ideal and $s \in \mathbb{N}_0$ is large enough so that $(X_1^{q^s}, \ldots, X_n^{q^s}) \subseteq I$, then*

$$\alpha(\mathrm{Wu}(I)) = \mathrm{Wu}(I^{\mathrm{fit}}) \cdot \alpha\big(\mathrm{Wu}(X_1^{q^s}, \ldots, X_n^{q^s})\big)$$

and

$$\alpha(\chi\mathrm{Wu}(I)) = \chi\mathrm{Wu}(I^{\mathrm{fit}}) \cdot \alpha\big(\chi\mathrm{Wu}(X_1^{q^s}, \ldots, X_n^{q^s})\big)$$

in $\mathbb{F}_q[z_1, \ldots, z_n]/I^{\mathrm{fit}}$. □

Following S. Mitchell and R. E. Stong [61] we note that if $\mathbb{F}_q[X_1, \ldots, X_n]$ is an unstable algebra over the Steenrod algebra $\mathscr{P}^*$ of the Galois field $\mathbb{F}_q$ with $q = p^\nu$ elements, $p \in \mathbb{N}$ a prime, then Corollary VI.6.4 implies

$$\begin{aligned}\mathrm{Wu}\big((X_1^{q^{s+1}}, \ldots, X_n^{q^{s+1}})\big) &= \mathrm{Wu}\big((X_1^{q}, \ldots, X_n^{q})\big)^{q^s + \cdots q + 1}\\ &= \mathrm{Wu}\big((X_1^{q}, \ldots, X_n^{q})\big)^{q^s} \cdot \mathrm{Wu}\big((X_1^{q^s}, \ldots, X_n^{q^s})\big)\end{aligned}$$

in $\mathbb{F}_q[X_1, \ldots, X_n]/(X_1^{q^{s+1}}, \ldots, X_n^{q^{s+1}})$. Next observe that the quotient map

$$\mathbb{F}_q[X_1, \ldots, X_n]_i \longrightarrow \big(\mathbb{F}_q[X_1, \ldots, X_n]/(X_1^{q^{s+1}}, \ldots, X_n^{q^{s+1}})\big)_i$$

is an isomorphism for $i < d(q^s - 1)$ where $d = \min\{\deg(X_1), \ldots, \deg(X_n)\}$. Therefore, we may regard $\mathrm{Wu}_i\big((X_1^{q^s}, \ldots, X_n^{q^s})\big)$ for $i \leq q^s - 1$ as an element of $\mathbb{F}_q[X_1, \ldots, X_n]$. Doing so, it therefore follows from the preceding formula that $\mathrm{Wu}_i\big((X_1^{q^{s+1}}, \ldots, X_n^{q^{s+1}})\big) = \mathrm{Wu}_i\big((X_1^{q^s}, \ldots, X_n^{q^s})\big)$ as elements of $\mathbb{F}_q[X_1, \ldots, X_n]$ for $i < q^s$. This allows us to make the following definition (cf. [61] Section 3).

DEFINITION: *Suppose $\mathbb{F}_q[X_1, \ldots, X_n]$ is an unstable algebra over the Steenrod algebra $\mathscr{P}^*$ of the Galois field $\mathbb{F}_q$ with $q = p^\nu$ elements, $p \in \mathbb{N}$ a prime. The total* **Mitchell–Stong element** $\mathrm{ms}(\mathbb{F}_q[X_1, \ldots, X_n])$ *and the total* **conjugate Mitchell–Stong element** $\chi\mathrm{ms}(\mathbb{F}_q[X_1, \ldots, X_n])$ *are the inhomogeneous elements of $\mathbb{F}_q[X_1, \ldots, X_n]$*

$$1 + \mathrm{ms}_1(\mathbb{F}_q[X_1, \ldots, X_n]) + \cdots + \mathrm{ms}_i(\mathbb{F}_q[X_1, \ldots, X_n]) + \cdots$$

and

$$1 + \chi\mathrm{ms}_1(\mathbb{F}_q[X_1, \ldots, X_n]) + \cdots + \chi\mathrm{ms}_i(\mathbb{F}_q[X_1, \ldots, X_n]) + \cdots$$

respectively, that are defined by the requirement that they satisfy

$$\mathrm{ms}_i(\mathbb{F}_q[X_1, \ldots, X_n]) \equiv \mathrm{Wu}_i\big((X_1^{q^s}, \ldots, X_n^{q^s})\big) \mod (X_1^{q^s}, \ldots, X_n^{q^s})$$

respectively,

$$\chi\mathrm{ms}_i(\mathbb{F}_q[X_1, \ldots, X_n]) \equiv \chi\mathrm{Wu}_i\big((X_1^{q^s}, \ldots, X_n^{q^s})\big) \mod (X_1^{q^s}, \ldots, X_n^{q^s}),$$

for $i < q^s$ in $\mathbb{F}_q[X_1, \ldots, X_n]_{i(q-1)}$.

The following result taken from [61] Section 3 provides us with an estimate of the Frobenius power needed to compute the conjugate Mitchell–Stong element of an unstable polynomial algebra over the Steenrod algebra. We include a proof as much for the reader's convenience as to complete the one in [61] which is given only for $q = 2$.

PROPOSITION VI.7.2 (Mitchell–Stong): *Let $\mathbb{F}_q[X_1, \ldots, X_n]$ be an unstable polynomial algebra over the Steenrod algebra $\mathscr{P}^*$ of the Galois field $\mathbb{F}_q$ with $q = p^\nu$ elements, $p \in \mathbb{N}$ a prime. Then $\chi\mathrm{ms}_i(\mathbb{F}_q[X_1, \ldots, X_n]) = 0$ if $i > \sum_{i=1}^n(\deg(X_i) - 1)$.*

PROOF: According to the preceding discussion we need to show that the i-th conjugate Wu class of the Poincaré duality algebras $\frac{\mathbb{F}_q[X_1, \ldots, X_n]}{(X_1^{q^s}, \ldots, X_n^{q^s})}$ is zero if $i > \sum_{i=1}^n(\deg(X_i) - 1)$ and $i < q^s$. Let $\mathbb{F}_q[z_1, \ldots, z_n]$ be the unstable polynomial algebra over $\mathscr{P}^*$ with $\deg(z_i) = 1$ for $i = 1, \ldots, n$. Choose an embedding $\alpha : \mathbb{F}_q[X_1, \ldots, X_n] \hookrightarrow \mathbb{F}[z_1, \ldots, z_n]$ of algebras over the Steenrod algebra: this is possible by [2] and [66]. Note that α splits as a map of $\mathbb{F}_q[X_1, \ldots, X_n]$-modules: this is because the extension is finite (loc. cit.) and hence $\alpha(X_1), \ldots, \alpha(X_n)$ is a system of parameters in $\mathbb{F}[z_1, \ldots, z_n]$, hence a regular sequence, and therefore $\mathbb{F}[z_1, \ldots, z_n]$ is a free $\mathbb{F}[X_1, \ldots, X_n]$-module.

For $s \in \mathbb{N}_0$ we have the coexact sequence

$$\mathbb{F}_q \longrightarrow \frac{\mathbb{F}_q[X_1, \ldots, X_n]}{(X_1^{q^s}, \ldots, X_n^{q^s})} \xrightarrow{\bar{\alpha}} \frac{\mathbb{F}_q[z_1, \ldots, z_n]}{(X_1^{q^s}, \ldots, X_n^{q^s})^{\text{ex}}} \xrightarrow{\pi} \frac{\mathbb{F}_q[z_1, \ldots, z_n]}{(X_1, \ldots, X_n)^{\text{ex}}} \longrightarrow \mathbb{F}_q$$

of Poincaré duality algebras. An $\mathbb{F}[X_1, \ldots, X_n]$-splitting of α induces an $\frac{\mathbb{F}_q[X_1,\ldots,X_n]}{(X_1^{q^s},\ldots,X_n^{q^s})}$-splitting of $\overline{\alpha}$ and by Proposition VI.4.12

$$\text{f-dim}\left(\frac{\mathbb{F}_q[z_1, \ldots, z_n]}{(X_1^{q^s}, \ldots, X_n^{q^s})^{\text{ex}}}\right) = \text{f-dim}\left(\frac{\mathbb{F}_q[X_1, \ldots, X_n]}{(X_1^{q^s}, \ldots, X_n^{q^s})}\right) + \text{f-dim}\left(\frac{\mathbb{F}_q[z_1, \ldots, z_n]}{(X_1, \ldots, X_n)^{\text{ex}}}\right).$$

Choose an element $a \in \mathbb{F}_q[z_1, \ldots, z_n]$ representing a fundamantal class of $\frac{\mathbb{F}_q[z_1,\ldots,z_n]}{(X_1,\ldots,X_n)^{\text{ex}}}$. Let $b_s = X_1^{q^s-1} \cdots X_n^{q^s-1} \in \mathbb{F}_q[X_1, \ldots, X_n]$ which represents a fundamental class of $\frac{\mathbb{F}_q[X_1,\ldots,X_n]}{(X_1^{q^s},\ldots,X_n^{q^s})}$. Then $\overline{\alpha}(b_s) \cdot a \in \mathbb{F}[z_1, \ldots, z_n]$ represents a fundamental class of $\frac{\mathbb{F}_q[z_1,\ldots,z_n]}{(X_1^{q^s},\ldots,X_n^{q^s})^{\text{ex}}}$ by Corollary I.2.5. Let

$$\chi\text{Wu}_i = \chi\text{Wu}_i\left(\frac{\mathbb{F}_q[z_1, \ldots, z_n]}{(X_1^{q^s}, \ldots, X_n^{q^s})^{\text{ex}}}\right).$$

If χWu_i is nonzero, then there is an element

$$c \in \frac{\mathbb{F}_q[z_1, \ldots, z_n]}{(X_1^{q^s}, \ldots, X_n^{q^s})^{\text{ex}}}$$

such that

$$< c \cdot \chi\text{Wu}_i \mid b_s > \neq 0.$$

Note that $\frac{\mathbb{F}[z_1,\ldots,z_n]}{(X_1^{q^s},\ldots,X_n^{q^s})^{\text{ex}}}$ has trivial Wu classes by Mitchell's theorem (see e.g. Theorem IV.2.3, alternatively by Proposition IV.1.3 and Corollary III.6.5), so by Proposition III.3.2

$$\begin{aligned} < c \cdot \chi\text{Wu}_i \mid b_s > &= < \chi(\mathcal{P}^i)(c) \mid b_s > = < \chi(\mathcal{P}^i)(c) \cdot a \mid \overline{\alpha}(b_s) \cdot a > \\ &= < c \cdot \mathcal{P}^i(a) \mid \overline{\alpha}(b_s) \cdot a >. \end{aligned}$$

The largest value of i for which $\mathcal{P}^i(a)$ could be nonzero is

$$\deg(a) = \text{f-dim}\left(\frac{\mathbb{F}_q[z_1, \ldots, z_n]}{(X_1, \ldots, X_n)^{\text{ex}}}\right) = \sum_{i=1}^{n} (\deg(X_i) - 1)$$

and the result follows. □

As an example we compute the conjugate Mitchell–Stong element for the mod 2 Dickson algebra $\mathbf{D}(2)$.

EXAMPLE 1 (Mitchell–Stong): Consider the Dickson algebra over the field $\mathbb{F}_2$. By [20] we have $\mathbf{D}(2) = \mathbb{F}_2[\mathbf{d}_{2,0}, \mathbf{d}_{2,1}]$ where $\deg(\mathbf{d}_{2,0}) = 3$ and $\deg(\mathbf{d}_{2,1}) = 2$. Hence by Proposition VI.7.2 $\chi\text{ms}_i(\mathbf{D}(2)) = 0$ for $i > 3$.

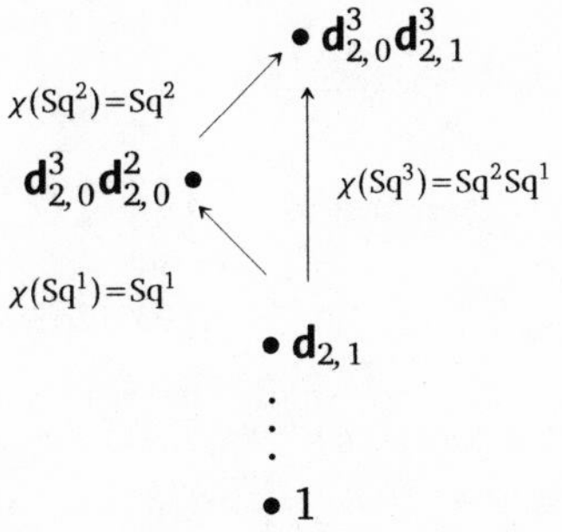

The algebra $\mathbf{D}(2)/(\mathbf{d}_{2,0}^4, \mathbf{d}_{2,1}^4)$

The conjugate Mitchell–Stong classes of $\mathbf{D}(2)$ coincide with the conjugate Wu classes of the quotient algebras $\mathbf{D}(2)/(\overline{\mathbf{D}(2)})^{[2^n]}$ for n satisfying $2^n > 3$. Hence it suffices to compute the conjugate Wu classes of the quotient algebra $\mathbf{D}(2)/(\overline{\mathbf{D}(2)})^{[4]}$. To this end we note that direct computation shows that the pattern of conjugate squaring operations into the top degree of the algebra $\mathbf{D}(2)/(\mathbf{d}_{2,0}^4, \mathbf{d}_{2,1}^4)$ is as in the adjacent graphic. In particular $\chi(\mathrm{Sq}^1)$ is zero into the top degree, as is $\chi(\mathrm{Sq}^4) = \mathrm{Sq}^4 + \mathrm{Sq}^1\mathrm{Sq}^2\mathrm{Sq}^1$, and $\chi(\mathrm{Sq}^5) = \chi(\mathrm{Sq}^4)\chi(\mathrm{Sq}^1)$. One reads off this diagram and these relations that $\chi\mathrm{Wu}(\mathbf{D}(2)/(\mathbf{d}_{2,0}^4, \mathbf{d}_{2,1}^4))$ is equal to $1 + \mathbf{d}_{2,1} + \mathbf{d}_{2,0}$ and therefore we have shown that $\chi\mathrm{ms}(\mathbf{D}(2)) = 1 + \mathbf{d}_{2,1} + \mathbf{d}_{2,0}$.

Using the Mitchell–Stong element we may restate Proposition VI.7.1 in the following way.

PROPOSITION VI.7.3: *Let $\alpha : \mathbb{F}_q[X_1, \ldots, X_n] \hookrightarrow \mathbb{F}_q[z_1, \ldots, z_n]$ be a finite extension of unstable polynomial algebras over the Steenrod algebra, where $z_1, \ldots, z_n$ all have degree 1. If $I \subset \mathbb{F}_q[X_1, \ldots, X_n]$ is an $\mathfrak{m}$-primary irreducible $\mathscr{P}^*$-invariant ideal then*

$$\alpha(\mathrm{Wu}(I)) = \mathrm{Wu}(I^{\mathrm{fit}}) \cdot \alpha(\mathrm{ms}(\mathbb{F}_q[X_1, \ldots, X_n]))$$

and

$$\alpha(\chi\mathrm{Wu}(I)) = \chi\mathrm{Wu}(I^{\mathrm{fit}}) \cdot \alpha(\chi\mathrm{ms}(\mathbb{F}_q[X_1, \ldots, X_n]))$$

in $\mathbb{F}_q[z_1, \ldots, z_n]/I^{\mathrm{fit}}$. □

There are two extreme cases of Proposition VI.7.3 of particular interest.

COROLLARY VI.7.4: *Let $\alpha : \mathbb{F}_q[X_1, \ldots, X_n] \hookrightarrow \mathbb{F}_q[z_1, \ldots, z_n]$ be a finite extension of unstable polynomial algebras over the Steenrod algebra, where $z_1, \ldots, z_n$ all have degree 1. Suppose that $I \subset \mathbb{F}_q[X_1, \ldots, X_n]$ is an $\mathfrak{m}$-primary irreducible $\mathscr{P}^*$-invariant ideal and $I^{\mathrm{fit}} \subset \mathbb{F}_q[z_1, \ldots, z_n]$ its fit extension to $\mathbb{F}_q[z_1, \ldots, z_n]$. If*

$$\mathrm{Wu}(I^{\mathrm{fit}}) \cdot \alpha\big(\mathrm{ms}(\mathbb{F}_q[X_1, \ldots, X_n])\big) = 1$$

or, equivalently,

$$\chi\mathrm{Wu}(I^{\mathrm{fit}}) \cdot \alpha\big(\chi\mathrm{ms}(\mathbb{F}_q[X_1, \ldots, X_n])\big) = 1$$

in $\mathbb{F}_q[z_1, \ldots, z_n]/I^{\mathrm{fit}}$ then $\mathbb{F}_q[X_1, \ldots, X_n]/I$ has trivial Wu classes.

PROOF: This follows from the fact that

$$\alpha : \mathbb{F}_q[X_1, \ldots, X_n]/I \longrightarrow \mathbb{F}_q[z_1, \ldots, z_n]/I^{\mathrm{fit}}$$

is monic and Proposition VI.7.3. □

The same proof yields in addition the following.

COROLLARY VI.7.5: *Let $\alpha : \mathbb{F}_q[X_1, \ldots, X_n] \hookrightarrow \mathbb{F}_q[z_1, \ldots, z_n]$ be a finite extension of unstable polynomial algebras over the Steenrod algebra, where $z_1, \ldots, z_n$ all have degree 1. Suppose that $I \subset \mathbb{F}_q[X_1, \ldots, X_n]$ is an $\mathfrak{m}$-primary irreducible $\mathscr{P}^*$-invariant ideal and $I^{\mathrm{fit}} \subset \mathbb{F}_q[z_1, \ldots, z_n]$ its fit extension to $\mathbb{F}_q[z_1, \ldots, z_n]$. If I^{fit} has trivial Wu classes, then $\mathrm{Wu}(I) = \mathrm{ms}(\mathbb{F}_q[X_1, \ldots, X_n])$ as well as $\chi\mathrm{Wu}(I) = \chi\mathrm{ms}(\mathbb{F}_q[X_1, \ldots, X_n])$ in $\mathbb{F}_q[X_1, \ldots, X_n]/I$.* □

Example 1 of Section III.3 falls nicely under Corollary VI.7.4. In that example we saw that the total Wu class of $\mathbb{F}_2[x, y]/(\mathbf{d}_{2,0}, \mathbf{d}_{2,1}^2)$ turned out to be $1 + \mathbf{d}_{2,1}$. The Mitchell–Stong element of the Dickson algebra $\mathbf{D}(2) = \mathbb{F}_2[x, y]^{\mathrm{GL}(2, \mathbb{F}_2)}$ was computed in Example 1 and we found $\mathrm{ms}(\mathbf{D}(2)) = 1 + \mathbf{d}_{2,1} + \mathbf{d}_{2,0}$. Hence for the total Wu class of the quotient algebra $\mathbf{D}(2)/(\mathbf{d}_{2,0}, \mathbf{d}_{2,1}^2)$ Corollary VI.7.4 yields $1 + \mathbf{d}_{2,1}^2 = 1$, which is pretty obvious as the accompanying graphic shows.

$\bullet\mathbf{d}_{2,1}$

$\bullet 1$

The algebra $\mathbf{D}(2)/(\mathbf{d}_{2,0}, \mathbf{d}_{2,1}^2)$

VI.8 The Hit Problem for the Dickson and other algebras

We begin this last section with a generalization of Proposition III.5.3 which allows us to connect up the *Hit Problem* for $\mathscr{P}^*$-unstable polynomial algebras with Wu classes of Poincaré duality quotients.

PROPOSITION VI.8.1: *Let $\mathbb{F}_q[X_1, \ldots, X_n]$ be a $\mathscr{P}^*$-unstable polynomial algebra, $I \subset \mathbb{F}_q[X_1, \ldots, X_n]$ a $\mathscr{P}^*$-invariant $\mathfrak{m}$-primary irreducible ideal. Assume that $\mathbb{F}_q[X_1, \ldots, X_n] \xrightarrow{\pi} \mathbb{F}_q[X_1, \ldots, X_n]/I$, the corresponding $\mathscr{P}^*$-Poincaré duality quotient, has trivial Wu classes. Regard a generator $\theta \in I^{-1} \subset \mathbb{F}_q[X_1^{-1}, \ldots, X_n^{-1}]$ of the inverse principal system as a linear form on $\mathbb{F}[X_1, \ldots, X_n]_d$, where $d = \mathrm{f\text{-}dim}(\mathbb{F}_q[X_1, \ldots, X_n]/I)$. If a monomial X^D represents a fundamental class of $\mathbb{F}_q[X_1, \ldots, X_n]/I$ then X^D is $\mathscr{P}^*$-indecomposable. Furthermore, if X^E is $\mathscr{P}^*$-equivalent to X^D then X^E also represents a fundamental class. Therefore $\mathrm{supp}(\theta)$ is a union of $\mathscr{P}^*$-equivalence classes of $\mathscr{P}^*$-indecomposable monomials.*

PROOF: Translate the proof of Proposition III.5.3 from the language of dual systems to that of inverse systems using the results of Sections VI.1 and VI.6. □

We examine the consequences for the Dickson algebra. Note that a regular[15] ideal $I \subset \mathbf{D}(n)$ is closed under the action of the Steenrod algebra if and only if the extended ideal $I^{\text{ex}} \subset \mathbb{F}_q[z_1, \ldots, z_n]$ is closed under the action of the Steenrod algebra. This means that in the special case $q = 2$ we have directly from Proposition V.4.1 the following complete list of $\mathscr{A}^*$-invariant ideals in $\mathbf{D}(n)$ generated by powers of Dickson polynomials.

PROPOSITION VI.8.2: *Let the ground field be $\mathbb{F}_2$, and let $n, a_0, \ldots, a_{n-1} \in \mathbb{N}$. For $k = 0, \ldots, n-1$ write $a_k = b_k \cdot 2^{s_k}$ with b_k odd. Then the ideal*

$$\mathfrak{d}(a_0, \ldots, a_{n-1}) = (\mathbf{d}_{n,0}^{a_0}, \mathbf{d}_{n,1}^{a_1}, \ldots, \mathbf{d}_{n,n-1}^{a_{n-1}}) \subset \mathbf{D}(n)$$

is $\mathscr{A}^*$-invariant if and only if $2^{s_k} \geq a_{k-1}$ for $k = 1, \ldots, n-1$. □

Let $\mathfrak{d}(a_0, \ldots, a_{n-1}) = (\mathbf{d}_{n,0}^{a_0}, \ldots, \mathbf{d}_{n,n-1}^{a_{n-1}}) \subset \mathbf{D}(n)$ be invariant under the action of the Steenrod algebra $\mathscr{P}^*$. Example 1 in Section VI.6 shows that it need not be the case that $\mathbf{D}(n)/\mathfrak{d}(a_0, \ldots, a_{n-1})$ has trivial Wu classes if $\mathbb{F}_q[z_1, \ldots, z_n]/\mathfrak{d}(a_0, \ldots, a_{n-1})$ does. To the contrary, Corollary VI.7.5 shows that there is a correction term needed, namely the conjugate Mitchell–Stong element of $\mathbf{D}(n)$. According to [61] this correction term (see also Example 1 in Section VI.7) for $q = 2$ is

$$\chi\text{ms}(\mathbf{D}(n)) = (1 + \mathbf{d}_{n,0} + \cdots + \mathbf{d}_{n,n-1})^{n-1} \in \mathbb{F}_2[z_1, \ldots, z_n].$$

This leads to a shift in the indices $(a_0, \ldots, a_{n-1})$ for which the quotient $\mathbf{D}(n)/\mathfrak{d}(a_0, \ldots, a_{n-1})$ has trivial Wu classes as opposed to those for which the corresponding quotient $\mathbb{F}_q[z_1, \ldots, z_n]/\mathfrak{d}(a_0, \ldots, a_{n-1})$ has trivial Wu classes. For $q = 2$ and a small number of variables we may combine this with Corollaries V.1.7 and V.3.6 to determine the $\mathscr{A}^*$-invariant ideals with trivial Wu classes. This includes the solution to the *Hit Problem* for the Dickson algebras $\mathbf{D}(2)$ and $\mathbf{D}(3)$ (see [81] Lemmas 9.2 through 9.5 for $\mathbf{D}(2)$ and [33] Theorem 2.6 for $\mathbf{D}(3)$).

---

15 If $A \hookrightarrow B$ is an extension of unstable algebras over the Steenrod algebra and $I \subset A$ is a $\mathscr{P}^*$-invariant ideal, then so is the ideal $I^{\text{ex}} \subset B$. The converse is however problematic: here is why. If $J \subset A$ is an ideal and $J^{\text{ex}} \subset B$ is $\mathscr{P}^*$-invariant, then of course so is $J^{\text{ex}} \cap A$. However, it may happen that $J \subsetneq J^{\text{ex}} \cap A$. In other words, $J^{\text{ex}} \cap A$ may well contain elements not present in J, which a priori, could be of the form $\mathscr{P}^i(a)$ for some $a \in J$. In the case A and B are polynomial algebras, and J is $\overline{A}$-primary and regular, then $J^{\text{ex}} = J^{\text{fit}}$, so J^{ex} lies over J, and therefore $J = J^{\text{ex}} \cap A$ is $\mathscr{P}^*$-invariant.

THEOREM VI.8.3: *An ideal $\mathfrak{d}(a, b) = (\mathbf{d}_{2,0}^{a}, \mathbf{d}_{2,1}^{b}) \subset \mathbf{D}(2)$ is invariant under the action of the mod 2 Steenrod algebra $\mathscr{A}^*$ if and only if when we write $b = 2^\ell \cdot c$ with c odd, then $a \le 2^\ell$. The quotient algebra $\mathbf{D}(2)/\mathfrak{d}(a, b)$ has trivial Wu classes if and only if $a = 1$ and $b = 2^s$. Hence the images of the monomials $\mathbf{d}_{2,1}^{2^s-1}$ for $s \ge 0$ in $\mathbb{F}_2 \otimes_{\mathscr{A}^*} \mathbf{D}(2)$ provide an $\mathbb{F}_2$-vector space basis for the $\mathscr{A}^*$-indecomposable elements.*

PROOF: By Proposition VI.8.2, if we write $b = 2^\ell \cdot c$ with c odd, then $a \le 2^\ell$. In addition, if we choose $m \in \mathbb{N}_0$ such that $2^m \le c \le 2^{m+1} - 1$ then

$$\chi\mathrm{Wu}(\mathfrak{d}(a, b)^{\mathrm{ex}}) = (1 + \mathbf{d}_{2,1} + \mathbf{d}_{2,0})^{2^\ell - a} \cdot (1 + \mathbf{d}_{2,1}^{2^\ell})^{2^{m+1}-(c+1)}$$

by Corollary V.1.7. Since $\mathfrak{d}(a, b) \subset \mathbf{D}(2)$ is a regular ideal its fit and ordinary extensions coincide by Corollary VI.3.2 and Construction VI.4.4. By Proposition VI.7.1 and Example 1 in Section VI.7 we therefore obtain

$$\begin{aligned} \alpha\big(\chi\mathrm{Wu}(\mathfrak{d}(a, b))\big) &= \chi\mathrm{Wu}(\mathfrak{d}(a, b)^{\mathrm{fit}}) \cdot \alpha\big(\chi\mathrm{ms}(\mathbf{D}(2))\big) \\ &= (1 + \mathbf{d}_{2,1} + \mathbf{d}_{2,0})^{2^\ell - a + 1} \cdot (1 + \mathbf{d}_{2,1}^{2^\ell})^{2^{m+1}-(c+1)}, \end{aligned}$$

where $\alpha : \mathbf{D}(2) \hookrightarrow \mathbb{F}_2[x, y]$ is the canonical inclusion. Hence $\mathfrak{d}(a, b) \subset \mathbf{D}(2)$ has trivial Wu classes if and only if

$$(\because) \qquad 1 = (1 + \mathbf{d}_{2,1} + \mathbf{d}_{2,0})^{2^\ell - a + 1} \cdot (1 + \mathbf{d}_{2,1}^{2^\ell})^{2^{m+1}-(c+1)}.$$

In order that the expansion of the first term contribute no nonzero term of the form $\mathbf{d}_{2,0}^{i}$ it must be the case that the exponent $2^\ell - a + 1$ is power of 2 times an odd integer, say

$$2^\ell - a + 1 = 2^t \cdot k$$

for some $k, t \in \mathbb{N}_0$, k odd. If $2^t < 2^\ell$ then

$$2^t < 2^\ell \le 2^\ell \cdot c = b.$$

So the right hand side of $(\because)$ contains the term $\mathbf{d}_{2,1}^{2^t}$ which does not belong to $\mathfrak{d}(a, b)$ and in this case the ideal $\mathfrak{d}(a, b) \subset \mathbf{D}(2)$ cannot have trivial conjugate Wu classes. Therefore $t = \ell$ and so $a = 1 = k$.

If we substitute $a = 1$ back into the formula $(\because)$ for the conjugate Wu classes to be trivial we find

$$1 = (1 + \mathbf{d}_{2,1}^{2^\ell})^{2^{m+1}-c}.$$

Since c is odd the expansion of the right hand side contains the term $\mathbf{d}_{2,1}^{2^\ell}$ which belongs to $(\mathbf{d}_{2,0}, \mathbf{d}_{2,1}^{b}) \subset \mathbf{D}(2)$ if and only if $2^\ell \ge b$, i.e., if and only if $c = 1$. So if an ideal $\mathfrak{d}(a, b)$ has trivial Wu classes in $\mathbf{D}(2)$ then $a = 1$ and $b = 2^\ell$.

On the other hand if $a = 1$ and $b = 2^\ell$ then we may choose $m = 0$ and formula (·⁝·)

$$\begin{aligned}\chi\mathrm{Wu}(\mathfrak{d}(1, 2^{\ell})) &= (1 + \mathbf{d}_{2,1} + \mathbf{d}_{2,0})^{2^{\ell}} \cdot (1 + \mathbf{d}_{2,1})^{2-2} \\ &= (1 + \mathbf{d}_{2,1} + \mathbf{d}_{2,0})^{2^{\ell}} = 1 + \mathbf{d}_{2,1}^{2^{\ell}} + \mathbf{d}_{2,0}^{2^{\ell}} \\ &= 1 \bmod \mathfrak{d}(1, 2^{\ell})\end{aligned}$$

completing the proof. □

THEOREM VI.8.4: *Let $a_0, a_1, a_2 \in \mathbb{N}$ and write $a_i = 2^{s_i} \cdot b_i$ with b_i odd, $i = 0, 1, 2$. The ideal $\mathfrak{d}(a_0, a_1, a_2) = (\mathbf{d}_{3,0}^{a_0}, \mathbf{d}_{3,1}^{a_1}, \mathbf{d}_{3,2}^{a_2}) \subset \mathbf{D}(3)$ is $\mathscr{A}^*$-invariant if and only if $2^{s_1} \geq a_0$ and $2^{s_2} \geq a_1$. The Poincaré duality algebra $\mathbf{D}(3)/\mathfrak{d}(a_0, a_1, a_2)$ has trivial Wu classes if and only if*

(i) $(a_0, a_1, a_2) = (1, 1, 2^t)$ for some $t \in \mathbb{N}_0$, or

(ii) $(a_0, a_1, a_2) = (2, 2^s, 2^r - 2^s)$ for some $r > s \in \mathbb{N}$.

The image in $\mathbf{D}(3)_{\mathscr{A}^*}$ of the monomials $\mathbf{d}_{3,2}^{2^t-1}$, $\mathbf{d}_{3,0} \cdot \mathbf{d}_{3,1}^{2^s-1} \cdot \mathbf{d}_{3,2}^{2^r-2^s-1}$ such that $t \in \mathbb{N}_0$ and $r > s \in \mathbb{N}$ are an $\mathbb{F}_2$ basis.

PROOF: Either by direct computation as in Example 1 of Section VI.7 or from [61] we have that

$$\chi\mathrm{ms}(\mathbf{D}(3)) = (1 + \mathbf{d}_{3,2} + \mathbf{d}_{3,1} + \mathbf{d}_{3,0})^2.$$

Since the ideal $\mathfrak{d}(a_0, a_1, a_2) \subset \mathbf{D}(3)$ is $\mathscr{A}^*$-invariant by Proposition VI.8.2 we have

$$a_0 \leq 2^{s_1} \text{ and } a_1 \leq 2^{s_2},$$

where $a_i = 2^{s_i} \cdot b_i$ with b_i odd for $i = 0, 1, 2$. Moreover, since $\mathfrak{d}(a_0, a_1, a_2)$ is a regular ideal its fit and usual extension to $\mathbb{F}_2[x, y, z]$ coincide by Corollary VI.3.2 and Construction VI.4.4.

Choose a large power of 2, say 2^r. By Corollary V.3.6

$$\begin{aligned}&\chi\mathrm{Wu}(\mathbf{d}_{3,0}^{a_0}, \mathbf{d}_{3,1}^{a_1}, \mathbf{d}_{3,2}^{a_2}) \\ &\quad= (1 + \mathbf{d}_{3,2} + \mathbf{d}_{3,1} + \mathbf{d}_{3,0})^{2^r - a_0} \cdot (1 + \mathbf{d}_{3,2} + \mathbf{d}_{3,1})^{2^r - a_1} \cdot (1 + \mathbf{d}_{3,2})^{2^r - a_2},\end{aligned}$$

so if we apply Proposition VI.7.1 we obtain that

$$(\maltese)\quad (1 + \mathbf{d}_{3,2} + \mathbf{d}_{3,1} + \mathbf{d}_{3,0})^{2^r - a_0 + 2} \cdot (1 + \mathbf{d}_{3,2} + \mathbf{d}_{3,1})^{2^r - a_1} \cdot (1 + \mathbf{d}_{3,2})^{2^r - a_2}$$

is the image of the total conjugate Wu class of the ideal $\mathfrak{d}(a_0, a_1, a_2) \subset \mathbf{D}(3)$ regarded as an element of $\mathbb{F}_2[x, y, z]$ under the natural inclusion $\mathbf{D}(3) \hookrightarrow \mathbb{F}_2[x, y, z]$. This product must yield 1 in the quotient algebra $\mathbb{F}_2[x, y, z]/\mathfrak{d}(a_0, a_1, a_2)$ for the ideal $\mathfrak{d}(a_0, a_1, a_2) \subset \mathbf{D}(3)$ to have trivial Wu classes.

Reasoning as in the proof of Theorem VI.8.3 the expansion of (✠) contains a power of $\mathbf{d}_{3,0}$ unless $2^r - a_0 + 2$ is an odd multiple of a power of 2, say

$2^r - a_0 + 2 = 2^{t_0} \cdot k_0$, with $2^{t_0} \geq a_0$ and k_0 odd. Since r is large it follows that 2^{t_0} divides 2^r and hence also $a_0 - 2$. If $a_0 \neq 2$, since $2^{t_0} \geq a_0$ it must be the case $t_0 = 0$. If $t_0 = 0$ then the expansion of $(\maltese)$ will contain the term $\mathbf{d}_{3,0}$, so for $\mathfrak{d}(a_0, a_1, a_2) \subset \mathbf{D}(3)$ to have trivial Wu classes forces $a_0 = 1$.

To summarize: if $\mathfrak{d}(a_0, a_1, a_2) \subset \mathbf{D}(3)$ has trivial Wu classes then $a_0 = 1$ or $a_0 = 2$.

CASE: $a_0 = 1$. In this situation the formula for the conjugate Wu classes simplifies to

$$(1 + \mathbf{d}_{3,2} + \mathbf{d}_{3,1})^{2^{r+1}+1-a_1} \cdot (1 + \mathbf{d}_{3,2})^{2^r - a_2}.$$

In order that the expansion not contain nonzero powers of $\mathbf{d}_{3,1}$ we need that

$$2^{r+1} + 1 - a_1 = 2^{t_1} \cdot k_1$$

with $k_1 \in \mathbb{N}$ odd and $2^{t_1} \geq a_1$. Since 2^{t_1} divides 2^{r+1} it must divide $a_1 - 1$. As $2^{t_1} \geq a_1$ this is only possible if $a_1 = 1$. This means the formula for the conjugate Wu classes further simplifies to

$$(1 + \mathbf{d}_{3,2})^{2^{r+1}+2^r-a_2}$$

since both a_0 and a_1 are 1. Reasoning as before, this expansion will contain nonzero powers of $\mathbf{d}_{3,2}$ unless

$$2^{r+1} + 2^r - a_2 = 2^{t_2} \cdot k_2$$

with $k_2 \in \mathbb{N}$ odd and $2^{t_2} \geq a_2$. Since 2^{t_2} divides 2^{r+1} and 2^r it must also divide a_2. As $2^{t_2} \geq a_2$ this can only happen if $a_2 = 2^{t_2}$. Therefore an ideal $\mathfrak{d}(a_0, a_1, a_2) \subset \mathbf{D}(2)$ with trivial Wu classes and $a_0 = 1$ must be as in statement (i) of the theorem. As the preceding analysis shows these ideals indeed have trivial conjugate Wu classes.

CASE: $a_0 = 2$. The formula for the conjugate Wu classes of the ideal $\mathfrak{d}(2, a_1, a_2)$ simplifies to

$$(1 + \mathbf{d}_{3,2} + \mathbf{d}_{3,1})^{2^{r+1}-a_1} \cdot (1 + \mathbf{d}_{3,2})^{2^r - a_2}.$$

As before, in order that no nonzero powers of $\mathbf{d}_{3,1}$ appear in the expansion we must have

$$2^{r+1} - a_1 = 2^{t_1} \cdot k_1$$

where $k_1 \in \mathbb{N}$ is odd and $2^{t_1} \geq a_1$. Using that 2^{t_1} divides 2^{r+1} we see that 2^{t_1} also divides a_1 so in fact $a_1 = 2^{t_1}$. Given this, the formula for the conjugate Wu classes simplifies to

$$(1 + \mathbf{d}_{3,2})^{2^{r+1}+2^r-2^{t_1}-a_2}.$$

For the conjugate Wu classes to be trivial we must have

$$2^r - 2^{t_1} - a_2 = 2^{t_2} \cdot k_2$$

with $k_2 \in \mathbb{N}$ odd and $2^{t_2} \geq a_2$. Then 2^{t_2} must divide $2^{t_1} + a_2$, and since $2^{t_1} = a_1 \leq 2^{s_2} \leq a_2$ this can only happen if $2^{t_1} + a_2 = 2^{t_2}$, i.e., $a_2 = 2^{t_2} - 2^{t_1}$. Thus those ideals $\mathfrak{d}(a_0, a_1, a_2) \subset \mathbf{D}(3)$ with $a_0 = 2$ and trivial Wu classes are exactly those listed in statement (ii) of the theorem. □

We have found no *simple* construction on ideals in $\mathbf{D}(n)$ analogous to the Frobenius power operator that produces new $\mathscr{A}^*$-invariant ideals with trivial Wu classes from old ones. In the general case one needs to multiply the formula of Theorem V.4.3 with the conjugate Mitchell–Stong element and redo the computations leading up to the proof of Theorem V.4.7.

In a completely analogous fashion we could obtain results for the *Hit Problem* for $\mathbb{F}_2[w_2, w_3, w_4] = H^*(B\mathbb{SO}(4); \mathbb{F}_2)$ from the results of Section V.5. In this manner one would obtain alternative proofs of results in [109] and [36].

References

[1] J. F. Adams, On Formulae of Thom and Wu, *Proc. London Math. Soc.(3)* **11** (1961), 741–752.

[2] J. F. Adams and C. W. Wilkerson, Finite H-spaces and Algebras over the Steenrod Algebra, *Ann. Math. (2)* **111** (1980), 95–143.

[3] M. A. Alghmadi, M. C. Crabb and J. R. Hubbuck, Representations of the Homology of BV and the Steenrod Algebra I, in: *Adams Memorial Symposium on Algebraic Topology (Manchester 1990),* Ed. N. Ray and G. Walker, LMS Lecture Notes 176, Cambridge: Cambridge University Press, 1992, pp. 217–234.

[4] S. Balcerzyk and T. Józefiak, *Commutative Noetherian and Krull Rings,* Warsaw: Polish Scientific Publishers, 1989.

[5] S. Balcerzyk and T. Józefiak, *Commutative Rings: dimension, multiplicity and homological methods,* Warsaw: Polish Scientific Publishers, 1989.

[6] H. Bass, On the Ubiquity of Gorenstein Rings, *Math. Z.* **82** (1963), 8–28.

[7] T. P. Bisson and A. Joyal, Q-Rings and the Homology of the Symmetric Groups, *Operads: Proc. Renaissance Conferences (Hartford, CT/Luminy 1995),* Ed. J.-L. Loday, J. D. Stasheff and A. A. Voronov, *Contemp. Math.* **202** (1997), 235–286.

[8] J. M. Boardman, Modular Representations on the Homology of Powers of Real Projective Space, *Algebraic Topology: Oaxtepec 1991,* Ed. M. C. Tangora, *Contemp. Math.* **146** (1993), 49–70.

[9] N. Bourbaki, *Groupes et Algèbres de Lie, Chapitres* 4, 5, *et* 6, Paris: Masson, 1981.

[10] C. Broto, L. Smith and R. E. Stong, Thom Modules and Pseudoreflection Groups, *J. Pure Appl. Algebra* **60** (1989), 1–20.

[11] W. Bruns and J. Herzog, *Cohen–Macaulay Rings,* Cambridge Studies in Advanced Math., 39, Cambridge: Cambridge University Press, 1993.

[12] D. P. Carlisle and R. M. W. Wood, The Boundedness Conjecture for the Action of the Steenrod Algebra on Polynomials, in: *Adams Memorial Symposium on Algebraic Topology (Manchester 1990),* Ed. N. Ray and G. Walker, LMS Lecture Notes 176, Cambridge: Cambridge University Press, 1992, pp. 203–216.

[13] H. Cartan, *Algèbres d'Eilenberg–Maclane et Homotopie,* Seminaire H. Cartan 7eme année: 1954/1955.

[14] H. Cartan, Quotient d'un Éspace Analytique par un Groupe d'Automorphismes, in: *Algebraic Geometry and Topology, A Symposium in Honor of S. Lefschetz,* Ed. R. H. Fox, D. C. Spencer and A. W. Tucker, Princeton, NJ: Princeton University Press, 1957.

[15] H. Cartan and S. Eilenberg, *Homological Algebra,* Princeton, NJ: Princeton University Press, 1956.

[16] M. C. Crabb and J. R. Hubbuck, Representations of the Homology of BV and the Steenrod Algebra II, in: *Algebraic Topology (New Trends in Localization and Periodicity, St Feliu de Guixols 1994),* Ed. C. Broto, C. Casacuberta and G. Mislin, Progr. Math. 136, Basel: Birkhäuser Verlag, 1996, pp. 143–154.

[17] M. D. Crossley, $\mathcal{A}^*(p)$-annihilated Elements in $H_*(\mathbb{CP}(\infty)\times\mathbb{CP}(\infty))$, *Math. Proc. Cambridge Philos. Soc.* **120** (1996), 441–453.

[18] C. W. Curtis and I. Reiner, *Representations of Finite Groups and Associative Algebras,* Pure and Applied Math XI, New York: Interscience, 1962.

[19] M. Demazure, Invariants Symétriques Entiers des Groupes de Weyl et Torsion, *Inventiones Math.* **21** (1973), 287–301.

[20] L. E. Dickson, A Fundamental System of Invariants of the General Modular Linear Group with a Solution of the Form Problem, *Trans. Am. Math. Soc.* **12** (1911), 75–98.

[21] D. Eisenbud, *Commutative Algebra,* Heidelberg: Springer-Verlag, 1995.

[22] R. M. Fossum, One Dimensional Formal Group Actions, *Invariant Theory (Denton, TX 1986),* Ed. R. Fossum, W. Haboush, M. Hochster and V. Lakshmibai, *Contemp. Math.* **88** (1989), 227–397.

[23] P. Gabriel, Objets injectifs dans les catégories abéliennes, Séminaire P. Dubriel (12eme année 1958/59), Exposé 17, 17-01–17-32.

[24] A. Geramita, Inverse Systems of Fat Points: Waring's Problem, Secant Varieties of Veronese Varieties, and Parameter Spaces for Gorenstein Ideals, *The Curves Seminar at Queen's, Vol. X (Kingston, ON, 1995), Queen's Papers Pure Appl. Math.* **102** (1996), 2–114.

[25] D. Glassbrenner, The Cohen–Macaulay Property and F-Rationality in Certain Rings of Invariants, *J. Algebra* **176** (1995), 824–860.

[26] O. E. Glenn, Modular Invariant Processes, *Bull. Am. Math. Soc.* **21** (1914–1915), 167–173.

[27] J. P. C. Greenlees and L. Smith, A Local Cohomology Interpretation of Macaulay's Inverse Systems, (in preparation).

[28] W. Gröbner, Über irreduzible Ideale in kommutativen Ringen, *Math. Ann.* **110** (1934), 187–222.

[29] H. Hasse and F. K. Schmidt, Noch eine Begründung der Theorie des höheren Differentialquotienten in einem algebraischen Funktionkörper in einer Unbestimmten, *J. Reine Angew. Math.* **177** (1937), 215–237.

[30] M. Hochster, Contracted Ideals from Integral Extensions of Regular Rings, *Nagoya Math. J.* **51** (1973), 25–43.

[31] M. Hochster and J. A. Eagon, Cohen–Macaulay Rings, Invariant Theory, and the Generic Perfection of Determinantal Loci, *Am. J. Math.* **93** (1971), 1020–1058.

[32] M. Hochster and C. Huneke, Tight Closure, Invariant Theory, and the Briançon–Skoda Theorem, *J. Am. Math. Soc.* **3** (1990), 31–116.

[33] N. H. V. Hung and F. P. Peterson, $\mathscr{A}$-Generators for the Dickson Algebra, *Trans. Am. Math. Soc.* **347** (1995), 4687–4728.

[34] A. Iarrobino and V. Kanev, *Power Sums, Gorenstein Ideals, and Determinantal Loci,* Lecture Notes in Math. 1721, Heidelberg: Springer-Verlag, 1999.

[35] N. Jacobson, *Lectures on Abstract Algebra II, Linear Algebra,* Princeton, NJ: D. van Nostrand, 1953.

[36] A. S. Janfada and R. M. W. Wood, The Hit Problem for Symmetric Polynomials over the Steenrod Algebra, *Math. Proc. Cambridge Philos. Soc.* **133** (2002), 295–303.

[37] A. S. Janfada and R. M. W. Wood, Generating $H^*(B\mathbb{SO}(3);\mathbb{F}_2)$ as a Module over the Steenrod Algebra, *Math. Proc. Cambridge Philos. Soc.* **134** (2003), 239–258.

[38] D. L. Johnson, The Group of Formal Power Series under Substitution, *J. Aust. Math. Soc.* **45** (1988), 296–302.

[39] M. Kameko, *Products of Projective Spaces as Steenrod Modules, Thesis,* Johns Hopkins University, Baltimore, MD, 1990.

[40] R. Kane, Poincaré Duality and the Ring of Coinvariants, *Can. Math. Bull.* **37** (1994), 82–88.

[41] R. Kane, *Reflection Groups and Invariant Theory,* CMS Books in Math., New York: Springer-Verlag, 2001.

[42] A. Kerber, *Algebraic Combinatorics via Finite Group Actions,* Mannheim: Bibliographisches Institut & F. A. Brockhaus AG, 1991.

[43] W. Krull, *Idealtheorie,* zweite ergänzte Auflage, Erg. Math. Bd. 46, Heidelberg: Springer-Verlag, 1968.

[44] K. Kuhnigk, *Poincarédualitätsalgebren, Koinvarianten, und Wu Klassen, Doktorarbeit*, Universität Göttingen, 2003.

[45] E. Kunz, Characterizations of Regular Local Rings for Characteristic p, *Am. J. Math.* **91** (1969), 772–784.

[46] G. I. Lehrer, A New Proof of Steinberg's Fixed-Point Theorem, *Int. Math. Res. Notices* **28** (2004), 1407–1411.

[47] T.-C. Lin, *Coinvariants of Pseudoreflection Groups,* Göttingen: Cuvillier Verlag, 2003.

[48] F. S. Macaulay, On the Resolution of a given Modular System into Primary Systems, *Math. Ann.* **74** (1913), 66–121.

[49] F. S. Macaulay, *The Algebraic Theory of Modular Systems,* Camb. Math. Lib., Cambridge: Cambridge University Press, 1916 (reissued with an introduction by P. Roberts 1994).

[50] F. S. Macaulay, Modern Algebra and Polynomial Ideals, *Proc. Cambridge Philos. Soc.* **30** (1934), 27–46.

[51] W. S. Massey and F. P. Peterson, The Cohomology Structure of Certain Fibre Spaces I, *Topology* **4** (1965), 47–65.

[52] B. H. Matzat, *Differential Galois Theory in Positive Characteristic,* Notes written by J. Hartmann, IWR Universität Heidelberg, Preprint 2001-35.

[53] D. M. Meyer, Injective Objects in Categories of Unstable K-modules, *Bonn. Math. Schriften* **316** (1999).

[54] D. M. Meyer, Hit Polynomials and Excess in the Mod p Steenrod Algebra, *Proc. Edinburgh Math. Soc. (2)* **44** (2001), 323–350.

[55] D. M. Meyer, Stripping and Conjugation in the Mod p Steenrod Algebra and its Dual, *Homology, Homotopy, and Appl.* **2** (2000), 1–16 (electronic).

[56] D. M. Meyer and L. Smith, Realization and Nonrealization of Poincaré Duality Quotients of $\mathbb{F}_2[x, y]$ as Topological Spaces, *Fund. Math.* **177** (2003), 241–250.

[57] E. Miller and B. Sturmfels, Combinatorial Commutative Algebra, Graduate Texts in Math. 227, New York: Springer-Verlag, 2005.

[58] J. W. Milnor, The Steenrod Algebra and its Dual, *Ann. Math. (2)* **67** (1958), 150–171.

[59] J. W. Milnor and J. C. Moore, On the Structure of Hopf Algebras, *Ann. Math. (2)* **81** (1965) , 211–264.

[60] S. A. Mitchell, Finite Complexes with $\mathscr{A}(n)$ Free Cohomology, *Topology* **24** (1985), 227–246.

[61] S. A. Mitchell and R. E. Stong, An Adjoint Representation for Polynomial Algebras, *Proc. Am. Math. Soc.* **101** (1987), 161–167.

[62] J.C. Moore and L. Smith, Hopf Algebras and Multiplicative Fibrations I, *Am. J. Math.* **90** (1968), 752–780.

[63] K. Morita and H. Tachikawa, Character Modules, Submodules of Free Modules, and Quasi-Frobenius Rings, *Math. Z.* **65** (1956), 414–428.

[64] M. Nagata, *Local Rings,* New York: John Wiley & Sons, 1962.

[65] Nam Trần Ngọc, Solution Générique du Problème “Hit” pour l’Algèbre Polynomiale, LAGA, Univ. de Paris-Nord, Preprint 2002-25.

[66] M. D. Neusel, *Inverse Invariant Theory and Steenrod Operations,* Memoirs of the Am. Math. Soc. 146 (2000), No. 692, Providence, RI: American Mathematical Society 2000.

[67] M. D. Neusel and L. Smith, The Lasker–Noether Theorem for $\mathscr{P}^*$-invariant Ideals, *Forum Math.* **10** (1998), 1–18.

[68] M. D. Neusel and L. Smith, *Invariant Theory of Finite Groups,* Math. Surveys and Monographs 94, Providence, RI: American Mathematical Society, 2002.

[69] E. Noether, Idealtheorie in Ringbereichen, *Math. Ann.* **83** (1921), 24–66.

[70] D. G. Northcott, Injective Envelopes and Inverse Polynomials, *J. London Math. Soc.(2)* **8** (1974), 290–296.

[71] N. Nossem, *On the Perfect Closure of Commutative Noetherian Rings of Positive Prime Characteristic, PhD Thesis,* University of Sheffield, 2002.

[72] K. Okugawa, Basic Properties of Differential Fields in arbitrary Characteristic and the Picard–Vesoit Theory, *Mem. Coll. Sci. Kyoto Univ. A,* **2** (1962/1963), 295–322.

[73] C. Peskine and L. Szpiro, Dimension Projective finie et Cohomologie Locale, *Pub. Math. I.H.E.S.* **42** (1973), 47–119.

[74] F. P. Peterson, Generators of $H^*(\mathbb{RP}(\infty) \wedge \mathbb{RP}(\infty))$ as a Module over the Steenrod Algebra, Abstr. Am. Math. Soc. (1987), 833-55-89.

[75] F. P. Peterson, *A*-Generators for Certain Polynomial Algebras, *Math. Proc. Cambridge Philos. Soc.* **105** (1989), 311–312.

[76] L. Schwartz, *Unstable Modules over the Steenrod Algebra and Sullivan’s Fixed Point Set Conjecture,* Chicago Lectures in Math, Chicago, IL: University of Chicago Press, 1994.

[77] H. Seifert and W. Threlfall, *Lehrbuch der Topologie,* New York: Chelsea Publ. Co., 1947.

[78] J.–P. Serre, Sur la dimension cohomologique des groupes profinis, *Topology* **3** (1964), 413–420.

[79] C. Shengmin and S. Xinyao, On the Action of Steenrod Powers on Polynomial Algebras, in: *Algebraic Topology: Homotopy and Group Cohomology (St. Feliu de Guixols 1990),* Ed. J. Aguadé, M. Castellet and F. R. Cohen, Lecture Notes in Math. 1509, Heidelberg: Springer-Verlag, 1991.

[80] G. C. Shephard and J. A. Todd, Finite Unitary Reflection Groups, *Can. J. Math.* **6** (1954), 274–304.

[81] W. M. Singer, The Iterated Transfer in Homological Algebra, *Math. Z.* **202** (1989), 493–523.

[82] W. M. Singer, On the Action of Steenrod Squares on Polynomial Algebras, *Proc. Am. Math. Soc.* **111** (1991), 577–583.

[83] B. Singh, Invariants of Finite Groups Acting on Local Unique Factorization Domains, *J. Indian Math. Soc. (N. S.)* **34** (1970), 31–38.

[84] L. Smith, Homological Algebra and the Eilenberg–Moore Spectral Sequence, *Trans. Am. Math. Soc.* **129** (1967), 58–93.

[85] L. Smith, Split Extensions of Hopf Algebras and Semi Tensor Products, *Math. Scand.* **26** (1970), 17–41.

[86] L. Smith, A Note on the Realization of Complete Intersection Algebras by the Cohomology of a Space, *Q. J. Math. Oxford Ser. (2)* **33** (1982), 379–384.

[87] L. Smith, *Polynomial Invariants of Finite Groups,* Wellesley, MA: A.K. Peters, Ltd., 1995, second printing 1997.

[88] L. Smith, *Linear Algebra,* New York: Springer-Verlag, 1998, 3rd edn.

[89] L. Smith, Cohomology Automorphisms over Galois Fields and Group Like Elements in the Steenrod Algebra, AG-Invariantentheorie Göttingen, Preprint 2000.

[90] L. Smith, An Algebraic Introduction to the Steenrod Algebra, Course Notes from the Summer School *Interactions Between Invariant Theory and Algebraic Topology, Ioannina 2000.*

[91] L. Smith, Invariants and Coinvariants of Finite Pseudoreflection Groups, Jacobian Determinants and Steenrod Operations, *Proc. Edinburgh Math. Soc. (2)* **44** (2001), 597–611.

[92] L. Smith, On a Theorem of R. Steinberg on Rings of Coinvariants, *Proc. Am. Math. Soc.* **131** (2003), 1043–1048.

[93] L. Smith, On Alternating Invariants and Hilbert Ideals, *J. Algebra* **280** (2004), 488–499.

[94] L. Smith and R. E. Stong, On the Invariant Theory of Finite Groups: Orbit Polynomials, Chern Classes and Splitting Principles, *J. Algebra* **109** (1987), 134–157.

[95] R. P. Stanley, Hilbert Functions of Graded Algebras, *Adv. Math.* **28** (1978), 57–83.

[96] R. P. Stanley, *Enumerative Combinatorics,* Cambridge Studies in Advanced Mathematics 49 and 62, Cambridge: Cambridge University Press, 1997/1999, volumes I and II.

[97] N. E. Steenrod and D. B. A. Epstein, *Cohomology Operations, Lectures by N.E. Steenrod written and revised by D. B. A. Epstein,* Annals of Math. Studies 50, Princeton, NJ: Princeton University Press, 1962.

[98] R. Steinberg, Differential Equations Invariant under Finite Reflection Groups, *Trans. Am. Math. Soc.* **112** (1964), 392–400.

[99] R. Steinberg, On Dickson's Theorem on Invariants, *J. Fac. Sci. Univ. Tokyo Sect. 1A Math.* **34** (1987), 699–707.

[100] H. Toda, Cohomology mod 3 of the Classifying Space BF_4 of the Exceptional Group F_4, *J. Math. Kyoto. Univ.* **13** (1973), 97–115.

[101] W. N. Traves, Tight Closure and Differential Simplicity, *J. Algebra* **228** (2000), 457–476.

[102] W. V. Vasconcelos, Ideals Generated by R-Sequences, *J. Algebra* **6** (1967), 309–316.

[103] K. Watanabe, A Note on Gorenstein Rings of Embedding Dimension Three, *Nagoya Math. J.* **50** (1973), 227–232.

[104] K. Watanabe, Certain Invariant Subrings are Gorenstein, *Osaka J. Math.* **11** (1974), 1–8.

[105] H. Wiebe, Über homologische Invarianten lokaler Ringe, *Math. Ann.* **179** (1969), 257–274.

[106] C. W. Wilkerson, Classifying Spaces, Steenrod Operations, and Algebraic Closure, *Topology* **16** (1977), 227–237.

[107] R. M. W. Wood, Steenrod Squares of Polynomials and the Peterson Conjecture, *Math. Proc. Cambridge Philos. Soc.* **105** (1989), 307–309.

[108] R. Wood, Problems in the Steenrod Algebra, *Bull. London Math. Soc.* **30** (1998), 449–517.

[109] R. Wood, *Hit Problems and the Steenrod Algebra,* Course Notes from the Summer School *Interactions Between Invariant Theory and Algebraic Topology, Ioannina 2000.*

[110] R. M. W. Wood, *Invariant Theory and the Steenrod Algebra,* Course Notes from the Lecture Series at the Conference *Invariant Theory and its Interactions with related Fields, Göttingen 2003.*

[111] Wu Wen-Tsün, Classes Characteristiques et i-carré d'une Varieté, *C. R. Acad. Sci. Paris* **230** (1950), 508–511.

[112] Wu Wen-Tsün, *Sur les puissances de Steenrod,* Colloque de Topologie de Strasbourg, 1951, No. IX, La Bibliothèque Nationale et Universitaire de Strasbourg, 1952.

[113] I. O. York, The Exponent of certain p-Groups, *Proc. Edinburgh Math. Soc. (2)* **33** (1990), 483–490.

[114] O. Zariski and P. Samuel, *Commutative Algebra,* Graduate Texts in Math. 28, 29, Berlin: Springer-Verlag, 1975, volumes I and II.

Notation

General

$\square$ indicates the end of a proof, or that the proof of what has been stated is left to the reader

$\delta_{E,F}$ Kronecker delta symbol, i.e., 1 if $E = F$ and 0 otherwise

$|X|$ number of elements in the set X

$|a| = \deg(a)$ the degree of the homogeneous element a in a graded object

$a \overset{\circ}{=} b$ means a and b are nonzero multiples of each other

$\nu(n)$, $\omega(n)$ for $n \in \mathbb{N}$ are defined by $n = 2^{\nu(n)} \cdot \omega(n)$ where $\nu(n), \omega(n) \in \mathbb{N}$ and $\omega(n)$ is odd

$\alpha_i(n)$ is the i-th digit in the binary expansion of n so $n = \sum_{i=0}^{\infty} \alpha_i(n) 2^i$

– an arbitrary object of a specified category

Set Theory

$\subset$ subset

$\subseteq$ subset or equal

$\subsetneqq$ proper subset

$\hookrightarrow$ an injective map

Rings, Fields, and Vector Spaces

$\mathbb{N} = \{1, 2, 3, \ldots, \}$, the natural numbers

$\mathbb{N}_0 = \{0, 1, 2, \ldots, \}$, the nonnegative integers

$\mathbb{Z} = \{, \ldots, -2, -1, 0, 1, 2, \ldots, \}$, the ring of integers

$\mathbb{Z}/m$ the integers modulo m, cyclic group of order m

$\mathbb{Q}$ the field of rational numbers

$\mathbb{R}$ the field of real numbers

$\mathbb{C}$ the field of complex numbers

$\mathbb{F}_q$ the field with q elements ($q = p^\nu$, $p \in \mathbb{N}$ a prime)

$\mathbb{F}^\times$ the group invertible elements in the field $\mathbb{F}$ with respect to multiplication

V^* dual vector space of V, if $\mathbb{F}$ is the ground field then $V^* = \mathrm{Hom}_{\mathbb{F}}(V, \mathbb{F})$
$\mathbb{PF}(n)$ projective space of dimension n, i.e., the set of 1-dimensional subspaces of $\mathbb{F}^{n+1}$
$\mathbb{P}(V)$ projective space of V, i.e., the set of 1-dimensional subspaces of V
$\Phi(-)$ is the graded object $-$ regraded so $\Phi(-)_{kp} = -_k$, and 0 otherwise, where p is the characteristic of the ground field

Groups
$\mathrm{GL}(n, \mathbb{F})$ the group of $n \times n$ invertible matrices over $\mathbb{F}$
$\mathrm{SL}(n, \mathbb{F})$ subgroup of $\mathrm{GL}(n, \mathbb{F})$ of matrices $\mathbf{A}$ with $\det \mathbf{A} = 1$
$\mathrm{Uni}(n, \mathbb{F})$ unipotent subgroup of $\mathrm{GL}(n, \mathbb{F})$
Σ_n the symmetric group of the n-element set
$\mathrm{sgn} : \Sigma_n \longrightarrow \mathbb{Z}/2$ the signum character of Σ_n
$\mathrm{sgn}(\sigma)$ the sign of the permutation $\sigma \in \Sigma_n$
$|G|$ order of the group G
$|G : H|$ the index of the subgroup H in G
$\mathrm{GL}(V)_I$ for an ideal $I \subset \mathbb{F}[V]$ is the subgroup $\mathrm{GL}(V)_I = \{\mathbf{T} \in \mathrm{GL}(V) \mid \mathbf{T}(I) = I\}$

Algebras: Their Elements and Ideals
I^{ex} is the extended ideal of $I \subset A$ to an overring $B \supseteq A$
$I^{[q^s]}$ q^s-th Frobenius power of the ideal I
$\mathbb{F}[V]$ algebra of homogeneous polynomial functions on the vector space V
$S(V)$ symmetric algebra on the vector space V
$\Gamma(V)$ divided power algebra on the vector space V
γ_k divided power operation on $\Gamma(V)$
$f \cap \gamma$ the action of $f \in \mathbb{F}[V]$ on $\gamma \in \Gamma(V)$
$M(\gamma)$ for $\gamma \in \Gamma(V)$ is the $\mathbb{F}[V]$-submodule generated by γ
$I^{\perp}$ dual system in $\Gamma(V)$ of an ideal $I \subset \mathbb{F}[V]$
I^{-1} inverse system in $\mathbb{F}[X_1^{-1}, \ldots, X_n^{-1}]$ of an ideal $I \subset \mathbb{F}[X_1, \ldots, X_n]$
θ_I generator for the dual principal system of an $\overline{\mathbb{F}[V]}$-primary irreducible ideal $I \subset \mathbb{F}[V]$
$\mathbf{Cat}_\theta(i, j)$ is the (i, j)-th catalecticant matrix associated to $\theta \in \mathbb{F}[X_1, \ldots, X_n]$
$z^E = z_1^{e_1} \cdot z_2^{e_2} \cdots z_n^{e_n}$ a monomial in $z_1, \ldots, z_n$, E is called the exponent sequence
γ_E element of $\Gamma(V)$ dual to z^E with respect to the monomial basis
e_i i-th elementary symmetric polynomial
w_i i-th Stiefel–Whitney class
$\mathbf{D}(n)$ Dickson algebra in n variables over a Galois field
$\mathbf{d}_{n,i}$ i-th Dickson polynomial in n variables over a finite field
$\mathbf{L}_n$ Dickson–Euler class, characterized by $\mathbf{L}_n^{q-1} = \mathbf{d}_{n,0} \in \mathbb{F}_q[z_1, \ldots, z_n]$
$\mathfrak{d}(k_0, \ldots, k_n)$ ideal of $\mathbb{F}_q[z_1, \ldots, z_n]$ generated by $\mathbf{d}_{n,0}^{k_0}, \ldots, \mathbf{d}_{n,n-1}^{k_{n-1}}$
$D_n = (\deg(\mathbf{d}_{n,0}), \ldots, \deg(\mathbf{d}_{n,n-1})) \in \mathbb{N}^n$

$\mathfrak{D}_n$ the Σ_n orbit of $D_n \in \mathbb{N}^n$ under the permutation action
$\overline{A}$ augmentation ideal of the algebra A
$\mathfrak{m}$ the maximal ideal of the algebra under discussion
$A^{\geq k}$ denotes the ideal of all elements in A whose degree is greater than or equal to k
$B//A = \mathbb{F} \otimes_A B$ where B is an algebra over the algebra A
$\mathbb{F}[V]^G$ algebra of invariants of $G \hookrightarrow \mathrm{GL}(V)$, $V = \mathbb{F}^n$
$\mathfrak{h}(G)$ Hilbert ideal of G, viz., ideal in $\mathbb{F}[V]$ generated by $\overline{\mathbb{F}[V]^G}$
$\mathbb{F}[V]_G$ algebra of coinvariants of G, viz., $\mathbb{F}[V]/\mathfrak{h}(G) = \mathbb{F} \otimes_{\mathbb{F}[V]^G} \mathbb{F}[V]$
$\mathrm{Ann}_X(Y)$ the annhilator in X of Y
over(I) the set of ideals J that contain I
Ξ the self map of over(I) defined by $\Xi(J) = (I : J)$, an involution if I is $\overline{A}$-primary and irreducible
$\mathfrak{a}(W)$ the little ancestor ideal of W for $W \subset A_d$ of codimension one
$\mathfrak{A}(W) = \mathfrak{a}(W) + A^{\geq d+1}$ the big ancestor ideal of W
$H' \# H''$ connected sum of the Poincaré duality algebras H' and H''
$I(\gamma)$ for $0 \neq \gamma : \mathbb{F}[V]_d \longrightarrow \mathbb{F}$ is the big ancestor ideal of $\ker(\gamma)$
$\mathrm{supp}(\theta)$ for $\theta \in \Gamma(V)$ is $\{z^D \mid |D| = \deg(\theta) \text{ and } \theta(z^D) \neq 0\}$
$\mathbf{P}$ periodicity operator defined on $\mathbb{F}[z_1, \ldots, z_n]$ by $\mathbf{P}(f) = (z_1 \cdots z_n)^{q-1} f^q$
$\mathbf{P}$ periodicity operator defined on $\Gamma(V)$ by $\mathbf{P}(\theta) = \gamma_{q-1}(u_1) \cdots \gamma_{q-1}(u_n)\Upsilon_q(\theta)$

Steenrod Algebra and Wu Classes

$\mathscr{P}^*$ Steenrod algebra over a Galois field
$\mathscr{P}^{**} = \prod_{i=0}^{\infty} \mathscr{P}^i$
$\mathscr{A}^*$ Steenrod algebra over the field $\mathbb{F}_2$
$\mathscr{A}^{**} = \prod_{i=0}^{\infty} \mathscr{A}^i$
$\mathcal{P}^i$ i-th Steenrod reduced power operation over a Galois field
Sq^i i-th Steenrod squaring operation over the field $\mathbb{F}_2$
$\mathcal{P} = 1 + \mathcal{P}^1 + \cdots + \mathcal{P}^k + \cdots \in \mathscr{P}^{**}$
$\mathbf{Sq} = 1 + \mathrm{Sq}^1 + \cdots + \mathrm{Sq}^k + \cdots \in \mathscr{A}^{**}$
$\mathbf{Q}_i = \mathbf{Sq}(\mathbf{d}_{n,i})$
$= [(1 + \mathbf{d}_{n,n-1} + \cdots + \mathbf{d}_{n,i+1})(\mathbf{d}_{n,i-1} + \cdots + \mathbf{d}_{n,0}) + \mathbf{d}_{n,i}(1 + \mathbf{d}_{n,n-1} + \cdots + \mathbf{d}_{n,i})]$
$\chi : \mathscr{P}^* \longrightarrow \mathscr{P}^*$ the canonical anti-automorphism of $\mathscr{P}^*$
$\mathrm{Wu}_k(H)$, $\chi\mathrm{Wu}(H)$ k-th Wu class respectively k-th conjugate Wu class of the $\mathscr{P}^*$-Poincaré duality algebra H
$\mathrm{Wu}(H)$, $\chi\mathrm{Wu}(H)$ total Wu class respectively total conjugate Wu class of the $\mathscr{P}^*$-Poincaré duality algebra H
$\mathrm{Wu}_k(I)$, $\chi\mathrm{Wu}(I)$ for a $\mathscr{P}^*$-invariant $\overline{\mathbb{F}_q[V]}$-primary irreducible ideal $I \subset \mathbb{F}_q[V]$ the k-th Wu class respectively k-th conjugate Wu class of the $\mathscr{P}^*$-Poincaré duality algebra $\mathbb{F}_q[V]/I$
$\mathrm{Wu}(I)$, $\chi\mathrm{Wu}(I)$ for a $\mathscr{P}^*$-invariant $\overline{\mathbb{F}_q[V]}$-primary irreducible ideal $I \subset \mathbb{F}_q[V]$ the total Wu class respectively total conjugate Wu class of the

$\mathscr{P}^*$-Poincaré duality algebra $\mathbb{F}_q[V]/I$

$\mathrm{ms} = 1 + \mathrm{ms}_1 + \cdots$ total Mitchell–Stong element of an unstable polynomial algebra over $\mathscr{P}^*$

$\chi\mathrm{ms} = 1 + \chi\mathrm{ms}_1 + \cdots$ total conjugate Mitchell–Stong element of an unstable polynomial algebra over $\mathscr{P}^*$

$\mathbb{F}[V]_{\mathscr{P}^*}$ module of $\mathscr{P}^*$-indecomposable elements

$\Gamma(V)^{\mathscr{P}^*}$ subalgebra of $\mathscr{P}^*$-invariants in $\Gamma(V)$

Topology

$H^*(X;\mathbb{F})$ denotes the cohomology of the space X with coefficients in $\mathbb{F}$

S^d the d-dimensional sphere

$\mathbb{RP}(k)$ the k-dimensional real projective space

$\mathbb{CP}(k)$ the k-dimensional complex projective space, a real $2k$-dimensional manifold

Index

BOLD face page entries indicate definitions, statement of theorems, etc, or clarification of notations. Index entries for the most important symbols appear on this page: the alphabetized index begins overleaf.

$(\mathbf{d}_{n,0}, \ldots, \mathbf{d}_{n,n-1})^{\perp}$ 88
2-adic expansion **102**
2-adic valuation 111–112
$\cap$-product 35, **36**
χ 56, **58**
$\chi\mathrm{Wu}_k$ **61**
$\Delta_{s,t}$ **101**
$\mathscr{D}$ 75–76
γ-operations **33**, 71
$\bar{A}$-primary 19
$\Phi(-)$ 50
$\wp$ **56**
Ξ **14**, 15–16, 41, 95
D_n **88**, 89
G-set 25
p-adic expansion 34
Wu_k **60**
$\mathscr{A}^*$ **54**
$\mathscr{A}^*$-equivalent **75**
$\mathscr{A}^*$-indecomposable 6, 73, 93, 139
$\mathscr{A}^*$-invariant 64, 97, 105, 109
$\mathscr{P}^*$ **54**
$\mathscr{P}^*$ Double Annihilator Theorem. *See* Double Annihilator Theorem, $\mathscr{P}^*$-version
$\mathscr{P}^*$-absorbing **67**
$\mathscr{P}^*$-decomposable **65**
$\mathscr{P}^*$-equivalence class 68
$\mathscr{P}^*$-equivalent **65**
$\mathscr{P}^*$-indecomposable **65**, 170–171
$\mathscr{P}^*$-invariant **65**
$\mathscr{P}^*$-invariant ideal. *See* ideal, $\mathscr{P}^*$-invariant
$\mathscr{P}^*$-invariant subalgebra. *See* subalgebra, $\mathscr{P}^*$-invariant
$\mathscr{P}^{**}$ 56
$\wp^k(\mathbf{d}_{n,i})$ 82
Sq **64**, 94
$\mathbf{Sq}(\mathbf{d}_{2,0})$ 94, 96
$\mathbf{Sq}(\mathbf{d}_{2,1})$ 94, 96–97
$\mathbf{Sq}(\mathbf{d}_{n,k})$ 109
$\mathfrak{D}_n$ **88**, 89
$\mathfrak{m}$-primary ideal 51
$\mathbf{D}(2)$ 160, 168, 170, 172
$\mathbf{D}(3)$ 173
$\mathbf{D}(n)$ 81, 85, 160, 171
$\mathbf{d}_{2,0}$ 41–42, 63–64, 94, 99, 160
$\mathbf{d}_{2,1}$ 41–42, 63–64, 94, 99, 160
$\mathbf{d}_{n,0}$ 83
$\mathbf{d}_{n,i}$ 81, **82**
$\mathbf{L}_n$ 83, 85
P **49**, 71, 74, 77

absorbing subspace 67–68
Adem–Wu relations 109
ancestor ideal 17–18, 38–39, 133
ancestor ideal, big **17**, 18, 21, 76, 137
ancestor ideal, little **17**, 137–138
augmentation ideal **9**, 42

big ancestor ideal. *See* ancestor ideal, big
bilinear pairing 10
binomial coefficient 98
block parameters **118**, 121
boundedness conjecture 74

canonical anti-automorphism 56, **58**
Cartan formula 86
catalecticant 39, 133, **136**, 137–140, 155
catalecticant equations **137**
catalecticant matrix 7, 76, 78, **136**
Cauchy–Frobenius lemma 23, **24**
change of rings 7, 155, 157, 162
characteristic classes 60
cocommutative 65
cocontraction pairing **155**
codimension one subspace 18–19, 158
coexact **50**, 51, 86–87, 151–153, 157
coexact sequence. *See* coexact
cofactor matrix 41, 84, 91
cofibration 152
Cohen–Macaulay 145
Cohen–Macaulay algebra 12
cohomology 95
coinvariants 81, 86
complete intersection 2, 11, 141
comultiplication 32
conjugacy class 27
conjugate Mitchell–Stong element **167**
conjugate Wu class, total **61**, 64, 117
conjugate Wu classes **61**, 62, 97, 108–109, 167
conjugate Wu classes, k-th **61**
conjugate Wu classes, trivial **61**, 62
connected sum **11**, 49, 69, 138, 141
contraction pairing **36**, 67
coproduct 32
Cramer's rule 43
cup-product 53
cyclic 48

degree one 17
diagonal 32
Dickson algebra 7, 81, 85, 87, 160, 168, 170–173
Dickson algebra, fractal of 87
Dickson coinvariants 6, 83, 85
Dickson polynomial 6, 41, 63, 81–82, 85, 94–96, 100, 105
Dickson polynomial, formulae for **82**
Dickson polynomial, power of 93, 95–96, 105, 107–109
Dickson-Euler class 83
differential operator 53
divided power algebra 5, **32**
divided power operation 32, **33**, 38
divided powers. *See* divided power operation
double annihilator 12
Double Annihilator Theorem 5, 37, 58, **65**, 134–135
Double Annihilator Theorem, $\mathscr{P}^*$-version 59, 67
double duality 10
dual principal system 5, **37**, 40, 42–44, 49, 62, 89, 135
dual system 5, 12, 31, 35, **37**, 47–49

elementary symmetric polynomial 42–43, 81
expansion, dyadic 120

finite extension 146, 148, 150–151, 155–156, 158, 165, 169–170
finite type 1
fit **149**
fit extension **149**, 150–151, 156, 165, 169–170

fitness 154
flat 44–45
formal dimension **1**, 9, 17–18, 23
fractal 87
Frobenius automorphism 158
Frobenius functor 44, 51
Frobenius homomorphism 5, 45, 53
Frobenius operator 72
Frobenius periodicity 96, 99
Frobenius power 31, 41, **44**, 64, 77, 97, 164
fundamental class **1**, 6, 9–10, 15, 17–19, 44, 47, 49, 66, 68, 77–78, 85–86, 90, 93–94, 100, 131, 152, 170

Galois field 5, 23, 41–42, 53–54, 56, 90
Gorenstein algebra 5, 9, 12, **13**, 16, 57, 146, 148, 150–151, 155

Hall subgroup 29
Hasse–Schmidt differentials 53
Hilbert ideal 81, 91
Hit Problem 4–7, 54, 65, 74, 93, 100, 123, 160
homogeneous component 94–95
Hopf algebra 5, 12, 31–32, 35, 65, 135, 162

ideal quotient 45
ideal, $\mathscr{A}^*$-invariant 93, 95–97, 99, 105–110, 112–113, 117, 128, 131, 171
ideal, $\mathscr{P}^*$-invariant **54**, 56, 62–63, 162–163, 165, 169–170
ideal, ancestor. *See* ancestor ideal
ideal, irreducible 4–5, 7, 9, **12**, 13, 15–19, 31–32, 35, 40, 42, 46, 49–50, 53–54, 56, 62–63, 90, 133, 135–136, 139, 146, 148, 150–151, 155–156, 158, 163, 169–170
ideal, over. *See* over ideal
ideal, parameter. *See* parameter ideal
ideal, principal 13, 15
ideal, regular 7, **12**, 142, 171
ideals, decreasing chain 97
indecomposable Poincaré duality algebra **11**
index sequence 34
injective hull 135
instability condition 161
integral domain 147
integrally closed 147
inverse polynomial 31, 133–134, 137, 158
inverse polynomial algebra **134**
inverse principal system 135
inverse system 5, 31, **135**, 156, 158, 163, 170
involution 16, 27, 38, 41, 95
irreducible ideal. *See* ideal, irreducible
isomorphism classes 19
isotropy subgroup 22

Jordan form 24

Krull dimension 146

lighthouse 95, 100
linear form 57
little ancestor ideal. *See* ancestor ideal, little
lying over 7, 145–146, 148

Macaulay dual 6, **37**, 41–42, 47–48, 85
Macaulay dual of the ideal $(\mathbf{d}_{2,0}^{2^r}\mathbf{d}_{2,1}^{2^r})$ 101
Macaulay dual of the ideal $(\mathbf{d}_{2,0}^{2^t}, \mathbf{d}_{2,1}^{2^s-2^t})$ 101–102
Macaulay inverse **135**
Macaulay's double annihilator correspondence 37
Macaulay's Double Annihilator Theorem. *See* Double Annihilator Theorem
Macaulay's dual principal system. *See* dual principal system
Macaulay's inverse system **135**
map of degree one 17
middle associative 10
middle dimension 100

Mitchell–Stong element **167**
Mitchell–Stong element, conjugate. *See* conjugate Mitchell–Stong element
modular case 2
module of $\mathscr{P}^*$-indecomposables **65**
moduli space 19, 23
Molien's theorem 27
monomial 66, 68, 70, 72, 75, 78, 89, 93–94, 138
monomial action 70
monomial ideal 37, 42

nil radical 32
node 94
Noether isomorphism theorem 44
Noetherian algebra 15
nondecreasing sequence 100
nondegenerate 10
number theoretic function 111

obstruction 109
orbit 22, 26, 39, 88
orbit space 19, **23**
over ideal 14, 41

palindromic 37, 137
paradigm, $K \subset L$ 5, 31, **40**, 88–89, 113
parameter ideal **12**, 13, 16, 41, 47, 55, 57, 145, 148–149, 151, 159
parameter ideal, $\mathscr{P}^*$-invariant 55
periodicity map 74
periodicity operator **49**, 71, 74, 77, 100
Poincaré dual 15–17, 35, 131, 152, 154
Poincaré duality 17–23, 26–29, 31, 35, 37–40, 44, 49, 51, 53–54, 57–58, 60–63, 65–66, 68–70, 73–74, 76–79, 81, 83, 85–86, 90, 93–94, 99–100, 105, 123, 131, 146–148, 151–154, 158, 160, 164, 168
Poincaré duality algebra **1**, 2, 5, 9, 11, 17–23, 26–29, 31, 35, 37–40, 44, 49–51, 53–54, 57–58, 60–63, 65–66, 68–70, 73–74, 76–79, 81, 83, 85–86, 90, 93–94, 99–100, 105, 123, 131, 146–148, 151–154, 158, 160, 164, 168, 173
Poincaré duality quotient 6, 17–18, 21, 23, 39, 49, 53–54, 63, 65–66, 70, 73, 77, 79, 90, 93, 95, 100, 170
Poincaré duality quotient of $\mathbb{F}[V]$ **12**
Poincaré polynomial 95
Poincaré series 20, 27–28, 99, 137, 141
polynomial algebra 155–156
prime ideal 7, 146
primitive element, Milnor 83
principal 47
principal ideal. *See* ideal, principal
projective space **19**
pseudoreflection group 2

rank **12**
rank of a Poincaré duality quotient of $\mathbb{F}[V]$ 12
rank sequence 138
regular ideal. *See* ideal, regular
regular sequence 12, 49, 66, 81, 86–87, 94, 142, 145, 148
relative transfer 148
ring extension, finite. *See* finite extension
ring of invariants 81
row-echelon form 137

spike 6, 93
split 50–51, 147–148, 151–152
Steenrod algebra 4–7, 53–54, 58, 66, 79, 100, 160, 163, 169–170
Steenrod operation, total **56**, 57, 71, 105, 109
Steenrod operations 7, 54, 66, 71
Steenrod square, total **64**
Steinberg, R. 2
Stiefel–Whitney class, power of 93
Stiefel–Whitney classes 62, 93, 123
Stiefel-Whitney classes 6
Stong–Tamagawa formulae 82
stripping operation 35
subalgebra, $\mathscr{P}^*$-invariant 86–87

subspace, codimension one. *See* codimension one subspace
support **36**, 70, 78
Sylow subgroup 29
symmetric 10
symmetric coinvariants 6
system of parameters 12, 42, 146, 148

Thom class **56**, 57, 63–64, 66, 95–96, 105, 109, 165
Thom–Wu formulae 4
topological space 95
total Steenrod operation. *See* Steenrod operation, total
total Steenrod square. *See* Steenrod square, total
totalization 13
transition element **14**, 15, 41, 43, 142, 149, 151
transition invariant **14**
transitive 155
transversal 27
trivial conjugate Wu classes. *See* conjugate Wu classes, trivial
trivial Wu classes. *See* Wu classes, trivial
type, finite 1

unstable algebra (over $\mathscr{P}^*$) **54**
unstable algebra 53–54, 56, 163
unstable Poincaré duality algebra **54**
unstable polynomial agebra 162, 169–170
upper triangular 141

Wu class 5–6
Wu class, total 60, 64, 170
Wu class, total conjugate **61**
Wu classes 3, **60**, 64, 66, 68, 71, 77, 83, 86, 90, 165
Wu classes, k-th 60
Wu classes, conjugate. *See* conjugate Wu classes
Wu classes, nontrivial 99
Wu classes, trivial 54, **60**, 61, 63, 65, 68–70, 73, 79, 86, 93–99, 105–109, 121, 123, 125, 128, 131, 169–170, 173
Wu Wen Tsün 54
Wu's formula 124–125